Behavior and Neurodynamics for Auditory Communication
KANWAL and EHRET

How do animals produce and process sounds for communication? How do their brains encode the large amounts of sensory information so rapidly and how do they use this to cope with their environment? These questions not only concern the evolution of sound communication systems but also aim to understand how arousal, motivation, emotion, and behavioral contexts are vocally expressed and how important sound attributes are recognized and perceived. This book highlights auditory communication in several species from four perspectives: actual sound communication, auditory and vocal adaptations, neural adaptations for sound processing and representation in higher auditory brain centers, and emotional and cognitive adaptations in sound signaling and processing. Broad in scope and geared towards graduate students and researchers within the fields of auditory communication and cognition, this book will appeal to auditory neurobiologists, speech, hearing, and communication scientists and engineers, students of animal behavior, and neuroethologists at large.

Jagmeet S. Kanwal is Associate Professor of Physiology and Biophysics at Georgetown University Medical Center in Washington, DC, USA.

Günter Ehret is Professor of Neurobiology at the University of Ulm in Germany.

Behavior and Neurodynamics for Auditory Communication

Edited by

JAGMEET S. KANWAL

and

GÜNTER EHRET

CAMBRIDGE UNIVERSITY PRESS

Cambridge, New York, Melbourne, Madrid, Cape Town, Singapore, São Paulo

Cambridge University Press
The Edinburgh Building, Cambridge CB2 2RU, UK

Published in the United States of America by Cambridge
University Press, New York

www.cambridge.org
Information on this title: www.cambridge.org/9780521829182
© Cambridge University Press 2006

This publication is in copyright. Subject to statutory exception and to the provisions of relevant collective licensing agreements, no reproduction of any part may take place without the written permission of Cambridge University Press.

First published 2006

Printed in the United Kingdom at the University Press, Cambridge

A catalogue record for this publication is available from the British Library

ISBN-13 978-0-521-82918-2 hardback
ISBN-10 0-521-82918-6 hardback

Cambridge University Press has no responsibility for the persistence or accuracy of URLs for external or third-party internet websites referred to in this book, and does not guarantee that any content on such websites is, or will remain, accurate or appropriate.

About the cover

Cover illustration: Wahlberg's epauletted fruit bat: *Epomophorus wahlbergi*, female and young in fig tree © Oxford Scientific. Neural trace from Medvedev AV, Chiao F and Kanwal JS (2002) Modeling complex sound perception: grouping harmonics with combination-sensitive neurons. *Biological Cybernetics* 86: 497–505. With kind permission of Springer Science and Business Media.

Wahlberg's epauletted fruit bat, *Epomophorus wahlbergi*, is a megabat native to Africa. During the breeding season, males of this species congregate at traditional sites, where they puff up their white shoulder patches, puff out their cheek pouches, fan their wings and make repeated gonglike calls to attract a passing female (Nowak, 1994). These calls combine four short chirps, and are one second in duration. The cheek pouches are used for collecting food and also as inflatable sacs that act as a resonance chamber to amplify their calls. These fruit bats recognize their young ones through vocalizations and olfaction. Other species, such as spear-nosed fruit bats, are capable of audiovocal learning during mother–infant interactions as well as with changing group compositions (Esser, 1994; Boughman, 1998). The line traces above (time on the horizontal axis and voltage on the vertical axis) are simulated excitatory post-synaptic potentials and facilitated spiking in a combination-sensitive auditory neuron in response to a complex sound, such as a communication call (Medvedev *et al.*, 2002).

Nonlinear facilitation (solid yellow line) results in a higher firing rate compared to a weak response (dashed blue line) to a simple sound, such as a single tone. Combination-sensitivity is considered to be important for the neural processing and recognition of communication calls in many species of animals.

REFERENCES

Boughman JW (1998) Vocal learning by greater spear-nosed bats. *Proc. Biol. Sci.* 265: 227–333.

Esser KH (1994) Audio-vocal learning in a non-human mammal: the lesser spear-nosed bat Phyllostomus discolor. *Neuroreport.* 5: 1718–1720.

Medvedev AV, Chiao F, Kanwal JS (2002) Modeling complex sound perception: grouping harmonics with combination-sensitive neurons. *Biological Cybernetics* 86: 497–505.

Nowak, R (1994) *Walker's Bats of the world.* Baltimore: Johns Hopkins University Press. 66pp.

Wickler W, Seibt U (1976) Field studies of the African fruit bat *Epomophorus wahlbergi*, with special reference to male calling. *Z. Tierpsychol.* 40: 345–376.

Contents

Preface   *xvii*
Contributors   *xix*

PART 1   BEHAVIORAL AND ANATOMIC/ PHYSIOLOGIC ADAPTATIONS

1   Vocal mechanisms for avian communication   **3**
    1.1   Introduction   3
    1.2   Avian vocal systems   3
        1.2.1   The syrinx   3
        1.2.2   The labia   5
        1.2.3   Syringeal membranes   6
        1.2.4   Integration with respiration   7
    1.3   Vocal learning   9
    1.4   Controlled and intrinsic modulation of the vocal source   10
        1.4.1   Amplitude modulation   11
        1.4.2   Frequency modulation   13
            1.4.2.1   Syringeal muscles   13
            1.4.2.2   Air sac pressure   13
            1.4.2.3   Other mechanisms   13
    1.5   Lateral independence in the bipartite syrinx   14
        1.5.1   Unilateral dominance versus bilateral parity   16
        1.5.2   Two-voice vocalizations   17
        1.5.3   Patterns of song lateralization and acoustic specialization   19
    1.6   Nonlinear dynamics in vocalizations   20
    1.7   Vocal tract filtering   24

1.8	Production, propagation, and perception	28
1.9	Future directions	29
References		30

2 The blind mole rat: an example of seismic communication via acoustic channels — 36

2.1	General overview	36
2.2	Properties of the vibratory signals	38
2.3	How are the seismic signals perceived?	40
2.4	Somatosensory or auditory?	41
2.5	Electrophysiologic experiments	42
2.6	Behavioral tests	43
2.7	Morphology	46
2.8	Discussion	49
2.9	Conclusion	52
2.10	Abbreviations	52

Acknowledgments — 53
References — 53

3 Audiovocal communication and social behavior in mustached bats — 57

3.1	Introduction	57
3.2	Social behaviors in captive free-flying bats	60
	3.2.1 Maintenance and recording procedures	60
	3.2.2 Scoring behaviors and calls	60
3.3	Results and discussion	63
	3.3.1 Roosting structure, activity patterns, and social interactions	63
	3.3.2 Behavioral postures and associated calls	64
	3.3.2.1 Crouching behavior	64
	3.3.2.2 Marking behavior	64
	3.3.2.3 Grooming, licking, and yawning behaviors	64
	3.3.2.4 Nipping behavior	65
	3.3.2.5 Wing-flicking behavior	65
	3.3.2.6 Boxing and poking behavior	65
	3.3.2.7 Wrestling and biting behavior	66
	3.3.2.8 Arching back and "kissing" behavior	66
	3.3.2.9 Inspection behavior	66
	3.3.2.10 Fly-by behavior	67

		3.3.3	Behavioral context of simple syllabic calls	67
		3.3.4	Inspection and appeasement	69
		3.3.5	Territoriality and social dominance	70
	3.4	Social interactions in caged bats		71
		3.4.1	Recording procedures	71
	3.5	Results and discussion		73
		3.5.1	Natural aggression	73
		3.5.2	Experimentally elicited aggression	74
	3.6	Call types and acoustic signal design		77
	3.7	Conclusions		81
Acknowledgments				81
References				82

4 Common rules of communication sound perception — **85**

    4.1 Introduction — 85
    4.2 The basis: common psychoacoustical measures and relations — 87
        4.2.1 Audiograms — 87
        4.2.2 Temporal summation — 87
        4.2.3 Frequency discrimination — 88
        4.2.4 Intensity discrimination — 88
        4.2.5 Duration discrimination — 88
        4.2.6 Spectral resolution, spectral summation — 89
        4.2.7 Pitch perception and discrimination — 89
    4.3 The six rules of communication sound perception — 90
        4.3.1 Rule 1: Perception is individualized to the level of matched groups — 90
        4.3.2 Rule 2: Perception of acoustic meanings or objects refers to JMDs between acoustic patterns — 90
        4.3.3 Rule 3: Perception of acoustically expressed levels of arousal and emotions refers to the variations in acoustic patterns — 92
        4.3.4 Rule 4: The audible parameter space is partitioned for the perception of three basic meanings — 93
            4.3.4.1 High-frequency group — 96
            4.3.4.2 Broadband group — 96
            4.3.4.3 Low-frequency group — 96

	4.3.5	Rule 5: The integrated extent of the ranges activated in the audible parameter space is related to the perception of the urgency of the response	99
	4.3.6	Rule 6: Perceptions of categories from continuums of acoustic patterns are built upon natural perceptual boundaries	100
		4.3.6.1 Spectral (formant) structure	102
		4.3.6.2 Direction of frequency sweeps (formant transitions)	104
		4.3.6.3 Duration of information-bearing elements or parameters	105
4.4	Conclusions		106
Acknowledgments			107
References			107

Part 1 Behavioral and Physiologic Adaptations: Summary and Discussion — 115

Anatomic and physiologic adaptations — 115
Behavioral adaptations and the perception of meaning — 117
References — 119

PART 2 NEURAL ADAPTATIONS AND PLASTICITY

5 Neural mechanisms of vocal communication: interfacing with neuroendocrine mechanisms — 123
 5.1 Introduction — 123
 5.2 Hormonal control of vocalizations — 123
 5.3 Vocal-auditory coupling — 126
 5.4 Conclusion and future prospects — 129
Acknowledgments — 130
References — 130

6 Processing of species-specific vocalizations in the auditory brainstem and midbrain of Mexican free-tailed bats (*Tadarida brasiliensis*) — 132
 6.1 Introduction — 132
 6.2 The importance of hearing for bats — 133
 6.3 The structure of the auditory brainstem and midbrain: a brief overview — 135
 6.4 Representation of communication sounds in lower auditory nuclei — 137

	6.5	Processing of communication sounds is much more complex in the inferior colliculus	143
	6.6	Communication sounds elicit a unique spatio-temporal pattern of activity in the ICC	147
	6.7	Conclusion	150
	Acknowledgments		151
	References		151

7 A distributed cortical representation of social communication calls — **156**

- 7.1 Introduction — 156
- 7.2 The mammalian AC — 157
    - 7.2.1 Organization of the AC — 158
    - 7.2.2 Response properties of primary auditory cortical neurons — 160
- 7.3 The acoustic structure of complex sounds in mustached bats — 161
    - 7.3.1 The structure of echolocation signals — 161
    - 7.3.2 The structure of social communication calls — 161
- 7.4 The AC in mustached bats — 162
    - 7.4.1 Functional organization for echolocation — 162
    - 7.4.2 Neural mechanisms for pulse–echo representation — 163
    - 7.4.3 Neural mechanisms for call representation — 164
- 7.5 Call responses in mustached bats — 165
    - 7.5.1 FM–CF combinations — 167
    - 7.5.2 FM–FM combinations: the time domain — 169
    - 7.5.3 CF/CF combinations: high-frequency domain — 172
    - 7.5.4 Harmonic complexity: low-frequency domain — 172
- 7.6 A hypothesis for the cortical representation of calls — 175
    - 7.6.1 Multiparametric distributed representations — 175
    - 7.6.2 Representation versus perception — 180
- 7.7 Conclusions — 182

Acknowledgments — 183
References — 183

8 Spatiotemporal processing in the guinea pig auditory cortex — **189**

- 8.1 Introduction — 189
- 8.2 Optical recording methods — 189
- 8.3 Tonotopic organization in auditory cortical fields — 191

8.4	Spatiotemporal representation of constant and frequency-modulated tones	193
8.5	Responses in AI to vocalized sounds	195
8.6	Functional significance of the spatiotemporal representation of sounds	195
8.7	Activity spreads to the higher auditory fields	197
8.8	Functional differences of the multiple auditory fields	198
8.9	Principal component analysis applied to optical signals	200
8.10	Conclusions	202
Acknowledgments		202
References		202

9 Hierarchic processing of communication sounds in primates — 205

9.1	Auditory communication as a pattern recognition problem	205
9.2	Early parallel processing in the auditory cortex	206
9.3	Processing of sounds with intermediate complexity	207
	9.3.1 Selectivity for BPN	207
	9.3.2 Selectivity for parameters of frequency sweeps	210
9.4	Responses to species-specific calls	214
	9.4.1 Nonlinear integration mechanisms	214
	9.4.2 MC and spatial selectivity	216
9.5	Auditory belt projections to prefrontal cortex	217
9.6	Human imaging studies	218
9.7	Conclusions	219
References		220

10 Synaptic mechanisms and sensitive periods for song learning — 223

10.1	Sensitive periods for learned behavior	223
10.2	Neuronal and behavioral sensitive periods: common themes	224
10.3	Sensitive periods for neural circuit plasticity: search for mechanisms	224
10.4	Birdsong development: sensitive periods for a learned behavior	227
10.5	The song system: a neural circuit for the learning and production of birdsong	232

10.6	What are the neural correlates of song plasticity during sensorimotor learning?	236
10.7	The nucleus RA is an important site to search for neuronal correlates of vocal plasticity	237
10.8	Neuronal codes for song in the nucleus RA	237
10.9	Extrinsic and intrinsic patterns of synaptic connectivity in the nucleus RA	238
10.10	The structure and function of RA subregions	239
10.11	A cellular analysis of the nucleus RA during sensorimotor learning	240
10.12	A functional synaptic analysis of the nucleus RA during sensorimotor learning	241
10.13	Molecular effectors of synaptic and vocal plasticity	244
10.14	Alternative possibilities and further tests of the model	249
10.15	Conclusions	252
References		253

11 Neuronal substrates of sensory processing for song perception and learning in songbirds: lessons from the mormyrid electric fish — **265**

11.1	Introduction	265
11.2	Song perceptual processing and song learning	267
11.3	Electrosensory processing in the mormyrid electric fish	271
11.4	The perceptual processing and memorization of song: in search of a neuronal substrate	276
11.5	Implementing feedback evaluation: what do we learn from comparing songbirds and the electric fish	282
11.6	Conclusion	285
Acknowledgments		286
References		286

12 Cortical plasticity and auditory communication — **294**

12.1	Introduction	294
12.2	Experience-dependent plasticity of the auditory cortex during early development	297
	12.2.1 Critical role of sensory experience in early cortical development	297
	12.2.2 Normal development of cortical IBEs/IBPs and tonotopic maps	298

	12.2.3 Cortical development in a distorted acoustic environment	299
	12.2.4 Impact of inner ear ablation on cortical development	300
12.3	Experience-dependent plasticity of the auditory cortex in adult animals	301
	12.3.1 Enhancement of cortical representation induced by acoustic signals alone	301
	12.3.2 Enhanced cortical representation as a result of learning	301
	12.3.3 Cortical plasticity after localized hearing loss	303
12.4	Cortex-oriented plasticity in the central auditory system: corticofugal modulation	304
	12.4.1 Frequency-specific modulation of frequency tuning and tonotopic maps	305
	12.4.2 Modulation of IBEs/IBPs in different domains	307
	12.4.3 Corticofugal modulation of response properties of combination-sensitive neurons	308
	12.4.4 Species-specific differences in corticofugal modulation	308
12.5	Perspective of cortical plasticity for cortical coding of vocalizations	309
12.6	Conclusion	313
References		313

13 Mesoscopic neurodynamics in auditory cortex during auditory concept learning — **319**

13.1	Introduction	319
	13.1.1 Traditional theoretical underpinnings of the study of category learning	319
	13.1.2 Eco-ethologic aspects of categorization	320
	13.1.3 Categorization and generalization	321
13.2	A new animal model of category learning	323
	13.2.1 Categorization of modulation direction in frequency-modulated tones	324
	13.2.1.1 Stimuli	324
	13.2.1.2 Apparatus and training paradigm	324
	13.2.1.3 Behavioral results	325
	13.2.2 Physiologic correlates of category learning	326

	13.2.2.1 General aspects of the search for physiologic correlates of learning	326
	13.2.2.2 Neuronal representation of utilized stimuli	327
	13.2.2.3 Single trial analysis of electrocorticograms	328
13.3	Conclusions	330
Acknowledgments		331
References		331

Part 2 Neural adaptations and plasticity: summary and discussion — 334

Auditory communication and hormones — 334
Auditory representations — 336
Learning, memory and plasticity — 339
References — 342

Appendix — 346
Basics of acoustic signal processing — 346

Index — 351

Preface

A major function of the auditory system is to provide and represent information about (1) the emotional or ascribed meaning of sounds for intraspecies communication, (2) the identity of animal senders or other sound sources, and (3) the spatial location of sound sources. This book highlights the first two points and approaches them from four perspectives: (1) the actual sound communication in a variety of species, (2) audio-vocal adaptations, (3) adaptations for sound processing and representation in higher auditory centers, and (4) emotional and cognitive adaptations in signaling and perception. In this way, it builds upon the topics of the symposium on "*Cracking the Code for Auditory Communication*" held at the *International Ethological Conference* in Tübingen (2001) and deepens them in scope. The readers will find descriptions of original data and treatments of theory at the cutting edge of present research in this field. Recent findings of the existence of many parallels between animal and human communication via the auditory channel indicate the importance of animal models for a basic understanding of the functioning of the auditory system in humans.

The book is arranged in two parts. In the first part, two amazing cases representing special adaptations in communication-sound production and perception, the blind mole rat and the mustached bat, are introduced and discussed in contrast to general adaptations of audiovocal performance in birds and of perceptual performance in mammals. In the second part, neural adaptations for perceiving communication sounds in the auditory system are examined from various perspectives, and conditions of adjustment for generating adaptive behavioral responses (i.e. complex sound recognition) are discussed. Both parts are summed up and discussed in additional chapters to facilitate rapid access to the essence of the book.

The book is geared towards graduate students and researchers in the fields of auditory communication and cognition (including researchers working in the speech and hearing sciences, communication sciences, and in engineering design), auditory neuroscience, psychoacoustics, animal behavior, and, last but not the least, neuroethology at large. We thank Cambridge University Press and its editors for their constant help and encouragement, and our authors for their fine contributions and their patience during the long process of editing this volume.

Jagmeet S. Kanwal and Günter Ehret

Contributors

Andrew H. Bass
Department of Neurobiology
and Behavior, Cornell University,
Seeley G. Mudd Hall, Ithaca,
NY 14853-3901, USA

Eric E. Bauer
Department of Zoology,
University of Texas at Austin,
Austin, TX 78712, USA

Gabriël J.L. Beckers
Medical Sciences, Jordan Hall,
Indiana University, Bloomington,
IN 47405, USA

Matthew J. Clement
Department of Physiology and
Biophysics, Georgetown University
Medical Center, 3900 Reservoir
Road, NW, Washington
DC 20057-1460, USA

Nicole Dietz
Department of Physiology and
Biophysics, Georgetown University
School of Medicine, 3900
Reservoir Road, NW, Washington
DC 20057, USA

Jos J. Eggermont
Department of Physiology and
Biophysics, University of Calgary,
Calgary, Alberta, Canada
Department of Psychology,
University of Calgary, Calgary,
Alberta, Canada
Neuroscience Research Group,
University of Calgary, Calgary,
Alberta, Canada

Günter Ehret
Department of Neurobiology,
University of Ulm,
Albert-Einstein-Allee 11,
D-89069 Ulm, Germany

Walter J. Freeman
University of California at Berkeley,
Berkeley, CA 94720-3200, USA

Punita Gupta
New England Medical Center,
750 Washington St, Box 299,
Boston, MA 02111, USA

Joshua T. Hanson
Section of Neurobiology, University
of Texas, Austin, TX 78712, USA

Andreas Hess
Leibniz Institute for Neurobiology,
Brenneckestrasse 6, 39118
Magdeburg, Germany Institute for
Pharmacology and Toxicology
FAU Erlangen, Fahrstrasse 17,
91054 Erlangen, Germany

Junsei Horikawa
Department of Knowledge-Based
Information Engineering, Toyohashi
University of Technology,
1-1 Hibarigaoka, Tempaku,
Toyohasi 441-8580, Japan

Yutaka Hosokawa
Department of Neurophysiology,
Medical Research Institute, Tokyo
Medical and Dental University,
2-3-10 Kanda-surugadai,
Chiyoda-ku, Tokyo 101-0062,
Japan

Jagmeet S. Kanwal
Department of Physiology and
Biophysics, Georgetown
University Medical Center, 3900
Reservoir Road, NW, Washington,
DC 20057-1460, USA
Department of Psychology,
Georgetown University, 37th and
O Streets, NW, Washington,
DC 20057, USA

Krasnow Institute for
Advanced Study
George Mason University
Fairfax, VA 22030, USA

J. Matthew Kittelberger
Department of Neurobiology and
Behavior, Cornell University, Ithaca,
NY 14853, USA

Achim Klug
Department of Neurobiology,
University of Munich,
Grosshaderner Strasse 2,
82152 Martinsried, Germany

Claudio V. Mello
Neurological Sciences Institute,
Oregon Health and Science
University, West Campus,
505 NW 185th Avenue,
Beaverton, OR 97006, USA

Richard Mooney
Department of Neurobiology,
Duke University Medical Center,
Durham, NC 27710, USA

Brian S. Nelson
Department of Biology, Indiana
University, Bloomington, IN 47405,
USA

Frank W. Ohl
Leibniz Institute for Neurobiology,
Brenneckestrasse 6, 39118
Magdeburg, Germany

George D. Pollak
Department of Zoology, University
of Texas at Austin, Austin,
TX 78712, USA

Rony Rado
Department of Zoology,
George S. Wise, Faculty of Life
Sciences, Tel Aviv University, Tel
Aviv 69978, Israel

Josef P. Rauschecker
Department of Physiology and
Biophysics, Georgetown University
Medical Center, 3900 Reservoir
Road, NW, Washington
DC 20057-1460, USA

Patrick D. Roberts
Neurological Sciences Institute,
Oregon Health and Science
University, West Campus,
505 NW 185th Avenue, Beaverton,
OR 97006, USA

Ronen S. Sadka
Department of Zoology,
George S. Wise, Faculty of Life
Sciences, Tel Aviv University, Tel
Aviv 69978, Israel

Henning Scheich
Leibniz Institute for Neurobiology,
Brenneckestrasse 6, 39118
Magdeburg, Germany

Roderick A. Suthers
Medical Sciences, Jordan Hall,
Indiana University, Bloomington,
IN 47405, USA
Department of Biology, Indiana
University, Bloomington,
IN 47405, USA
Program for Neural Science,
Indiana University, Bloomington,
IN 47405, USA

Ikuo Taniguchi
Department of Neurophysiology,
Medical Research Institute, Tokyo
Medical and Dental University,
2-3-10 Kanda-surugadai,
Chiyoda-ku, Tokyo 101-0062,
Japan

Biao Tian
Department of Physiology and
Biophysics, Georgetown University
School of Medicine, Washington
DC 20057-1460, USA

Zvi Wollberg
Department of Zoology,
George S. Wise, Faculty of Life
Sciences, Tel Aviv University, Tel
Aviv 69978, Israel

Jun Yan
Department of Physiology and
Biophysics, University of Calgary,
Calgary, Alberta, Canada
Neuroscience Research Group,
University of Calgary, Calgary,
Alberta, Canada

PART ONE

Behavioral and anatomic/physiologic adaptations

1
Vocal Mechanisms for Avian Communication

Roderick A. Suthers, Gabriël J.L. Beckers and Brian S. Nelson

1.1 Introduction

Vocal signals are an important means of avian communication. The variety and complexity of the vocal repertoire varies greatly between species, as does the morphology of the vocal organ, but the correlation between vocal anatomy and acoustic output is not always apparent. Recent years have seen important advances in understanding how birds sing, including new insights into sound generation, production constraints, and vocal tract filtering, as well as a growing interest in the nonlinear components of vocalizations. Here we attempt to provide an overview of the peripheral mechanisms by which both songbirds and non-songbirds control the acoustic properties of their vocalizations, in the hope that it will help strengthen the bridge between mechanistic and behavioral approaches to avian acoustic communication.

1.2 Avian vocal systems

1.2.1 The syrinx

The avian vocal organ is located in the interclavicular air sac, where it is formed by modified bronchial and/or tracheal cartilages, and contains structures that vibrate in airflow to generate sound. Syringeal anatomy varies significantly across different groups. At one extreme are the vultures, which have no syrinx. At the other extreme are various songbirds whose well-developed syrinx includes multiple pairs of muscles with specialized functions in sound production (Fig. 1.1). In between these extremes are a variety of non-oscine syrinxes controlled by an intermediate number of tracheal and/or syringeal muscles.

Syrinxes are often categorized anatomically as being tracheal, bronchial, or tracheobronchial, but the location of the sound-generating oscillators is of

4   1 Vocal mechanisms for avian communication

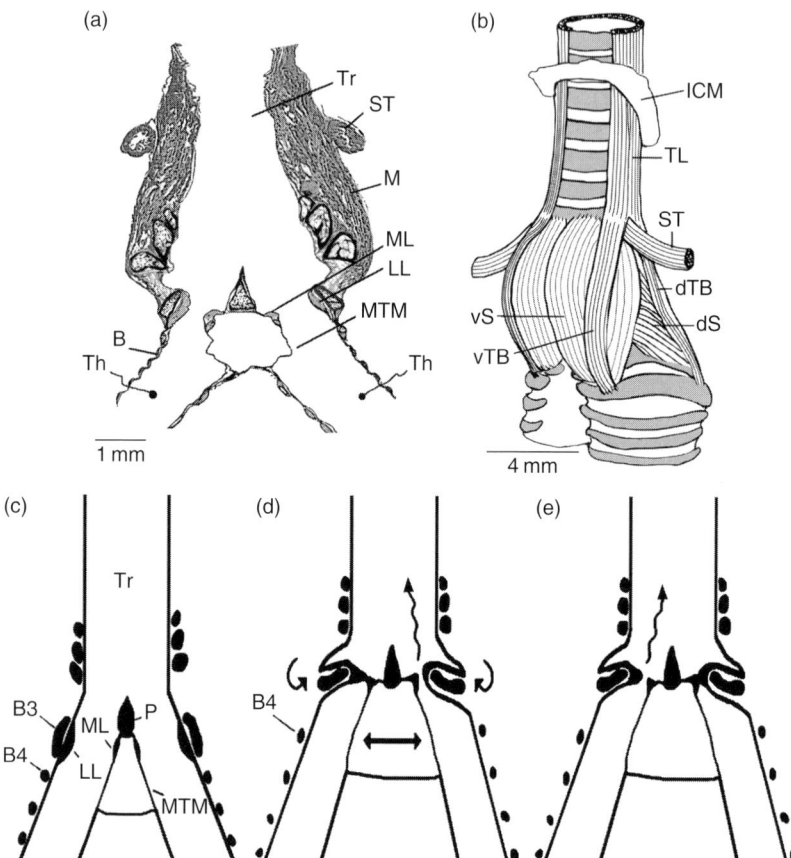

Figure 1.1. The songbird syrinx is a bipartite structure located at tracheobronchial junction. (a) Frontal section through the syrinx of a brown thrasher, showing the dual nature of the vocal organ and placement of microbead thermistors (Th) for recording airflow. (b) Ventrolateral external view of syrinx illustrating syringeal musculature. (c) Schematic ventral views of songbird syrinx during quiet respiration (d) phonation on the left side with labial valve closed on right side and (e) phonation of right side with left side closed. In preparation for phonation the syrinx moves rostrad. Contraction of the ipsilateral dorsal syringeal muscles (dS and dTB) rotates the bronchial cartilages (curved arrows) into the syringeal lumen, moving the labia into the airstream, where they are set into vibration, producing sound (wavy arrows). Phonation may be bilateral (not shown) or unilateral (shown). Tr: trachea; M: syringeal muscle; ML: medial labium; LL: lateral labium; B: bronchus; ICM: membrane of the interclavicular air sac; TL: m. tracheolateralis; ST: m. sternotrachealis; vS: m. syringealis ventralis; vTB: m. tracheobronchialis ventralis; dTB: m. tracheobronchialis dorsalis; dS: m. syringealis dorsalis; B3 and B4: third and fourth bronchial cartilages; P: pessulus (b: modified from Goller and Suthers, 1996a; c–e: modified from Suthers and Goller, 1997).

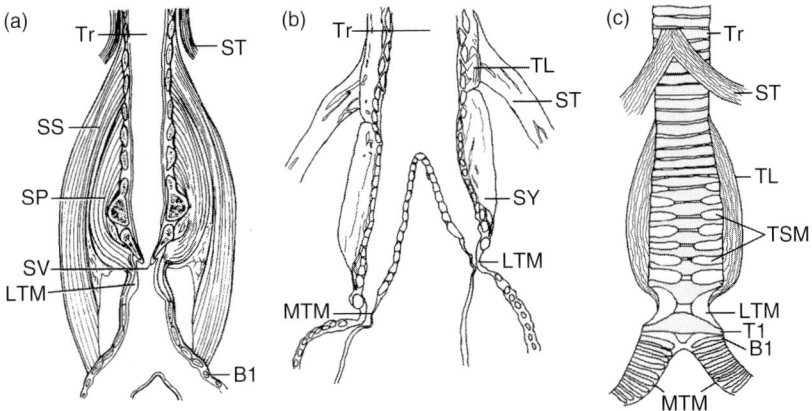

Figure 1.2. (a) Frontal section of monk parakeet syrinx showing tracheal location sound producing LTMs (modified from King, 1989). (b) Asymmetrical bronchial syrinx of the oilbird. Cranial end of each bronchus contributes a formant to the vocalization (Suthers, 1994; Fletcher and Tarnopolsky, 1999). Individual variation in degree of asymmetry endows different individuals in colony with unique formant frequencies (modified from Suthers and Hector, 1985). (c) Syrinx of collared dove. Sound is produced by LTM at base of trachea. Functional significance of the TSM is unknown (modified after Ballintijn and ten Cate, 1997). B1: first bronchial cartilage; T1: first tracheal ring; Tr: trachea; SS: superficial syringeal muscle; SP: deep syringeal muscle; ST: sternotrachealis muscle; SV: syringeal valve; TL: tracheolateralis muscle; TSM: tracheosyringeal membrane; SY: syringeal muscle; LTM: lateral tympaniform membrane; MTM: medial tympaniform membrane.

particular functional relevance. In storks, ibises, ducks, geese, some parrots (Fig. 1.2a), and several families of the suboscines, for example, these oscillators are located in the trachea where they share the same airstream during phonation and are likely to function together as a single sound source (though there may be exceptions). In most birds, including songbirds (order Passeriformes, suborder Oscine), sound is generated by separate bronchial airstreams across oscillating structures at the cranial end of each bronchus (Fig. 1.1). In a few taxa including penguins, nightjars, oilbirds (Fig. 1.2b), and some cuckoos the vocal organ is separated into two semi-syrinxes situated part way along the length of each bronchus.

1.2.2 The labia

Songbirds produce sound at the cranial end of each bronchus by vibration of two small masses of connective tissue, the medial and lateral labia that lie on opposite sides of the syringeal lumen (Fig. 1.1a). Phonation is initiated when syringeal adductor muscles move the labia into the expiratory airstream. Here they are caused to oscillate, presumably by Bernoulli forces acting in combination with the physical properties of the labial tissue in much the same way as occurs in the human vocal folds (Titze, 1994).

The motion of the labia has been observed with an endoscope during sounds elicited by brain stimulation of anesthetized northern cardinals (*Cardinalis cardinalis*) and brown thrashers (*Toxostoma rufum*), and during spontaneous vocalizations by an awake crow (*Corvus brachyrhynchus*) (Goller and Larsen, 1997b). In each case sound production was preceded by labial adduction to form a slit and accompanied by labial vibration. In other experiments, the dominant frequency of labial vibration in the hill myna (*Gracula religiosa*), measured with an optical vibration sensor, was found to agree with that in the vocalization (Larsen and Goller, 1999). Prior to these experiments by Goller and Larsen sound was assumed to be generated by vibration of the medial tympaniform membranes (MTMs) (Miskimen, 1951) that border the caudal edge of the medial labium. Destruction of the MTM has only a small effect on cardinal and zebra finch (*Taeniopygia guttata*) song, and its role in phonation needs further study (Goller and Larsen, 1997b).

1.2.3 Syringeal membranes

According to King (1989), the medial labium has not been reported in non-passerines. A lateral labium is present in some species, but whether it generates sound or acts only as a pneumatic valve is not known. Direct observation of oscillating syringeal structures during vocalization are available for only two non-passerines, neither of which has a lateral labium. In both of these species sound is produced by vibration of syringeal membranes. Endoscopic observations of the cockatiel syrinx indicate that in parrots sound is generated by oscillation of the lateral tympaniform membranes (LTMs) (Fig. 1.2a) as they are adducted into the syringeal lumen (Larsen and Goller, 1999). The syrinx of doves contains a thin MTM on each bronchus and thicker LTMs between the first and second tracheal rings where they extend around the dorsal side of the trachea (Fig. 1.2c) (Gaunt *et al.*, 1982). A third membrane, the tracheosyringeal membrane is also present in some species between narrow portions of tracheal rings along the dorsal side of the trachea (Ballintijn *et al.*, 1995). Sound was long assumed to be generated either by vibration of the MTMs (Warner, 1972; Ballintijn *et al.*, 1995) or possibly by an aerodynamic whistle mechanism (Gaunt *et al.*, 1982). The whistle mechanism is not supported by recent experiments (e.g. Ballintijn and ten Cate, 1998). Goller and Larsen (1997a) observed that when they induced phonation by blowing air through the trachea of a pigeon, the LTMs bulged into the syringeal lumen and vibrated. Stiffening the MTMs with tissue adhesive caused the coos to become "soft" and changed the spectral distribution of their energy, but the temporal and frequency patterns remained, indicating that the LTM is the primary sound source. A similar production mechanism presumably occurs in the closely related ring dove (*Streptopelia risoria*; a domesticated form of the African collared dove, *Streptopelia roseogrisea*, with which it is now thought to be conspecific) and the Eurasian collared dove (*Streptopelia decaocto*). Measures of the membrane vibration frequency in cockatiels and pigeons (Larsen and Goller, 1999) showed a dominant frequency similar to that of the fundamental in the vocalization (but see Beckers *et al.*, 2003b).

The LTM is absent in about half of the non-passerines (King, 1989) leaving other syringeal membranes, notably the MTM, as the presumptive sound source.

1.2.4 Integration with respiration

Since air flowing through the syrinx provides the energy for sound production, the control of respiratory ventilation is a critical component of vocalization. The temporal pattern of ventilation changes during song and sets the song's basic temporal pattern (Vicario, 1991). The avian respiratory system consists of a series of air sacs distributed in various parts of the body. Air sacs do not participate in gas exchange but act as bellows, moving air through the lungs and vocal tract as they are alternately expanded and compressed by inspiratory and expiratory respiratory muscles, respectively. The unpaired median interclavicular (= clavicular) air sac, which contains the syrinx (McLelland, 1989), also provides a pathway between the two sides and minimizes any pressure differences between the left and right bronchi (Wild *et al.*, 1998).

With only a few known exceptions, birds vocalize during expiratory airflow. Some zebra finches include inspiratory syllables in their song and these syllables which have a distinctively high fundamental frequency, are copied by juveniles (Goller and Daley, 2001). In all songbirds studied, singing is accompanied by a large increase in the amplitude of electromyograms (EMGs) in both the external oblique abdominal expiratory muscle (Hartley, 1990; Goller and Suthers, 1999) and thoracic inspiratory muscles (Wild *et al.*, 1998), which are alternately active during expiration and inspiration, respectively. Song requires precision coordination between syringeal and respiratory motor patterns. The robust nucleus of the arcopallium (RA) in the forebrain is likely to have a key role in this coordination, since it projects to premotor inputs of spinal respiratory motor neurons (Wild, 1997), as well as to syringeal motor neurons.

Different species vary greatly in the tempo of their songs and in the demand this makes on respiratory ventilation, which must continue to meet the needs for pulmonary gas exchange while supplying air to the syrinx at the appropriate pressure for phonation. This is usually accomplished by taking a small inspiration or minibreath (Calder, 1970) between syllables (Fig. 1.3a). In canaries,

8　　　　　　　　　　　*1 Vocal mechanisms for avian communication*

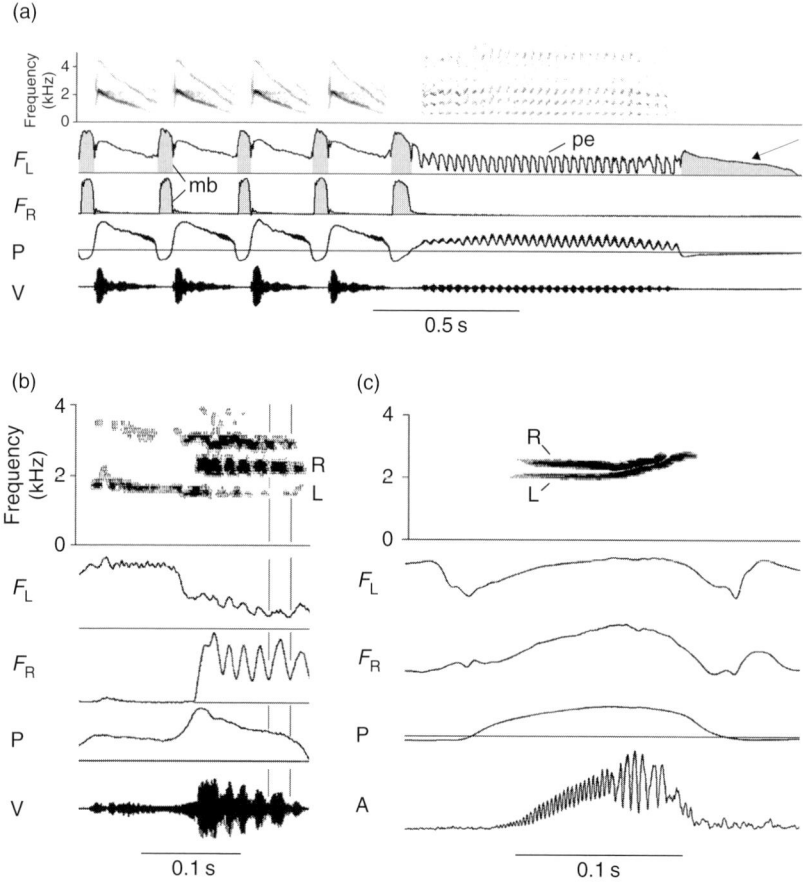

Figure 1.3. (a) Songbirds use different respiratory motor patterns during song, depending on syllable repetition rate. This segment of cardinal song includes the last 4 syllables of a phrase sung at about 2 syllables per second using a minibreath (mb) after each syllable. Note that both sides open for each mb during negative air sac pressure. These low-frequency syllables are sung entirely on the left side with the right side closed. This is followed by a trill at 30 syllables per second produced by pulsatile expiration in which the left side of the syrinx is repetitively opened while the right side remains closed and air sac pressure is positive during the entire phrase. Cardinals switch from a mb to a pulsatile pattern at about 16 syllables per second. Note the longer inspiration (arrow) immediately following the pulsatile portion of the trill. (b) A syllable from a catbird song. First half consists of fundamental and 2nd harmonic generated on left side of syrinx. Syllable becomes two-voice during last half when cyclically varying airflow controlled by syringeal muscles on right side produce a rapid AM. (c) Rapid amplitude modulation (A) generated by sustained simultaneous sounds in a brown thrasher two-voice syllable. F_L and F_R, rate of airflow through left and right sides of syrinx, airflow associated with positive pressure is expiratory; shaded flow is inspiratory (corresponds with

and probably many other birds, the volume of the minibreath is adjusted to match the volume of air exhaled to produce the syllable (Hartley and Suthers, 1989). This ensures that the bird does not run out of available air during long songs. There is, however, a physical limit on how quickly the respiratory movements necessary to produce a minibreath can occur. At some point, as syllable repetition rate increases, the interval between syllables becomes too short for a minibreath. Birds which exceed this repetition rate are forced to change to a different respiratory motor pattern of pulsatile expiration in which expiratory muscles maintain a positive subsyringeal pressure for the duration of a fast trilled phrase (Fig. 1.3a). The timing of each syllable within the phrase is controlled by the syringeal muscles that repetitively open and close the syrinx, and generate a syllable with each puff of escaping air.

1.3 Vocal learning

The vocal mechanisms of songbirds have received particular attention because they are one of the few groups capable of vocal learning, a capacity known to be shared only by humans, a few groups of birds, marine mammals, and perhaps some bats. Vocal learning in oscine songbirds involves a distinct system of interconnected song control nuclei in the brain (Nottebohm *et al.*, 1976; Nottebohm, 1991), which are absent in other passerines and non-passerines that lack vocal learning (Kroodsma and Konishi, 1991; Gahr *et al.*, 1993; Gahr, 2000). This song control system is usually best developed in males (Nottebohm and Arnold, 1976) and undergoes seasonal, hormonally controlled changes in size, being largest during the reproductive season (Brenowitz and Kroodsma, 1996). A detailed discussion of the song system is beyond the scope of this chapter. It receives input from the auditory system and includes two major circuits. One of these is the anterior forebrain pathway which is particularly important for song learning. The other circuit includes the motor pathway to syringeal muscles and a connection to respiratory premotor neurons. The brain stem, below the mesencephalon, motor pathway to the syrinx is generally similar for both vocal learners and non-learners (Wild, 1994), but songbirds have a major projection from the forebrain premotor nucleus, RA, to various

Caption for fig. 1.3. (cont.)
negative pressure). P: pressure in cranial thoracic air sac; V: oscillograph of vocalizations; A: amplitude envelope of rectified vocalization; R and L: sound generated on right or left side of syrinx, respectively; mb: minibreath; pe: pulsatile expiration. Horizontal lines indicate zero airflow and ambient pressure ((a) modified from Suthers and Zollinger, 2004; (b) modified after Suthers *et al.*, 1994; and (c) modified after Suthers and Goller, 1997).

song control brain stem nuclei and directly to syringeal motor neurons in the tracheosyringeal portion of the hypoglossal nucleus (nXIIts).

Vocal learning has also evolved in two other avian orders that are not closely related to songbirds: parrots (Psittaciformes) (Gramza, 1970; Todt, 1975; Pepperberg, 1981; Farabaugh et al., 1992; Farabaugh et al., 1994) and some hummingbirds (Trochiliformes) (Snow, 1968; Wiley, 1971; Snow, 1977; Baptista and Schuchmann, 1990).

Although vocal learning takes place in the brain, central plasticity in vocal motor programs is only useful if the peripheral vocal system can translate these motor patterns into sound. It has been suggested (Gaunt, 1983) that a peripheral prerequisite for vocal learning may be the possession of multiple pairs of syringeal muscles that increase the versatility of the oscine syrinx, and presumably increases the range of sounds that can be accurately reproduced during imitative learning. However, parrots also learn vocalizations, though they have only two pairs of syringeal muscles and a seemingly simple tracheal sound source. The details of psittacine vocal control are still poorly understood. Active modulation of the syringeal sound by the suprasyringeal vocal tract, for example by moving the fleshy tongue, may provide important added vocal versatility (Warren et al., 1996; Beckers et al., 2004). Central plasticity and peripheral vocal flexibility have apparently co-evolved in different ways as separate groups acquired vocal learning.

1.4 Controlled and intrinsic modulation of the vocal source

Bird vocalizations assume their complexity by modulating the amplitude and frequency of phonation. This is achieved through coordinated neuromuscular control of the respiratory, syringeal, and craniomandibular systems, but complex modulation can also arise spontaneously from intrinsic properties of the vocal production system. Songbirds have an additional way of achieving vocal complexity through independent motor control of the two separate sound sources in their duplex syrinx.

Neuromuscular systems are likely to be constrained due to limitations in how rapidly muscles are able to expand and contract. Thus, while most research has focused on how birds perceive relatively slow modulations, it is of interest that recent investigations have begun to suggest that avian auditory systems may also be well suited to perceiving the rapid fluctuations in both frequency and amplitude that may not always be under direct neuromuscular control (e.g. Langemann and Klump, 1992; Dooling et al., 2002).

1.4 Controlled and intrinsic modulation of the vocal source

Such fluctuations can be produced when relatively low-frequency spectral components in a harmonic signal do not radiate well or are filtered by the vocal tract (e.g. Dooling *et al.*, 2002), but can also be generated independently using mechanisms that remain unclear (e.g. Nowicki and Capranica, 1986a; Banta-Lavenex, 1999; Beckers and ten Cate, in press). For example, it is rather easy for humans to imitate the relatively slow frequency and amplitude modulation (FM and AM, respectively) patterns that exist in the eastern towhee's (*Pipilo erythrophthalmus*) onomatopoeic call (Nelson, 2004). Indeed, the relatively slow (<10 Hz) modulations that occur in this call are highly sinusoidal and are therefore easily whistled by humans. Unlike a human whistle, however, towhee calls are also modulated rapidly in frequency and amplitude (>450 Hz), and these rapid modulations give towhee calls a "rough" or "buzzing" acoustic quality that is not easily imitated when whistling. FM and AM span a wide range of periods (<2 ms to >1 s) in numerous vocalizations, and it is therefore interesting to consider how avian species are, in general, able to generate both relatively slow and surprisingly rapid modulations in their vocalizations.

1.4.1 Amplitude modulation

The amplitude of bird vocalization is modulated over different timescales and by different mechanisms. At the largest scale, syllables and notes or elements separated by silent intervals are produced through respiratory dynamics and syringeal motor action in both songbirds and non-songbirds (Brackenbury, 1978b, 1980; Hartley and Suthers, 1989; Beckers *et al.*, 2003a). In birdsong, notes are defined as the smallest units of continuous sound and repeated patterns of notes define a syllable. In the case of single-note syllables these terms are synonymous (Konishi, 1985).

Notes and syllables are produced by the coordinated action of respiratory and syringeal muscles (Brackenbury, 1978a; Gaunt, 1988; Suthers, 1997). Expiratory muscles provide the force to produce airflow through the syrinx (Hartley, 1990; Goller and Suthers, 1999) and the syringeal or tracheal muscles move the sound-generating membranes or labia into the stream of air flowing through the syringeal lumen (Suthers, 1990; Suthers *et al.*, 1999; Goller and Larsen, 2002). When a bird repeats a syllable the pressure pattern is also similar (Allan and Suthers, 1994; Suthers *et al.*, 1996) and different zebra finches use similar pressure patterns to produce similar syllables (Franz and Goller, 2002).

Non-oscines use a variety of mechanisms and syringeal structures to segment their vocalizations into notes or syllables. In swiftlets (*Aerodramus*) (Suthers and Hector, 1982) that lack intrinsic syringeal muscles and in oilbirds (*Steatornis carapensis*) that have but one intrinsic muscle (Suthers and Hector, 1985), this is accomplished by contraction of the sternotrachealis muscles, which stretch the trachea in a caudal direction and, by reducing tension on the syrinx, allow the external tympaniform membranes to fold into the lumen. Among sandpipers, the dunlin (*Calidris alpina*), has a prominent lateral labium, which Gaunt (1988) suggested might function as a syringeal valve and explain the pulsatile nature of the dunlin's call compared to the more continuous FM calls of most other members of the subfamily Calidrinae. In parrots, the syrinx of the monk parakeet (*Myopsitta monachus*) has a pair of flaps extending into the syringeal lumen just cranial to the LTM (Fig. 1.2a) that have been postulated to act as valves controlling airflow for phonation (Gaunt, 1988), but whose function is unknown. These flaps are absent in cockatiels where the antagonistic actions of a pair of syringeal muscles regulates the movement of the LTM into the syringeal lumen (Larsen and Goller, 2002).

In songbirds the labia are adducted into the syringeal lumen by contraction of the two dorsal syringeal muscles (Goller and Suthers, 1995, 1996b; Goller and Larsen, 2002). In addition to producing sound, the labia act as pneumatic valves that when strongly adducted close the ipsilateral bronchus and prevent phonation by stopping airflow (Fig. 1.1) (Suthers, 1990). At moderate or low syllable repetition rates adductor muscle tension is adjusted to let the high expiratory subsyringeal pressure push the labia slightly apart and force air through the slit between them. At the end of each syllable the adductors typically relax and the labia are withdrawn, opening the airway for a brief inspiration or minibreath (Hartley and Suthers, 1989; Goller and Suthers, 1996b).

Not all notes are produced this way, however. Separate notes can also be produced by active opening and closure of the syrinx against a relatively high and stable subsyringeal respiratory pressure, using the pattern of pulsatile expiration described above (Hartley and Suthers, 1989; Suthers, 1997; Suthers and Goller, 1997; Beckers *et al.*, 2003a). Such AM has been termed gating. By increasing the contraction of the adductor muscles (Goller and Suthers, 1996b) to completely close the syringeal lumen and stop airflow, phonation is interrupted to produce silent intervals between notes within a syllable or the short intervals between syllables sung at a high repetition rate. In gating, closure does not have to be complete, or happens only momentarily, which leads to a sound waveform that is not separated by silent intervals, but is continuous with a time-varying amplitude envelope. AM patterns attributable to syringeal regulation of airflow can reach 125 Hz in brown thrashers (Fig. 1.3b) (Suthers *et al.*, 1994; Goller and Suthers, 1996a).

In songbirds, such as the brown thrasher, AM is occasionally generated by the linear interaction, or beating, of tonal sounds having a similar amplitude but differing slightly in frequency, which originate from opposite sides of the syrinx (Fig. 1.3c) (Suthers and Goller, 1997).

Another mechanism that can cause AM is a nonlinear interaction between two or more vibrating structures in the syrinx. Such structures could be located in the two different halves of the duplex syrinx in songbirds (Nowicki and Capranica, 1986a, b), but interactions between two opposing membranes or labia in a single vocal structure may also lead to this type of AM (see Section 1.6). Some vocalizations of parrots (budgerigars) show evidence of AM by the nonlinear interaction of separate carrier and modulating frequencies (Banta-Lavenex, 1999), though it is not clear what oscillators are involved.

1.4.2 Frequency modulation

1.4.2.1 Syringeal muscles

The fundamental frequency in bird vocalization is determined by the resonance frequency of the vocal labia or membranes, which in turn is dependent on their geometry, density, and the tension applied to them (Fletcher, 1992). In songbirds (Miskimen, 1951; Goller and Suthers, 1996a; Larsen and Goller, 2002), and possibly parrots (Larsen and Goller, 2002), FM is achieved by the action of specialized musculature. The activity of the ventral syringeal muscles appears to provide the main mechanism for controlling the tension of the labia in brown thrashers, and accurately corresponds to FM in the sound (Fig. 1.4a and b) (Goller and Suthers, 1996a).

1.4.2.2 Air sac pressure

In ring doves, FM is of two types: continuous FM, characterized by gradual change of frequency in time, and instantaneous frequency jumps. Continuous FM shows a strong correlation with changes in air pressure in the interclavicular air sac, an air cavity in which the syrinx is located (Fig. 1.4c and d) (Beckers et al., 2003a). Presumably the changing air sac pressure directly modulates the tension in the LTMs, the sound source in doves.

1.4.2.3 Other mechanisms

Songbirds are also able to generate surprisingly rapid FM in excess of 500 Hz that often correspond with equally rapid AM (Fig. 1.5) (Nelson, 2004). Neither syringeal muscle contractions nor changes in air sac pressure are likely to produce these rapid modulations and, while the syringeal mechanisms that underlie these modulations remain unclear, they may be generated by the same nonlinear dynamics that function to generate rapid AM in vocalizations.

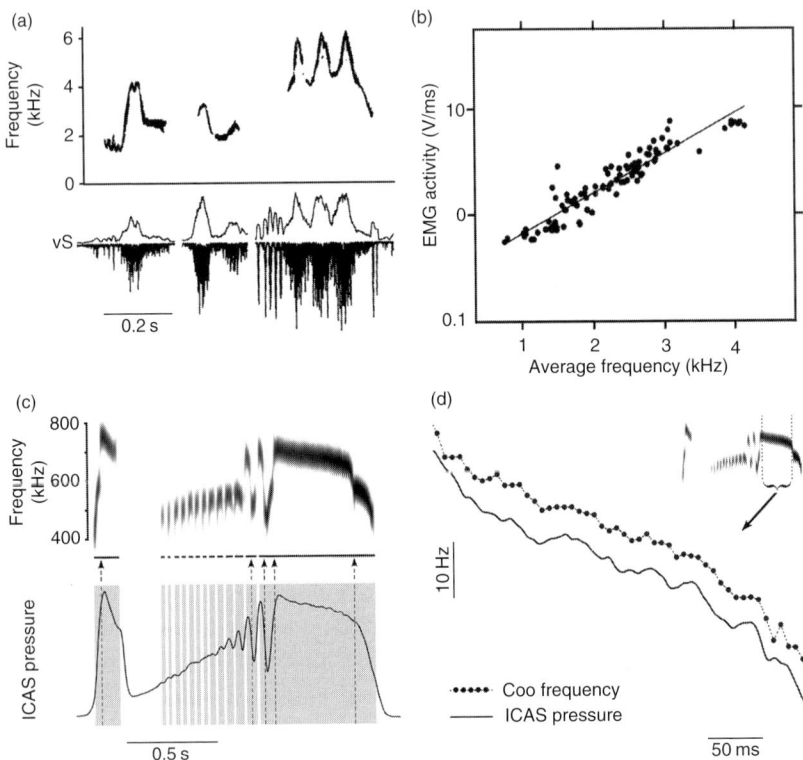

Figure 1.4. Control of fundamental frequency. (a) EMG amplitude of ipsilateral ventral syringeal muscle (vS) is positively correlated with FM of sounds generated on the ipsilateral side. EMG is shown integrated envelope (upward, time constant 5 ms) and rectified (downward). (b) Amplitude of EMG activity is correlated exponentially with fundamental frequency of ipsilaterally produced sounds. EMG activity was averaged over segments of syllables having a relatively constant frequency ((a) and (b) modified after Goller and Suthers, 1996a). (c) FM in the ring dove's coo is correlated with pressure changes in the interclavicular air sac, which surrounds the syrinx. This correlation does not hold for frequency jumps, indicated by vertical arrows and broken lines. (d) An expanded portion of coo showing detailed correlation of FM (dotted line) and interclavicular air sac pressure (smooth line). vS: m. syringealis ventralis; ICAS: interclavicular air sac (c) and (d) modified after Beckers *et al.*, 2003a).

1.5 Lateral independence in the bipartite syrinx

The temporal and spectral diversity of oscine song is in part due to the separate motor control of the left and right sides of the syrinx. Although both sides are exposed to a similar subsyringeal respiratory pressure, the muscles of each side

1.5 Lateral independence in the bipartite syrinx

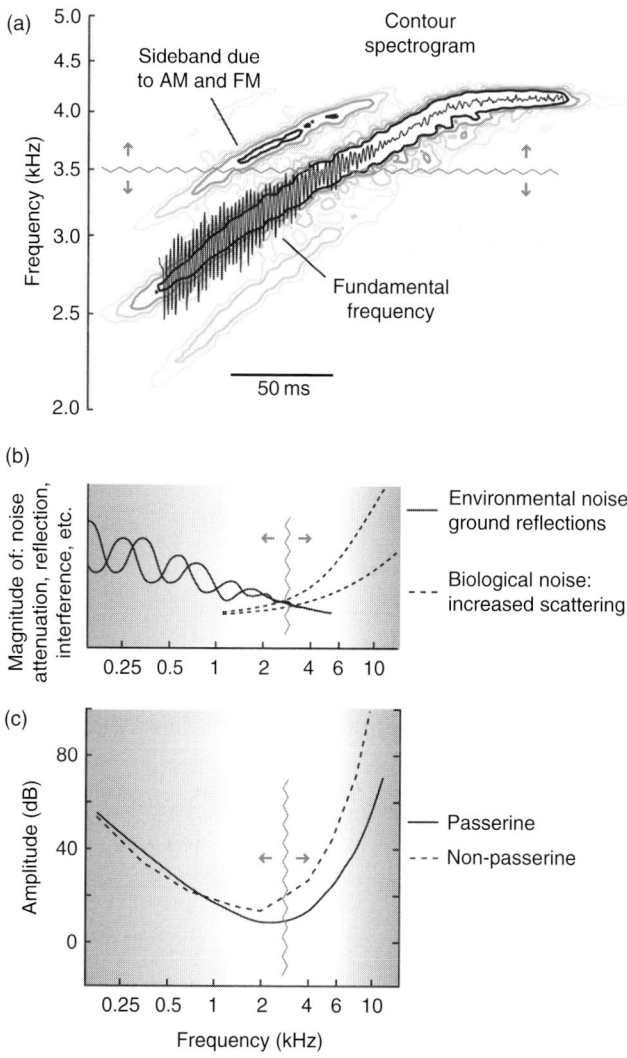

Figure 1.5. Vocalizations are often produced with sound frequencies that propagate well over distance (Wiley and Richards, 1978) and with frequencies that can be detected after attenuation (Wright et al., 2003). Birds may also often communicate within separate sound frequency channels, defined to each side of a mid-range frequency, if "sound windows" are bounded by different physical forces. (a) Contour spectrogram of an eastern towhee call overlaid with a trace tracking fundamental frequency. AM (not illustrated) and FM often have greater amplitude below ~3.5 kHz (solid lines; Nelson, 2004). (b) Schematic illustrating how relatively high sound frequencies often experience strong attenuation (dashed lines) while lower frequencies are often rippled due to ground reflections (Nelson, 2003). Background noise levels also often increase both below 1–2 kHz and above 4–5 kHz. (c) Mean audibility functions for passerines (20 species) and non-passerines (15 species) as reported by Dooling et al. (2000).

are innervated by motor neurons of the ipsilateral hypoglossal nucleus. Within the constraints of the underlying respiratory motor pattern, the songbird syrinx can thus potentially function as two independent vocal organs, driven by separate song motor programs. Sound production can be limited to one side or switched from side to side by controlling the valve-like action of the fully adducted labia which can silence one side of the syrinx by stopping air from flowing through it (Suthers, 1990, 1997).

Even though the vocal motor patterns to each side of the syrinx are often different, they are coordinated with temporal precision. The neural mechanisms responsible for coordination between the two sides of the syrinx are still poorly understood. Descending connections in the brain stem are predominantly ipsilateral, though a number of contralateral pathways exist. There is, however, no known cross connection between the telencephalic song nuclei (Wild, 1994, 1997) and no callosal connections between song nuclei on opposite sides of the brain (DeVoogd and Nottebohm, 1981; Nottebohm et al., 1982; DeVoogd et al., 1991). The strength of connections between RA and some contralateral nuclei including nXIIts and retroambigualis (RA) differ between species of songbirds in a way that might be related to differences in their bilateral syringeal motor patterns during song (Wild et al., 2000).

1.5.1 Unilateral dominance versus bilateral parity

One measure of lateral independence is the relative contribution each side of the syrinx makes to the total song repertoire; that is, its position along a continuum from strong unilateral dominance to bilateral parity. The most extreme example of functional lateralization in song production occurs in the Belgian waterslager strain of canary (*Serinus canaria*) that has been inbred for particular qualities of its song. Male waterslagers sing about 90% of their syllable repertoire with the left syrinx (Nottebohm and Nottebohm, 1976). This discovery together with similar, but less extreme, tendencies for left syringeal dominance in song production by several other species (e.g. Nottebohm, 1971; Nottebohm and Nottebohm, 1976; reviewed by Suthers, 1997) led to speculation that a left motor dominance might be a universal property of birdsong. However, subsequent studies on other songbirds have failed to find a similar overriding lateral dominance, even in the conspecific outbred domestic or common canary. In domestic canaries both sides of the syrinx produce approximately equal shares of the notes and syllables in the bird's repertoire. Some of these syllables are sung on the left, some on the right, and in some the bird switches phonation from one side to the other to produce syllables consisting of successive notes from opposite sides of the syrinx (Suthers et al., 2004b). Canaries thus provide evidence that selection for particular song characteristics can produce genetic strains with fundamentally different motor patterns of syringeal lateralization.

1.5.2 Two-voice vocalizations

Some birds sing separate sounds simultaneously on opposite sides of the syrinx that are harmonically unrelated and independently FM (Greenewalt, 1968; Stein, 1968). These two-voice syllables (Fig. 1.6a) often impart a dissonant quality to the song, but the behavioral significance of two-voice syllables in songbirds has not been studied.

In his analysis of recorded bird vocalizations, Greenewalt (1968) reported that he found two-voice vocalizations in "almost every avian family" including non-passerine species as well as songbirds. Many non-songbirds have a duplex syrinx, but the extent to which they can independently control sound production on each side of this organ is unclear. Aside from the reduction in number, or even total absence, of intrinsic syringeal muscles, the left and right tracheosyringeal nerves in at least some species of non-songbirds form a plexus so that each nerve innervates both sides of the syrinx. In Greenewalt's published examples of two voices by non-songbirds the overlap between "voices" is very short (about 7–30 ms) making it difficult to exclude the possibility that it is due to environmental reverberation at the end of the first tone that makes it appear to continue beyond the start of the second tone. Should this be the case, at least some examples assumed to be two simultaneous voices might represent a frequency jump by a single oscillator (see Section 1.6). There is also the possibility that two simultaneous fundamental frequencies could arise passively as a result of lateral anatomic asymmetries that produce different resonant frequencies for members of the pair of oscillators forming a single sound source (e.g. the labia of a single bronchus or LTMs of a tracheal syrinx), in which case they are examples of biphonation rather than two voices.

Penguins provide a notable exception to the paucity of information about two voices in non-oscines. Aubin et al. (2000) showed that the two-voice quality of display calls in emperor penguins (*Aptenodytes forsteri*) is important in parent–chick recognition. In this call, two independent fundamental frequencies are simultaneously produced, the interaction of which results in a beating amplitude pattern. Synthesized calls with only either one of the two components do not evoke any response from parents or chicks. Sound propagation experiments described in the same study demonstrated that the beating amplitude pattern is more robust to change when transmitted through a penguin colony than an AM signal. A two-voice system may thus have evolved in this species to enhance individual recognition over longer distances in a large crowd of conspecifics.

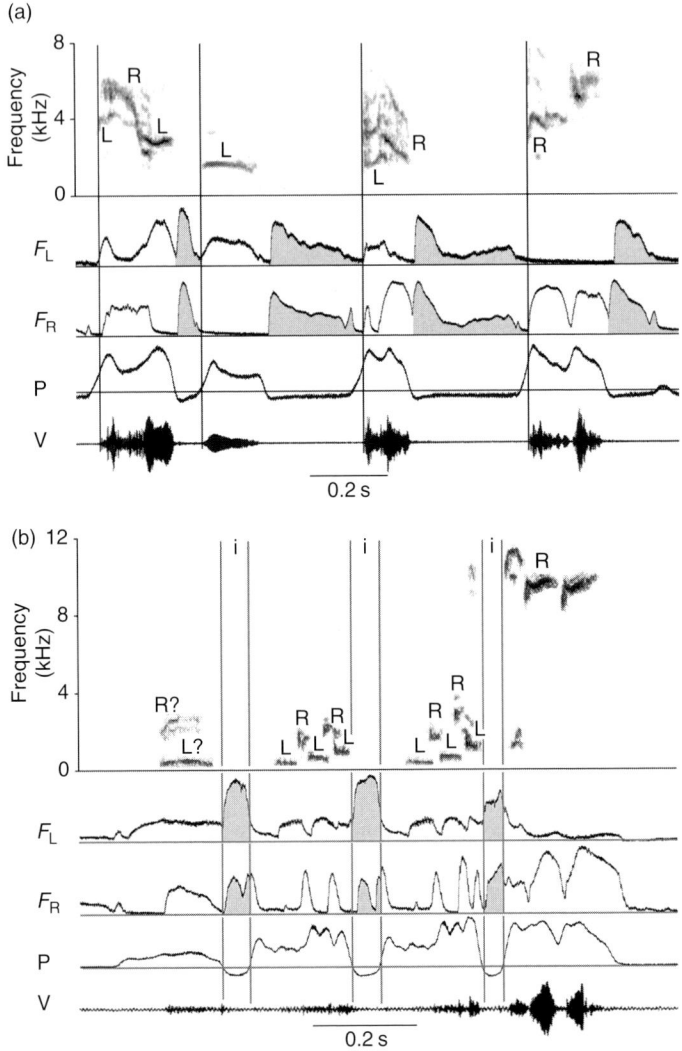

Figure 1.6. (a) The song of the gray catbird is highly variable with prominent FM, two-voice syllables in which each side of the syrinx produces independently modulated frequencies that are not harmonically related (first syllable) and frequent switching of sound production between sides of the syrinx. (b) Song of the brown-headed cowbird begins with three-note clusters separated by an inspiration. First note in each cluster is sung on left side of syrinx and following notes alternate between sides. By changing sides between notes the bird is able to achieve abrupt frequency changes. Last note cluster is followed by a final whistle, which is always sung on the right side. Songs from both of these species illustrate the general trend for the higher frequencies to be produced on the right side. Top panel is

1.5.3 Patterns of song lateralization and acoustic specialization

Even in species in which both sides of the syrinx contribute about equally to the total song repertoire, there is a lateralization of syringeal function such that each side is specialized for certain aspects of the acoustic performance. For example, in species that have been studied, the two sides of the syrinx are specialized to cover partially different ranges of fundamental frequency, with the right syrinx producing high frequencies and the left producing low frequencies, usually with considerable overlap for mid-range frequencies (Suthers, 1999). Certain aspects of syllable "morphology" may also be lateralized. In some species, such as brown thrashers, notes with rapid FM are usually produced by the right side of the syrinx. The extent to which these lateral differences in frequency range and the production of FM components are due to peripheral differences in the physical properties of the two sound generators or reflect central specializations expressed as lateral differences in motor output from the brain needs further study. In either case, this division of labor takes advantage of the duplex syrinx to increase the bandwidth and diversity of song.

Different species of songbirds have characteristic ways of using their bipartite syrinx to achieve their particular style of song. Two-voice syllables are common in the songs of gray catbirds (*Dumetella carolinensis*) and brown thrasher (Fig. 1.6a) (Suthers *et al.*, 1994, 1996). Brown-headed cowbirds (*Molothrus ater*) alternate sides of the syrinx during the note clusters at the beginning of their song but sing the high-frequency final whistle at the end entirely on the right side (Fig. 1.6b) (Allan and Suthers, 1994). Northern cardinals also often use the two sides of their syrinx in sequence, but connect the fundamental frequency from each side to achieve a wider bandwidth than could be produced by either side alone. In both upward and downward sweeping syllables, frequencies below about 3.5 kHz are sung on the left and higher frequencies are sung on the right side of the syrinx (Suthers, 1997, 1999).

Canaries take advantage of the acoustic potential of their duplex syrinx when producing a special class of high repetition rate, broad-band syllables, called A-syllables that are particularly effective in eliciting copulation solicitation displays from receptive females. Experiments have shown that the

Caption for fig. 1.6. (cont.)
spectrogram of song. F_L and F_R: rate of airflow through the left and right side of the syrinx, respectively; P: cranial thoracic air sac pressure; V: oscillograph of vocalization; R: right-side component or syllable; L: left-side component or syllable; i: inspiration. Horizontal lines indicate zero airflow and ambient pressure. Inspiratory airflow is shaded ((b) modified after Allan and Suthers, 1994).

perceptually important parameters of these "sexy" syllables include a high repetition rate ($>15\,s^{-1}$), a broad bandwidth including low frequencies below 2 kHz, and the presence of two notes (Vallet and Kreutzer, 1995; Vallet *et al.*, 1998; Draganoiu *et al.*, 2002). Domestic canaries achieve the required bandwidth, including the low frequencies, by rapidly switching phonation from one side to the other to generate an FM note on each side. Although the basis of female preference for A-syllables and their contribution to breeding success remain controversial, bilateral motor skills are clearly an important part of their production (Suthers, 2004).

There appears to be an obligatory relationship between the syringeal motor pattern and the acoustic characteristics of each species' song. When juvenile northern mockingbirds (*Mimus polyglottos*), a vocal mimic, are tutored with these songs and subsequently sing them as adults, they usually use the same pattern of syringeal lateralization as did the tutor. Furthermore, the fidelity of the acoustic copy is correlated with the degree of similarity between the motor patterns of the tutor and mimic. The fact that both the tutor and mimic independently arrive at a similar production mechanism through trial and error motor learning, suggests that different species have evolved their distinctive song motor patterns as the best way to achieve the species-specific acoustic effects (Zollinger and Suthers, 2004).

1.6 Nonlinear dynamics in vocalizations

In addition to the mechanisms described above for actively controlling phonation, there is another way in which birds can generate vocal complexity by modulating their syringeal sound source. The primary sound generators in birds, vibrating labia or membranes, are in fact *nonlinear* oscillators, and it has been known since the 1960s that coupled nonlinear oscillators in general can also exhibit complex dynamics *without* any external, complex control. Although the conceptual and mathematical tools of nonlinear systems theory have been very fruitful in a wide variety of studies in such diverse fields as ecology, physics, and economics, their relevance to bird vocalization has only been recognized recently (Fee *et al.*, 1998; Fletcher, 2000; Beckers and ten Cate, in press). The key point about nonlinear dynamics in vocalization is that a relatively simple system, such as two coupled vibrating membranes, can give rise to a series of different states of oscillatory behavior, some of which are very complex. Complexity emerges spontaneously from intrinsic properties of the sound source, and requires no external complex control.

Nonlinearities in the avian syrinx probably arise from syringeal pressure–flow relations, stress–strain relations in oscillating tissue, and the collision between

1.6 Nonlinear dynamics in vocalizations

vibrating labia or of membranes into each other. Systems consisting of two or more coupled, nonlinear oscillators are known to exhibit qualitatively different types of dynamics that correspond to three different attractor types in nonlinear dynamics theory: limit cycle, torus, and chaotic attractors. An attractor can be thought of as the state to which the dynamics of a system "settles" as time goes by.

In the simplest state, a limit cycle, the behavior of the dynamical system is periodic; that is, it repeats itself in time (Fig. 1.7a). This is a state in which the syrinx vibrates in a regular fashion; the resulting sound is called *harmonic sound*. Spectrographic representation of harmonic sounds shows acoustic energy in discrete bands, composed of a fundamental frequency that is inversely related to the vibration period, and zero or more integer multiples of this fundamental frequency. This is the type of phonation that is most often studied in birds. Sometimes, additional spectral components that are called subharmonics appear in the harmonic stack, usually at multiples of 1/2 or 1/3 of the fundamental frequency. Subharmonic regimes correspond to a different attractor type, a folded limit cycle, and are known to occur in mammalian vocal production when one

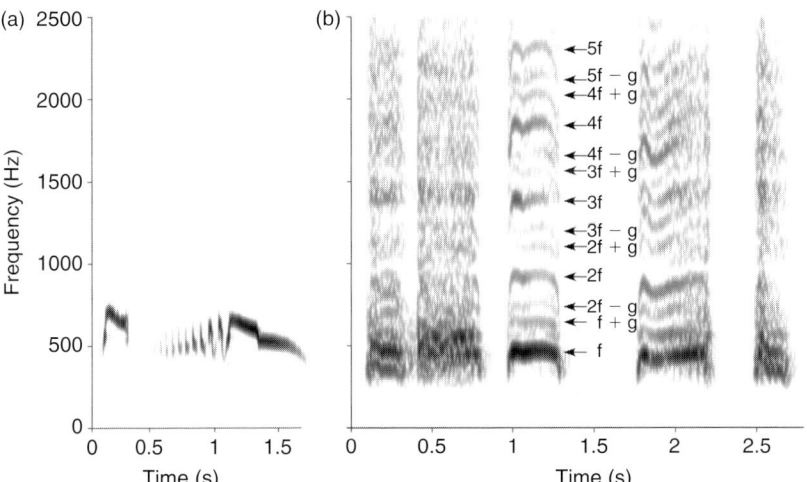

Figure 1.7. Tonal versus chaotic and biphonation vocalization in two species of turtle-dove. (a) Coo of the African collared dove, *S. roseogrisea*. (b) Part of a coo bout of the oriental turtle-dove, *S. orientalis*, in which syllables show both chaotic and biphonation regimes. In the third element, a harmonic signal (f, first five harmonics shown) is modulated by a lower-frequency component (g), which causes a side-band pattern around each harmonic of f. Note that in the fourth element, the regime changes from biphonation (or perhaps subharmonic) to chaotic. The very different types of phonation may be a manifestation of different attractor states of the same nonlinear dynamical system, and hence the large interspecific differences in phonation may be caused by small differences in underlying parameters.

of the two vocal folds oscillates at exactly half or one-third the frequency of the other (Wilden *et al.*, 1998).

In a more complex attractor, a torus, the dynamical system exhibits two oscillations that are not harmonically related, say f and g. In nonlinear systems, this also gives rise to linear combinations of these frequencies: $nf + mg$, where n and m are integer numbers (Fig. 1.7b). In the mammalian literature this has been termed biphonation (Wilden *et al.*, 1998). Biphonation can be identified in a spectrogram as simultaneous, non-parallel energy bands, and occurs for example when the two vocals folds that are normally entrained to vibrate harmonically at the same frequency, become desynchronized and vibrate at their own individual, different frequencies (Neubauer *et al.*, 2001). Some forms of AM in birdsong may be due to such dynamics. Biphonation should not be confused with "two-voice" sound production in birds. The former arises from the dynamics of a single sound source (one set of vibrating labia or membranes), while the latter arises from two independent sound sources (two independent sets of vibrating labia).

In a very complex attractor called a chaotic state, or "chaos", oscillations are very irregular. This results in very complex sounds, the spectrographic representation of which exhibits irregular patterns of energy over wide frequency bands, although some residual energy may still be concentrated in discrete bands. Chaotic sounds differ fundamentally from acoustic noise but they can resemble each other superficially in spectrograms. The complexity of chaotic sounds is rooted in the intrinsic behavior of a low-dimensional dynamical system, and is completely deterministic. Noise, in contrast, originates from a very high-dimensional system of random components. Noise is a normal component of human speech, and is not produced by vibrating vocal folds but by a vocal tract constriction that creates turbulence in the flow of air (Stevens, 1998). Deterministic chaos, in contrast, is produced by vibrating vocal folds, in pathologic human speech, and in the normal vocalizations of other mammals (Wilden *et al.*, 1998; Fitch *et al.*, 2002). Although it is not uncommon to see the quality of various bird vocalizations being described in the literature as "noisy", it remains to be demonstrated whether or not the avian syrinx or other vocal structures actually do produce noise sounds. Fletcher (2000) argues that syllables with a chaotic structure may be part of the vocal repertoire in many birds. Possibly, most if not all vocalizations described as noisy are in fact low-dimensional chaotic sounds.

Besides the typical attractor states, nonlinear dynamic systems can also exhibit another type of characteristic behavior, namely sudden transitions from one dynamical state to another, which are called *bifurcations*. Typical examples found in songbird and non-songbird vocalizations are the sudden

1.6 Non-linear dynamics in vocalizations

transition from harmonic to subharmonic regimes (period-doubling bifurcation), from harmonic to biphonation (secondary Hopf bifurcation), or from (sub)harmonic to chaotic regimes (Fee et al., 1998; Beckers and ten Cate, in press). Also sudden transitions within one type of attractor are bifurcations. Examples of this are mode-locking transitions, in which the harmonic vibrations instantaneously jump from one frequency to the other. Mode locking occurs when a nonlinear interaction constrains two oscillating components of a system to maintain a small integer ratio of frequencies. Mode-locking transitions may arise when the characteristic frequency of one component is changed relative to the other, and the oscillation frequency suddenly jumps to achieve a new stable integer ratio. Such a mechanism has been suggested as an explanation for frequency jumps in the song of zebra finches (Fee et al., 1998) and turtle-doves (Fig. 1.8) (Beckers et al., 2003a; Beckers and ten Cate, in press).

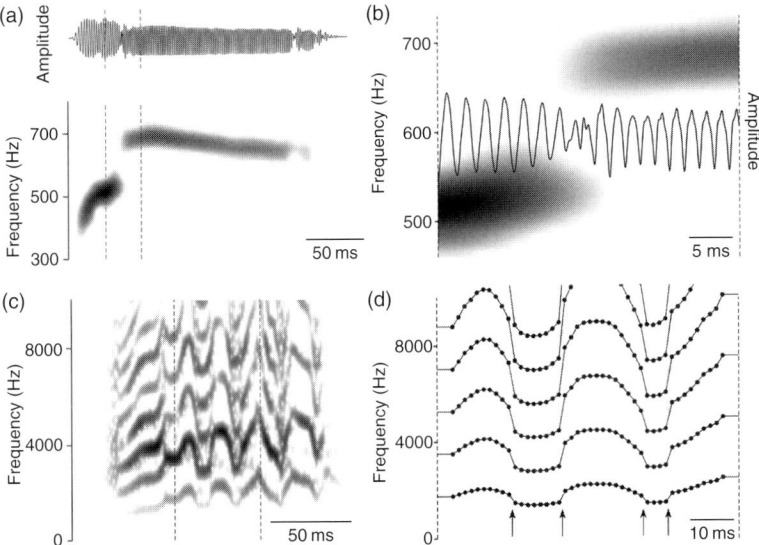

Figure 1.8. Examples of frequency jumps in bird vocalizations that may be caused by mode-locking transitions of nonlinear coupled oscillators. (a) Oscillogram (top) and spectrogram (bottom) of coo syllable of a ring dove, in which f_0 jumps from about 525–675 Hz within 5 ms. (b) A 5-ms segment including this jump (time frame is indicated by dashed lines in (a), in which the oscillogram is superimposed on the spectrogram. The oscillogram shows that sound is not interrupted during frequency jumps. (c) Multiple frequency jumps are present in a contact call of a monk parakeet. (d) Plot of the harmonic frequencies in a 60-ms segment of the contact call. Four frequency jumps are indicated by arrows. Note that in jumps downward, the 5th harmonic before the jump is continuous with the 6th harmonic after the jump, respectively. In jumps upward the 5th harmonic becomes continuous with the 4th, or the 6th with the 5th.

It should be noted that from a nonlinear dynamics point of view, "nonlinear phenomena" in bird vocalization, such as subharmonics and chaos, are not fundamentally different from, or special with respect to, "normal" pure-tone or multi-frequency harmonic vocalizations. All types of phonation, ranging from pure-tonal to chaotic, can be viewed as different states of the same nonlinear dynamical system. This has two very interesting implications. First, as even very simple nonlinear systems can exhibit very complex oscillatory behavior, part of the complexity in bird vocalization may be intrinsic to syringeal dynamics, and may thus not require any complex neuromuscular control (Fee *et al.*, 1998). Second, as one of the key features of nonlinear systems is that small and gradual changes in control parameters can cause large, sudden, and qualitative changes in dynamics, seemingly strong (inter- or intraspecific) differences in bird vocalizations may be a consequence of minor, quantitative changes in driving parameters or morphology. Acoustic transitions within zebra finch syllables are accompanied by rapid beak movements and large gapes, suggesting feedback between beak movements and the vibration of the syringeal labia (Goller *et al.*, 2004). Beckers and ten Cate (in press) recently hypothesized that such easy switching between dynamical states may play a role in the evolution of species-identifying coo vocalizations in turtle-doves (genus *Streptopelia*). Presence of bifurcations within species' song shows that changes in dynamics can be readily achieved, and may hence also be at the basis of vocal differences between turtle-dove species. Seemingly strong interspecific differences, for example, between the tonal coos of *S. risoria* and the chaotic ones of *S. orientalis* or *S. tranquebarica*, might thus have been a consequence of relatively small changes in underlying vocal mechanisms, without any need of large modifications in syringeal structure or control mechanisms (Fig. 1.7).

1.7 Vocal tract filtering

Vibratory dynamics of the syrinx are an important causal factor with respect to the characteristics of the sound that is produced. However, the sound that is generated by syringeal vibrations is also dependent on the specific acoustic impedance at the syrinx, which is determined by the resonance behavior of surrounding vocal tract air cavities such as the bronchi and trachea. Moreover, as sound propagates through the vocal tract it will be filtered by changes in tract diameter, which cause acoustic impedance changes, and the sound eventually radiates from the beak or through skin to the environment, which introduces additional changes in sound characteristics (Fletcher and Tarnopolsky, 1999). So the avian sound production system should be viewed as a source–filter system, in which the syrinx provides the necessary source vibrations and the filter determines the amplitudes of the sound frequencies produced.

Although acoustic filtering of syringeal source vibrations is physically inevitable, it has long been debated how important vocal tract filtering actually is in shaping specific time-frequency patterns in bird vocalization. Two questions in particular are related to this issue: is vocal tract filtering involved in pure-tone ("whistled") bird vocalization and does vocal tract filtering cause formant patterns that shape the amplitude spectrum of broad-frequency vocalizations?

Vocal tract filtering as a mechanism for pure-tone bird vocalization was originally suggested by Nowicki (1987), who tested the hypothesis that the syringeal acoustic source consisted of multi-frequency harmonic oscillations, which are filtered to a pure-tone by a vocal tract bandpass filter that responds only to one harmonic. In an elegant experiment he recorded several species of songbirds in a helium-enriched atmosphere. Resonant frequencies of air cavities shift upward in such an environment due to the increased velocity of sound, and he observed that harmonic overtones were added to song elements that appeared as pure-tones in normal atmosphere. Under the premise that the changed gas composition of the environment does not appreciably affect the dynamics of the syringeal source, this experiment provides evidence that resonant properties of air cavities of the vocal tract play an important role in pure-tone bird vocalizations. Brittan-Powell *et al.* (1997) concluded after similar experiments in budgerigars, however, that in this species vocal tract filtering only has a slight effect on the pure-tonality of its contact call vocalization, which suggests that its narrow-band frequency spectrum is produced as such by the syringeal source.

Two mechanisms have been proposed to explain the helium results in songbirds:

1. The syrinx acts as a vibrating valve that generates a multi-frequency harmonic source sound that is filtered to a pure-tone by a vocal tract (termed the "source–filter model").
2. Vocal tract resonances couple with the vibrating syrinx, suppressing the normal production of harmonic overtones at this source (termed the "soprano model" due to the analogy with mechanisms involved in human soprano singing (Nowicki and Marler, 1988)).

Recently, Beckers *et al.* (2003b) recorded sound signals close to the syringeal sound source during spontaneous, pure-tone vocalizations of two species of turtle-dove (non-songbirds), using tracheal thermistors and air sac cannulae. Their results show that in these species the syringeal sound source produces a multi-frequency harmonic sound, which is filtered to a pure-tone before it radiates from the animal.

The tonal, low-frequency vocalizations of some doves place special demands on vocal tract filters. Doves and pigeons coo with their beaks and nares closed (Gaunt *et al.*, 1982). Studies on ring doves (Riede *et al.*, 2004) show that during a coo air moves through the syrinx and inflates the esophagus, which becomes part of the suprasyringeal vocal tract (Fig. 1.9). The harmonics that are present in the syringeal sound are successively attenuated by the trachea and esophagus. Riede *et al.* hypothesized that tracheal resonance is tuned to the low fundamental frequency (*ca.* 500–600 Hz) of the coo by varying the opening of the glottis, located at the upper end of the trachea. The adducted syringeal tympaniform membranes at the lower end of the trachea cause it to

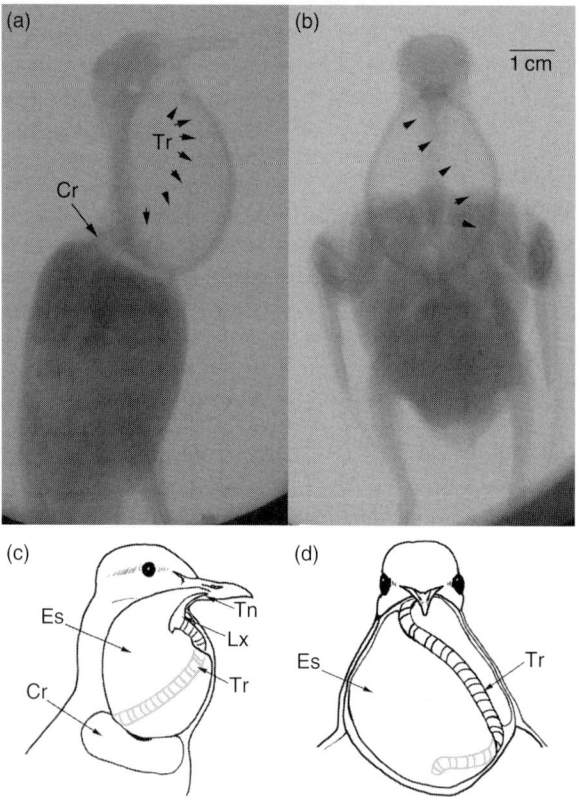

Figure 1.9. Cineradiograph of a cooing ring dove during maximum inflation of the esophagus: ((a) and (c)) lateral view and ((b) and (d)) frontal view. The course of the trachea around the esophagus is indicated by arrowheads. The crop is a separate cavity from the inflated upper esophagus. Cr: crop; Es: esophagus; Lx: larynx; Tn: tongue; Tr: trachea. Drawings by S.A. Zollinger. Modified after Riede *et al.* (2004).

act acoustically like a stopped tube. As sound from the trachea radiates into the environment through the wall of the inflated esophagus, and its thin layer of overlying skin, these elastic tissues form the vibrating mass of a Helmholtz resonator. A mathematical model of this system (Fletcher *et al.*, 2005) indicates that it may have evolved as a way of amplifying the very low-frequency fundamental.

Both the soprano model and the source–filter model require an adjustable vocal tract filter in birds that modulate their pure-tone vocalizations over more than an octave, as do many species of songbird. Changes in filter resonance frequency could in principle be achieved by changes in tracheal length, glottis opening, tongue position, expansion of the oropharyngeal cavity by hyoid movement, and beak gape. Indeed, beak gape appears to be positively correlated with fundamental frequency in tonal notes of swamp sparrows (*Melospiza georgiana*) and white-crowned sparrows, *Zonotrichia leucophrys* (Westneat *et al.*, 1993), as well as in zebra finches (Goller *et al.*, 2004). In northern cardinals this correlation is strongest below about 3.5 kHz (Suthers and Goller, 1997). Disruption of normal beak movements in these sparrows, canaries, or cardinals results in predicted frequency-dependent amplitude changes of their song (Suthers and Goller, 1997; Hoese *et al.*, 2000). This supports the idea that dynamic vocal tract motions are a necessary feature of song production in these species.

In the case of broad-frequency phonation vocal tract filtering causes emphasis in certain frequency ranges and de-emphasis in others. Regions of emphasis are called formants, and the pattern of formant frequency dispersion depends on the length of the trachea, and can in principle be used as an indicator of body size by receivers. Fitch (1999) hypothesized that this has lead to the evolution of tracheal elongation in a number of bird species, in order to exaggerate the apparent size of a vocalizing bird.

Interestingly, a vocal tract filter could be modulated independently from the syringeal source, and create a time-varying formant pattern in a broad-band source sound. This is analogous to the human speech production (Fant, 1970), and it has in fact been suggested that this is the mechanism by which parrots mimic human speech. Warren *et al.* (1996) used video imaging techniques (including infrared and X-ray radiography) to examine associations between formant frequencies and vocal tract movements in a gray parrot (*Psittacus erithacus*) that produces English speech. They found that formant frequencies are correlated with horizontal tongue position in the beak (and a number of other correlated variables of vocal tract movement, such as protracted trachea) in the two vowels /i/ and /a/. In a recent study, direct measurements in monk parakeets showed that their vocal tract has strong formant patterns that can be modulated significantly and in complex ways by tongue movements (Beckers *et al.*, 2004). Taken together, these studies offer strong support for a role of lingual articulation in parrot vocalization analogous to speech production.

1.8 Production, propagation, and perception

An understanding of how birds might use vocalizations when communicating requires an understanding of how vocalizations are perceived. In addition, birds often communicate over relatively long distances under noisy natural conditions and, as a consequence, it may also be important to consider how vocalizations propagate over distance.

The simplest illustration of how vocal production, propagation, and perception might coordinate with each other occurs when vocalizations are produced with energy concentrated within frequency bands that can best be detected (Wiley and Richards, 1978; Wright *et al.*, 2003). Many oscine species, for example, produce vocalizations with energy between 2 and 5 kHz (Wiley and Richards, 1982; Wiley, 1991), and birds in general are also most sensitive to these same sound frequencies (Dooling *et al.*, 2000). These same sound frequencies also generally propagate "well" over long distances in natural habitat (e.g. Wiley and Richards, 1978; Ryan and Brenowitz, 1985; Klump, 1996; Nelson, 2003) and, as a consequence, many vocalizations appear to be well adapted for long-distance communication (Fig. 1.5).

Environmental factors that influence sound propagation may limit the upper and lower frequencies that birds produce in vocalizations. Nevertheless, detection alone may often be insufficient for communication and vocal discrimination is expected to be a more difficult task (Lohr *et al.*, 2003). Furthermore, many vocalizations vary in both amplitude and acoustic structure (e.g. Cynx *et al.*, 1998; Manabe *et al.*, 1998; Nelson, 2000), and the modulations that are necessary for species or individual recognition cannot always propagate with maximum efficiency over distance.

Sound frequencies that propagate well over distance are sometimes described as sound windows. Sound windows are not bounded on all sides by the same physical forces. Wind noise and ground reflections, for example, can constrain communication below approximately 3.5 kHz while noise from insects and increased attenuation due to scattering can often limit communication above this frequency. These distinct interfering forces tend to increase to each side of $\sim$3.5 kHz. An alternative "strategy" for minimizing this interference may therefore be to produce vocalizations over a relatively wide range of frequencies so that the information in them can be discriminated over the widest range of environmental conditions (Nelson, 2002, 2003, 2004).

Whether birds concentrate energy in vocalizations so as to maximize communication distance or modulate vocalizations so that they span different sound frequency ranges remains unclear. However, at least several songbirds appear to produce sound frequencies to each side of 3–4 kHz using separate sides of their syrinx (see above). In addition, these two sound frequency ranges appear to radiate differently and may be filtered differently by the vocal tract (Larsen and Dabelsteen, 1990).

Many "quiet" vocalizations do not seem to be structured for maximum transmission distance (e.g. Dabelsteen *et al.*, 1998) and several species produce exceptionally low or high sound frequencies in their vocalizations (e.g. doves and cowbirds). These species do not appear to exploit sound windows but may still produce vocalizations that are adapted for communication within distinct sound frequency channels.

1.9 Future directions

Vocal production in birds involves multiple physiologic systems, both in the brain and periphery. Understanding vocal mechanisms therefore requires an integrated approach, in which not only the mechanisms of each separate vocal subsystem, but also the interactions between them and how central neural processes associated with phonation map onto the vocal periphery are studied.

It seems evident that a better understanding of the neural, syringeal, respiratory, and vocal tract systems that underlie vocal production in birds will be facilitated by combining EMG and measures of respiratory dynamics during song with relatively new measurement technologies, such as laser vibrometry, sonomicrometry, and magnetic resonance imaging. The importance of *in vivo* and *in vitro* physiologic measurements notwithstanding, it is also becoming evident that such work should be expanded to include additional avian taxa and combined with advanced signal analysis, biophysical modeling, and theoretic studies on dynamical systems. An approach that includes the new view of the vocal system as a nonlinear dynamical system should stimulate investigations into the production of vocalizations that do not behave very well in spectrographic representations. Chaotic vocalizations or vocalizations with other complex spectral patterns have been largely ignored, even though they seem to be common in many bird species of different phylogenetic groups.

Although the topic of bird vocalization has already had a long history in science, and considerable progress has been made in its understanding, it is probably fair to say that most of the work needed to arrive at a general understanding of bird song production is still ahead of us. Bird song is a biologic phenomenon, but it is very much physical in nature. A recent renewed interest of mathematicians and physicists in bird song has already provided interesting results, and we believe that future progress will benefit from true interdisciplinary cooperation between researchers in biology, neuroscience, physics, and mathematics.

REFERENCES

Allan SE, Suthers RA (1994) Lateralization and motor stereotypy of song production in the brown-headed cowbird. *J Neurobiol* 25: 1154–1166.

Aubin T, Jouventin P, Hildebrand C (2000) Penguins use the two-voice system to recognize each other. *Proc Roy Soc Lond (Biol)* 267: 1081–1087.

Ballintijn MR, ten Cate C (1997) Sex differences in the vocalization and syrinx of the collared dove (*Streptopelia decaocto*). *Auk* 114: 22–39.

Ballintijn MR, ten Cate C (1998) Sound production in the collared dove: a test of the "whistle" hypothesis. *J Exp Biol* 201: 1637–1649.

Ballintijn MR, ten Cate C, Nuijens FW, Berkhoudt H (1995) The syrinx of the collared dove (*Streptopelia decaocto*): structure, inter-individual variation and development. *Netherlands J Zool* 45: 455–479.

Banta-Lavenex P (1999) Vocal production mechanisms in the budgerigar (*Melopsittacus undulatus*): the presence and implications of amplitude modulation. *J Acoust Soc Am* 106: 491–505.

Baptista LF, Schuchmann KL (1990) Song learning in the anna hummingbird (*Calypte anna*). *Ethology* 84: 15–26.

Beckers GJL, ten Cate C (in press) Nonlinear phenomena and song evolution in *Streptopelia* doves. *Acta Zool Sinica*.

Beckers GJL, Suthers RA, ten Cate C (2003a) Mechanisms of frequency and amplitude modulation in ring dove song. *J Exp Biol* 206: 1833–1843.

Beckers GJL, Suthers RA, ten Cate C (2003b) Pure-tone birdsong by resonance filtering of harmonic overtones. *Proc Nat Acad Sci USA* 100: 7372–7376.

Beckers GJL, Nelson BS, Suthers RA (2004) Vocal-tract filtering by lingual articulation in a parrot. *Curr Biol* 14: 1592–1597.

Brackenbury J (1978a) A possible relationship between respiratory movements, syringeal movements and the production of song by skylarks *Alauda arvensis*. *Ibis* 120: 526–528.

Brackenbury J (1978b) Respiratory mechanics of sound production in chickens and geese. *J Exp Biol* 72: 229–250.

Brackenbury J (1980) Control of sound production in the syrinx of the fowl *Gallus gallus*. *J Exp Biol* 85: 239–251.

Brenowitz EA, Kroodsma DE (1996) The neuroethology of birdsong. In: Kroodsma DE, Miller EH (eds). *Ecology and Evolution of Acoustic Communication in Birds*, pp. 283–304. Comstock, Ithaca.

Brittan-Powell EF, Dooling RJ, Larsen ON, Heaton JT (1997) Mechanisms of vocal production in budgerigars (*Melopsittacus undulatus*). *J Acoust Soc Am* 101: 578–589.

Calder WA (1970) Respiration during song in the canary (*Serinus canaria*). *Comp Biochem Physiol* 32: 251–258.

Cynx J, Lewis R, Tavel B, Tse H (1998) Amplitude regulation on vocalizations in noise by a songbird, *Taeniopygia guttata*. *Anim Behav* 56: 107–113.

Dabelsteen T, McGregor PK, Lampe HM, Langmore NE, Holland J (1998) Quiet song in song birds: an overlooked phenomenon. *Bioacoustics* 9: 89–105.

DeVoogd TJ, Nottebohm F (1981) Sex differences in dendritic morphology of a song control nucleus in the canary: a quantitative Golgi study. *J Comp Neurol* 196: 309–316.

DeVoogd TJ, Pyskaty DJ, Nottebohm F (1991) Lateral asymmetries and testosterone-induced changes in the gross morphology of the hypoglossal nucleus in adult canaries. *J Comp Neurol* 307: 65–76.

Dooling RJ, Lohr B, Dent ML (2000) Hearing in birds and reptiles. In: Dooling RJ, Fay RR, Popper AN (eds). *Comparative Hearing: Birds and Reptiles*, pp. 308–359. Springer-Verlag, New York.

Dooling RJ, Leek MR, Gleich O, Dent ML (2002) Auditory temporal resolution in birds: discrimination of harmonic complexes. *J Acoust Soc Am* 112: 748–759.

Draganoiu TI, Nagle L, Kreutzer M (2002) Directional female preference for an exaggerated male trait in canary (*Serinus canaria*) song. *Proc Roy Soc Lond B* 269: 2525–2531.

Fant G (1970) *Acoustic Theory of Speech Production*, p. 328. Mouton, The Hague.

Farabaugh SM, Brown ED, Dooling RJ (1992) Analysis of warble song of the budgerigar. *Bioacoustics* 4: 111–130.

Farabaugh SM, Linzenhold A, Dooling RJ (1994) Vocal plasticity in budgerigars (*Melopsittacus undulatus*): evidence for social factors in the learning of contact cells. *J Comp Psychol* 108: 81–92.

Fee MS, Shraiman B, Pesaran B, Mitra PP (1998) The role of nonlinear dynamics of the syrinx in the vocalizations of a songbird. *Nature* 395: 67–71.

Fitch WT (1999) Acoustic exaggeration of size in birds via tracheal elongation: comparative and theoretic analyses. *J Zool* 248: 31–48.

Fitch WT, Neubauer J, Herzel H (2002) Calls out of chaos: the adaptive significance of nonlinear phenomena in mammalian vocal production. *Anim Behav* 63: 407–418.

Fletcher NH (1992) *Acoustic Systems in Biology*, p. 333. Oxford University Press, New York.

Fletcher NH (2000) A class of chaotic bird calls? *J Acoust Soc Am* 108: 821–826.

Fletcher NH, Tarnopolsky A (1999) Acoustics of the avian vocal tract. *J Acoust Soc Am* 105: 35–49.

Fletcher NH, Riede T, Beckers GJL, Suthers RA (2005) Vocal tract filtering and the "coo" of doves. *J Acoust Soc Am* 116: 3750–3756.

Franz M, Goller F (2002) Respiratory units of motor production and song imitation in the zebra finch. *J Neurobiol* 51: 129–141.

Gahr M (2000) Neural song control system of hummingbirds: comparison to swifts, vocal learning (songbirds) and nonlearning (suboscines) passerines, and vocal learning (budgerigars) and nonlearning (dove, owl, gull, quail, chicken) non-passerines. *J Comp Neurol* 426: 182–196.

Gahr M, Güttinger HR, Kroodsma DE (1993) Estrogen receptors in the avian brain: survey reveals general distribution and forebrain areas unique to songbirds. *J Comp Neurol* 327: 112–122.

Gaunt AS (1983) An hypothesis concerning the relationship of syringeal structure to vocal abilities. *Auk* 100: 853–862.

Gaunt AS (1988) Interaction of syringeal structure and airflow in avian phonation. In: Ouillet H (ed.), *Acta XIX Congress of International Ornithology*, pp. 915–924. University Ottawa Press, Ottawa.

Gaunt AS, Gaunt SLL, Casey RM (1982) Syringeal mechanics reassessed: evidence from *Streptopelia*. *Auk* 99: 474–494.

Goller F, Daley MA (2001) Novel motor gestures for phonation during inspiration enhance the acoustic complexity of birdsong. *Proc Roy Soc Lond B* 268: 2301–2305.

Goller F, Larsen ON (1997a) In situ biomechanics of the syrinx and sound generation in pigeons. *J Exp Biol* 200: 2165–2176.

Goller F, Larsen ON (1997b) A new mechanism of sound generation in songbirds. *Proc Natl Acad Sci USA* 94: 14787–14791.

Goller F, Larsen ON (2002) New perspectives on mechanisms of sound generation in songbirds. *J Comp Physiol A* 188: 841–850.

Goller F, Suthers RA (1995) Implications for lateralization of bird song from unilateral gating of bilateral motor patterns. *Nature* 373: 63–66.

Goller F, Suthers RA (1996a) Role of syringeal muscles in controlling the phonology of bird song. *J Neurophysiol* 76: 287–300.

Goller F, Suthers RA (1996b) Role of syringeal muscles in gating airflow and sound production in singing brown thrashers. *J Neurophysiol* 75: 867–876.

Goller F, Suthers RA (1999) Bilaterally symmetrical respiratory activity during lateralized birdsong. *J Neurobiol* 41: 513–523.

Goller F, Mallinckrodt MJ, Torti SD (2004) Beak gape dynamics during song in the zebra finch. *J Neurobiol* 59: 289–303.

Gramza AF (1970) Vocal mimicry in captive budgerigars (*Melopsittacus undulatus*). *Z Tierpsychol* 27: 971–983.

Greenewalt CH (1968) *Bird Song: Acoustics and Physiology*. Smithsonian Institution Press, Washington, DC.

Hartley RS (1990) Expiratory muscle activity during song production in the canary. *Respir Physiol* 81: 177–187.

Hartley RS, Suthers RA (1989) Airflow and pressure during canary song: evidence for mini-breaths. *J Comp Physiol A* 165: 15–26.

Hoese WJ, Podos J, Boetticher NC, Nowicki S (2000) Vocal tract function in birdsong production: experimental manipulation of beak movements. *J Exp Biol* 203: 1845–1855.

King AS (1989) Functional anatomy of the syrinx. In: King AS, McLelland J (eds), *Form and Function in Birds*, pp. 105–192. Academic Press, London.

Klump GM (1996) Bird communication in the noisy world. In: Kroodsma DE, Miller EH (eds), *Ecology and Evolution of Acoustic Communication in Birds*, pp. 321–338. Cornell University Press, Ithaca.

Konishi M (1985) Birdsong: from behavior to neuron. In: Cowan WM (ed.), *Annu Rev Neurosci*, pp. 125–170. Annual Reviews, Inc., Palo Alto.

Kroodsma DE, Konishi M (1991) A suboscine bird (eastern phoebe, *Sayornis phoebe*) develops normal song without auditory feedback. *Anim Behav* 42: 477–487.

Langemann U, Klump GM (1992) Frequency discrimination in the European starling (*Sturnis vulgaris*): a comparison of different methods. *Hearing Res* 63: 43–51.

Larsen OL, Dabelsteen T (1990) Directionality of blackbird vocalization, implications for vocal communication and its further study. *Ornis Scand* 21: 37–45.

Larsen ON, Goller F (1999) Role of syringeal vibrations in bird vocalizations. *Proc Roy Soc Lond* 266: 1609–1615.

Larsen ON, Goller F (2002) Direct observation of syringeal muscle function in songbirds and a parrot. *J Exp Biol* 205: 25–35.

Lohr B, Wright TF, Dooling RJ (2003) Detection and discrimination of natural call in masking noise by birds: estimating the active space of a signal. *Anim Behav* 65: 763–777.

Manabe K, Sadr EI, Dooling RJ (1998) Control of vocal intensity in budgerigars (*Melopsittacus undulatus*): differential reinforcement of vocal intensity and the Lombard effect. *J Acoust Soc Am* 103: 1190–1198.

McLelland J (1989) Anatomy of the lungs and air sacs. In: King AS, McLelland J (eds), *Form and Function in Birds*, pp. 221–279. Academic Press, New York.

Miskimen M (1951) Sound production in passerine birds. *Auk* 68: 493–504.

Nelson BS (2000) Avian dependence on sound–pressure level as an auditory distance cue. *Anim Behav* 59: 57–67.

Nelson BS (2002) Duplex auditory distance assessment in a small passerine bird (*Pipilo erythrophthalmus*). *Behav Ecol Sociobiol* 53: 42–50.

Nelson BS (2003) Reliability of sound attenuation in Florida scrub habitat and behavioral implications. *J Acoust Soc Am* 113: 2900–2910.

Nelson BS (2004) Dynamics of frequency and amplitude modulations in vocalizations produced by eastern towhees, *Pipilo erythrophthalmus*. *J Acoust Soc Am* 115: 1333–1344.

Neubauer J, Mergell P, Eyshholdt U, Herzel H (2001) Spatio-temporal analysis of irregular vocal fold oscillations: biphonation due to desynchronization of spatial modes. *J Acoust Soc Am* 110: 3179–3192.

Nottebohm F (1971) Neural lateralization of vocal control in a passerine bird. I. Song. *J Exp Zool* 177: 229–262.

Nottebohm F (1991) Reassessing the mechanisms and origins of vocal learning in birds. *TINS* 14: 206–211.

Nottebohm F, Arnold AP (1976) Sexual dimorphism in vocal control areas of the songbird brain. *Science* 194: 211–214.

Nottebohm F, Nottebohm ME (1976) Left hypoglossal dominance in the control of canary and white-crowned sparrow song. *J Comp Physiol* 108: 171–192.

Nottebohm F, Kelley DB, Paton JA (1982) Connections of vocal control nuclei in the canary telencephalon. *J Comp Neurol* 207: 344–357.

Nottebohm F, Stokes T, Leonard CM (1976) Central control of song in the canary, *Serinus canarius*. *Comp Neurol* 165: 457–486.

Nowicki S (1987) Vocal tract resonances in oscine bird sound production: evidence from birdsongs in a helium atmosphere. *Nature* 325: 53–55.

Nowicki S, Capranica RR (1986a) Bilateral syringeal coupling during phonation of a songbird. *J Neurosci* 6: 3595–3610.

Nowicki S, Capranica RR (1986b) Bilateral syringeal interaction in vocal production of an oscine bird sound. *Science* 231: 1297–1299.

Nowicki S, Marler P (1988) How do birds sing? *Music Percept* 5: 391–426.
Pepperberg IM (1981) Functional vocalizations by an African gray parrot. *Z Tierpsychol* 55: 139–160.
Riede T, Beckers GJL, Blevins W, Suthers RA (2004) Inflation of the esophagus and vocal tract filtering in ring doves. *J Exp Biol* 207: 4025–4036.
Ryan MJ, Brenowitz EA (1985) The role of body size, phylogeny, and ambient noise in the evolution of bird song. *Am Nat* 126: 87–100.
Snow D (1968) The singing assemblies of little hermits. *Living Bird* 7: 47–55.
Snow D (1977) Comparison of the leks of Guy's hermit hummingbird *Phaethornis guy* in Costa Rica and Trinidad. *Ibis* 119: 211–214.
Stein RC (1968) Modulation in bird sound. *Auk* 94: 229–243.
Stevens KN (1998) *Acoustic Phonetics*. MIT Press, Cambridge, Massachusetts.
Suthers RA (1990) Contributions to birdsong from the left and right sides of the intact syrinx. *Nature* 347: 473–477.
Suthers RA (1994) Variable asymmetry and resonance in the avian vocal tract: a structural basis for individually distinct vocalizations. *J Comp Physiol A* 175: 457–466.
Suthers RA (1997) Peripheral control and lateralization of birdsong. *J Neurobiol* 33: 632–652.
Suthers RA (1999) The motor basis of vocal performance in songbirds. In: Hauser M, Konishi M (eds), *The Design of Animal Communication*, pp. 37–62. MIT Press, Cambridge, MA.
Suthers RA (2004) How birds sing and why it matters. In: Marler P, Slabbekoorn H (ed), *Nature's Music. The Science of Birdsong.* pp. 272–295. Elsevier Academic Press, New York.
Suthers RA, Goller F (1997) Motor correlates of vocal diversity in songbirds. In: Nolan Jr V, Ketterson E, Thompson CF (eds), *Curr Ornithol*, pp. 235–288. Plenum Press, New York.
Suthers RA, Hector DH (1982) Mechanism for the production of echolocating clicks by the grey swiftlet, *Collocalia spodiopygia*. *J Comp Physiol A* 148: 457–470.
Suthers RA, Hector DH (1985) The physiology of vocalization by the echolocating oilbird, *Steatornis caripensis*. *J Comp Physiol A* 156: 243–266.
Suthers RA, Zollinger SA (2004) Producing song: the vocal apparatus. *Ann NY Acad Sci* 1016: 109–129.
Suthers RA, Goller F, Hartley RS (1994) Motor dynamics of song production by mimic thrushes. *J Neurobiol* 25: 917–936.
Suthers RA, Goller F, Hartley RS (1996) Motor stereotypy and diversity in songs of mimic thrushes. *J Neurobiol* 30: 231–245.
Suthers RA, Goller F, Pytte C (1999) The neuromuscular control of birdsong. *Philos Trans Roy Soc Lond B* 354: 927–939.
Suthers RA, Vallet EM, Tanvez A, Kreutzer M (2004b) Bilateral song production in domestic canaries. *J Neurobiol* 60: 381–393.
Titze IR (1994) *Principles of Voice Production*. Prentice Hall, Englewood Cliffs, NJ.
Todt D (1975) Social learning of vocal patterns and modes of their application in grey parrots (*Psittacus erithacus*). *Z Tierpsychol* 39: 178–188.
Vallet E, Kreutzer M (1995) Female canaries are sexually responsive to special song phrases. *Anim Behav* 49: 1603–1610.
Vallet E, Beme I, Kreutzer M (1998) Two-note syllables in canary songs elicit high levels of sexual display. *Anim Behav* 55: 291–297.

Vicario DS (1991) Neural mechanisms of vocal production in songbirds. *Curr Opin Neurobiol* 1: 595–600.
Warner RW (1972) The syrinx in the family Columbidae. *J Zool* 166: 385–390.
Warren DK, Patterson DK, Pepperberg IM (1996) Mechanisms of American English vowel production in a grey parrot (*Psittacus erithacus*). *Auk* 113: 41–58.
Westneat MW, Long J, John H, Hoese W, Nowicki S (1993) Kinematics of birdsong: functional correlation of cranial movements and acoustic features in sparrows. *J Exp Biol* 182: 147–171.
Wild JM (1994) The auditory–vocal–respiratory axis in birds. *Brain Behav Evol* 44: 192–209.
Wild JM (1997) Neural pathways for the control of birdsong production. *J Neurobiol* 33: 653–670.
Wild JM, Goller F, Suthers RA (1998) Inspiratory muscle activity during birdsong. *J Neurobiol* 36: 441–453.
Wild JM, Williams MN, Suthers RA (2000) Neural pathways for bilateral vocal control in songbirds. *J Comp Neurol* 423: 413–426.
Wilden I, Herzel H, Peters G, Tembrock G (1998) Subharmonics, biphonation, and deterministic chaos in mammal vocalization. *Bioacoustics* 8: 1–30.
Wiley HR (1971) Song groups in a singing assembly of little hermits. *Condor* 73: 28–35.
Wiley RH (1991) Associations of song properties with habitats for territorial oscine birds of eastern North America. *Am Nat* 138: 973–993.
Wiley RH, Richards DG (1978) Physical constraints on acoustic communication in the atmosphere: implications for the evolution of animal vocalizations. *Behav Ecol Sociobiol* 3: 69–94.
Wiley RH, Richards DG (1982) Adaptations for acoustic communication in birds: sound transmission and signal detection. In: Kroodsma DE, Miller EH (eds), *Acoustic Communication in Birds*, pp. 132–163. Academic Press, New York.
Wright TF, Cortopassi KA, Bradbury JW, Dooling R (2003) Hearing and vocalizations in the orange-fronted conure. *J Comp Psychol* 117: 87–95.
Zollinger SA, Suthers RA (2004) Motor mechanisms of a vocal mimic: implications for birdsong production. *Proc Roy Soc Lond B* 271: 483–491.

2
The Blind Mole Rat: An Example of Seismic Communication via Acoustic Channels

Zvi Wollberg, Rony Rado and Ronen S. Sadka

2.1 General overview

Various animal species across the entire animal kingdom, invertebrate and vertebrate, aquatic and terrestrial, use substrate-borne vibrational signals as a means for intraspecific communication, for prey detection and predator avoidance. They include arthropods such as spiders, scorpions and insects (e.g. Michelsen et al., 1982; Gogala, 1985; Schuch and Barth, 1985; Schmitt et al., 1994; Sturzl et al., 2000; Hill and Shadley, 2001) as well as amphibians (Lewis and Narins, 1985; Narins, 1990, 2001), reptiles (Gans and Wever, 1972) and various mammals (Narins et al., 1997; Randall and Lewis, 1997; Randall, 1997, 2000, 2001; Francescoli and Altuna, 1998; Francescoli, 2000; Goold and Clarke, 2000; Randall et al., 2000; O'Connell-Rodwell et al., 2000, 2001; Mason and Narins, 2001). Different modes are used by these animals to generate vibratory signals and various sensory organs and mechanisms are used for detecting them, mainly depending on the particular substrate and habitat in which these signals are generated, travel and are perceived. Subterranean rodents, in particular those that are blind and spend their entire life below ground, represent in this regard a unique group because seismic vibrations (primarily Rayleigh and Love surface waves) constitute their only effective means for long-distance communication and apparently also for other purposes such as food detection and avoiding obstacles (e.g. Mason and Narins, 2002; Kimchi and Terkel, 2003a, b). Our model animal for studying intraspecific seismic communication has been the blind mole rat (*Spalax ehrenbergi* known also as *Nanospalax ehrenbergi*), a solitary and territorial subterranean rodent that rarely leaves its tunnel system, which it digs to its own body width (Nevo, 1961). Spending essentially all its life in the underground environment, it has evolved striking behavioral, morphologic and physiologic adaptations to this habitat (Topachevskii, 1976; Arieli and Ar, 1981; Storier et al., 1981; Arieli et al., 1986), including limited hearing capabilities for air-borne sounds (Bruns *et al.*, 1988; Bronchti *et al.*, 1989; Heffner and Heffner, 1992) and vestigial eyes that are covered by skin and fur (Cei, 1946; Sanyal *et al.*, 1990; Bronchti *et al.*, 1991; Cooper *et al.*, 1993). Hence, although the blind mole rat shows some sensitivity to light (Rado *et al.*, 1992b), in the usual sense this rodent is essentially blind.

Being a solitary animal, encounters between individuals are limited to the mating season, to contacts between mother and pups, and to the incidental intrusion of one mole rat into the tunnel system of another. Such unwanted encounters between two individuals usually prove fatal for at least one of them. To avoid the undesired events and yet enable contact between males and females during the mating season, there is need for a communication system between individuals inhabiting neighboring tunnels. Such a system is believed to be provided by "seismic signaling". Indeed, by inserting a geophone into the soil close to mole rat mounds, one can pick up the periodic vibratory bursts that are generated by these animals. It is even possible to distinguish between the different individuals that inhabit different tunnel systems, generating alternating vibratory signals.

When a mole rat is introduced into a Plexiglas tube, it generates vibrations similar to those picked up in the field. It produces these signals by striking the roof of the tube with the flattened, bony, anterodorsal surface of its head. When two individuals are placed within a Plexiglas tube, separated by a barrier to avoid physical contact, or alternatively, in two separate but touching tubes (Rado *et al.*, 1987), the two mole rats commence a vibratory "dialogue" similar to those recorded in the field. However, when the two compartments are entirely separated, so that no vibrations can cross, even though the two animals can hear and smell each other, the dialogue ceases. Thus, it is evident that the vibratory (seismic) signals produced by the blind mole rat in nature and under experimental conditions, serve as means for intraspecific communication.

During the reproduction season, females give birth in an underground nest chamber, where they rear their altricial pups. Communication between mother and pups during this period is based primarily on air-borne vocalizations (Heth *et al.*, 1986, 1988). It has been shown (Rado *et al.*, 1991, 1992a) that at about 7 weeks of age, juveniles begin to dig their own tunnels as extensions of the maternal tunnel system, but keep the connection still open. This situation lasts for about 4 weeks, during which vibratory signaling commences but vocalizations are still the main means for communication. From the age of about 12 weeks the young become independent and solitary and after sealing off their connection to the maternal tunnel system, the communication shifts mainly to the seismic channel.

In the following sections, we describe the properties of these vibratory (seismic) signals and how they are perceived. Data were obtained from both field

2.2 Properties of the vibratory signals

Using several geophones in an area inhabited by blind mole rats, we picked up vibratory "dialogues" between individual animals located as far as 15 m apart. Figure 2.1a and b illustrates a sequence of vibratory bursts performed by an

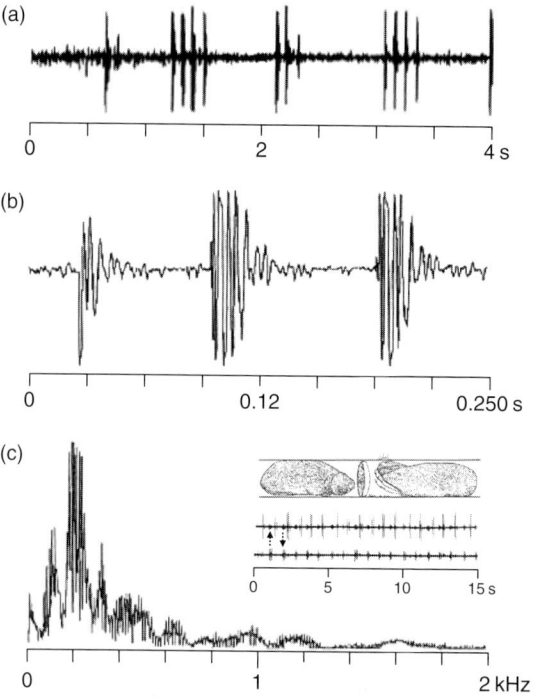

Figure 2.1. Temporal pattern of blind mole rat vibratory signals recorded in the field illustrating sequential bursts within a vibratory series (a) and vibrations within a single burst (b). (c) Spectral density function of vibratory signals (digitization rate: 4 kHz); fast Fourier transform (FFT) performed on consecutive 256 ms samples, at a rate of 1024 data point per sample. Inset in (c) shows two mole rats located within a Plexiglas tube, performing a seismic "dialogue", and two oscillograms of a dialogue between two mole rats recorded in the field. Recordings were made using two geophones located several meters apart, each within the territory of one of the two dialoguing mole rats. The vibrations from the animals can be differentiated by the different amplitudes in the two traces. Note (arrows), that while one mole rat is emitting its message the other one remains silent and vice versa. (After Rado et al., 1987.)

individual mole rat in nature. Figure 2.1c depicts a spectral density function of these vibrations, indicating that most of the energy is concentrated between 100 and 300 Hz with a peak at about 200 Hz and additional low-intensity harmonics (Heth et al., 1987; Rado et al., 1987).

Under laboratory conditions, when two mole rats are placed within a Plexiglas tube and "dialoguing" (inset in Fig. 2.1c), one often firmly presses its cheek and lower jaw against the vibrating wall of the tube and perceives the signals which the other one generates when tapping its head against the roof of the tube. Recordings from the field suggest that mole rats in their natural tunnel systems dialogue in a similar way. Due to the signals of the two remote animals were picked up by two separate geophones at different strengths, one can distinguish the mole rat emitting its message from the other remaining silent and vice versa (arrows).

An analysis of the patterns of the vibratory signals revealed that they were temporally generated in series, each one consisting of several consecutive periodic bursts with each burst composed of several vibratory waves. Our definition of a series (Rado et al., 1987) was based on the time interval between consecutive bursts. Sequential bursts with inter-burst intervals shorter than 2 s were defined as a series. Hence, successive series were separated from each other by intervals longer than 2 s. Recordings made during 10 min of dialoguing episodes revealed that the mean ($\pm$SD) number of bursts per series was 8.6 ± 8.2 (40 sequences from seven animals); the mean ($\pm$SD) duration of each burst was 0.31 ± 0.06 s (nine animals and 261 bursts) and the mean ($\pm$SD) number of vibrations per burst was 4.1 ± 0.9 (Rado et al., 1987).

To estimate the sensitivity of mole rats to the vibratory signals, we measured the acceleration (m/s^2) of the seismic waves close to the emitter and its attenuation while traveling towards the receiving mole rat. Measurements were obtained by using four geophones perpendicularly positioned between the emitter and the receiver, with the first geophone located about 1 m away from the emitter and the other three in a row 1 m apart. The results are depicted in Fig. 2.2. As can be seen, the best function fitting these experimental data is a polynomial inverse third order regression. The extrapolated acceleration of the vibrational signals at the receiver's site, 15 m away from the source, is about $0.013 \, m/s^2$ indicating a fairly high sensitivity of the blind mole rat to these vibrations (recalling anecdotal evidence of the extremely high sensitivity of animals to earthquakes). The velocity of the vibratory waves, calculated from traces such as in Fig. 2.2 (inset), is about 100 m/s, which is compatible with the velocity of Rayleigh surface waves in damp soil (Lewis and Narins, 1985).

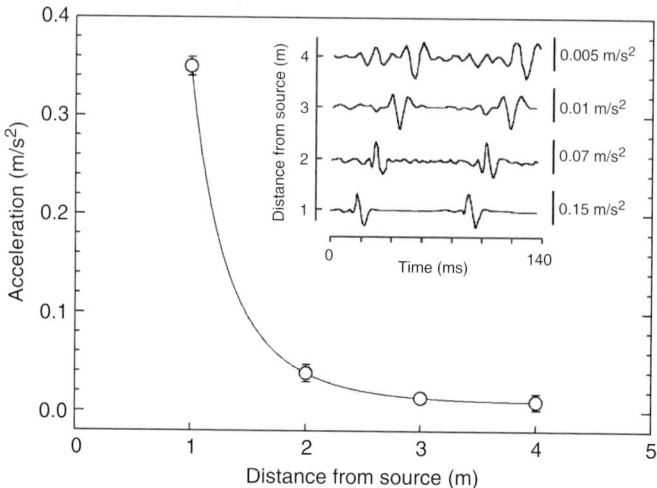

Figure 2.2. Decay of mole rat vibratory signals (mean ± SEM; eight measurements per point) as a function of distance from source. Measurements were made using four geophones, the first one positioned about 1 m away from the emitter and the other three in a row, 1 m apart. The curve represents a regression function that best describes the experimental data (a polynomial inverse third order regression: $Y = Y_0 + a/x + b/x^2 + c/x^3$, where $Y_0 = 0.016$, $a = -0.0415$, $b = -0.0283$, $c = 0.4041$; ordinate = acceleration, abscissa = distance). The inset shows a segment of a vibratory burst recorded at 1, 2, 3 and 4 m from source. Acceleration scales are indicated by the vertical bars at the right of the graphs (note the differences in the acceleration scales).

2.3 How are the seismic signals perceived?

Judging by our behavioral observations, it seems very likely that the vibratory signals generated by the blind mole rat serve for intraspecific communication. The receiver pressing its cheek and lower jaw against the vibrating wall of the tube while dialoguing with conspecifics suggests that signals are picked up by these parts of the body. This behavior occurs also in nature as revealed in a film made under semi-natural conditions by David Attenborough, BBC TV. To validate this assumption, we performed the following experiments. Two cylindrical Plexiglas tubes 80 and 30 cm long with a diameter of 7 cm were placed on separate tables with their openings facing each other at a distance of 2 cm so as to avoid vibrational cross-talk between the two tubes. Vibratory signals, produced by short clicks, were delivered through a mini-shaker placed under the longer tube at a distance of 60 cm from the animal's head. Event related responses (brainstem auditory evoked responses, BAER; and middle latency responses, MLR) elicited by the vibratory signals were recorded from the scalp of anesthetized animals under three different experimental settings:

1. The animal was placed inside the short Plexiglas tube with its head facing, but not touching, the long tube (Fig. 2.3A). Under this configuration the animal was exposed to the air-borne sounds from the mini-shaker while tapping the long tube, but not to the vibrations.
2. The animal's body was situated inside the long tube with its head inside the short tube (Fig. 2.3B), exposing the mole rat to the air-borne sounds from the mini-shaker, while its body, except for the head, was in physical contact with the vibrating substrate.
3. The animal's body was situated inside the short tube while its head was inside the long tube with the lower-jaw touching the Plexiglas floor of the tube (Fig. 2.3C). In this position the animal was exposed to the air-borne sounds produced by the vibrator and its lower jaw was in physical contact with the vibrating substrate.

The acceleration of the substrate and the sound pressure level produced by the mini-shaker, measured at the distal end of the long tube, were 7.82 m/s^2 and 70 dB SPL, respectively. Stimulation rate in all these experiments was 0.5 stimuli/s. The results of these experiments are presented in the left three panels of Fig. 2.3 (a–c). The air-borne sound produced by the mini-shaker did not elicit any detectable response (Fig. 2.3a), while the same stimulus causing vibration of the substrate in contact with either the body of the mole rat (Fig. 2.3b) or only its head (Fig. 2.3c), elicited clear MLRs. The strongest response was elicited when the head, with the jaw leaning on the floor of the tube, was exposed to the vibrations. This response was significantly stronger than when the entire body except the head was stimulated. We also noticed that the waveform and amplitudes of the responses to the vibratory stimuli were highly dependent on the position of the mole rat's lower jaw with respect to the vibrating floor of the tube. When the lower jaw was barely touching the substrate, the elicited response was weakest and the waveform complex. When the lower jaw was lying on the substrate the response was stronger and less complex. The strongest response was elicited when the lower jaw was gently pressed against the wall of the vibrating tube by means of a rubber coated screw that pushed the head in a dorso-ventral direction. Hence, the detection of vibratory signals by the mole rat is accomplished through a physical contact between the animal, especially its lower jaw, and the vibrating substrate (Rado et al., 1998).

2.4 Somatosensory or auditory?

Two sensory systems might be involved in the detection and analysis of these signals: the auditory system by means of bone conduction and the somatosensory

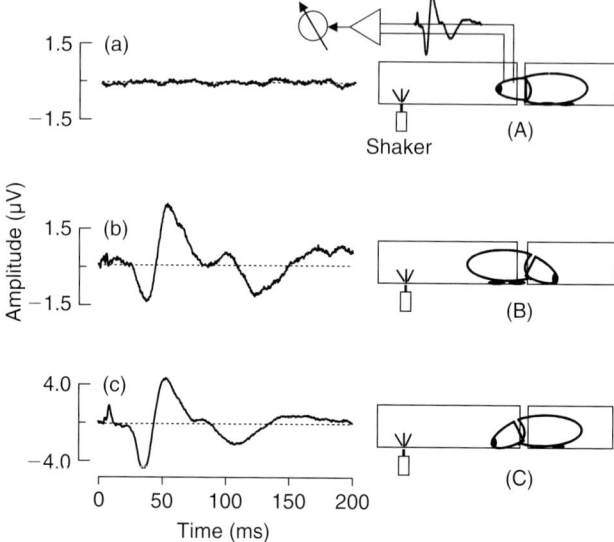

Figure 2.3. (a–c) MLRs recorded from the scalp of an anesthetized blind mole rat under three different experimental settings (A–C). (A) The animal is exposed to the air-borne sounds (70 dB SPL peak intensity) but not to the vibrations (7.8 m/s^2) produced by a mini-shaker vibrating the longer tube. (B) The animal is exposed to the sound produced by the mini-shaker while its body is in contact with the vibrating substrate. (C) The animal's head, especially its lower jaw, rests on the floor of the vibrating tube. Note that no response was elicited in setting A, whereas in the other two settings vibrations elicited prominent responses with a maximum response for setting (C) (note different amplitude scales in (b) and (c)) (from Rado et al., 1998).

system via mechanoreceptors. Studies on several animal species that use substrate-borne vibrations for communication indeed suggested that one or both of these mechanisms function in different animals (Francescoli, 2000; Mason, 2001; Narins, 2001). Based on electrophysiologic, behavioral and morphologic studies, we suggested that blind mole rats primarily use the auditory system for this purpose (Rado et al., 1989, 1998) rather than the somatosensory system, as suggested by others (Nevo et al., 1991).

2.5 Electrophysiologic experiments

As shown above (Fig. 2.3), vibratory stimuli picked up by the lower jaw elicited prominent responses, while a 70 dB (SPL peak intensity) air-borne sound produced by the same stimuli failed to elicit any noticeable response. However, extremely strong air-borne sound (120 dB peak intensity), compensating partially for the low sensitivity of blind mole rats to air-borne sounds (Heffner and Heffner, 1992), evoked BAERs and MLRs similar in shape to, although a great deal weaker than those evoked by the vibrations when being picked up through the jaw (Fig. 2.4a, b, d, e). This suggests that the responses elicited by the two modes of stimulation are processed by the auditory system though picked up differently. High-intensity air-borne sounds are detected and transmitted to the inner ear in the conventional way, whereas vibratory signals bypass the eardrum and are transmitted directly to the inner ear by bone conduction. To further corroborate this assumption we examined the responses of binaurally deafened mole rats to the same vibrations and air-borne clicks. Binaural deafness was achieved, under deep anesthesia, by mechanical destruction of the middle and inner ears (for details see Rado *et al.*, 1998). Deaf animals were tested shortly after behavioral recovery from surgery (2 weeks post-deafening for most animals) and again 3 weeks later. Substrate vibrations elicited in the deafened animals produced a considerably weaker MLR that also differed in shape and latency (Fig. 2.4c). This residual response in the deaf animals was essentially the same whether the lower jaw was pressed against the vibrating substrate or just barely touching it, probably representing a somatosensory response component usually obscured by the dominating auditory response. The high-intensity air-borne clicks failed to elicit any noticeable response in the deaf animals (Fig. 2.4f).

Next we tested the responses of intact mole rats to vibratory stimuli at three different accelerations, ranging between 7.8 and $0.9\,m/s^2$, with and without a stationary high intensity (120 dB SPL) broadband (0.02–20 kHz) background masking noise presented through the air. As in the previous experiments, animals were placed in a Plexiglas tube with their lower jaw pressed to the vibrating substrate. Figure 2.5 shows that an acceleration of $7.8\,m/s^2$ without back-ground noise elicited a marked response, the same acceleration in the presence of the noise, however, only a residual response. Weaker accelerations (1.3 and $0.9\,m/s^2$) without masking noise still elicited remarkable responses but with presence of noise failed to evoke a clear response.

2.6 Behavioral tests

Isolated mole rats occasionally perform spontaneous head drumming and jaw listening, suggesting that this is an innate behavior associated with exploring the environment and maintaining communication with their conspecifics (Zuri and Terkel, 1996). However, this activity fades with time if there is no response to the signals they emit. We used this finding for further assessment of the association between seismic communication and hearing. To that end, we examined

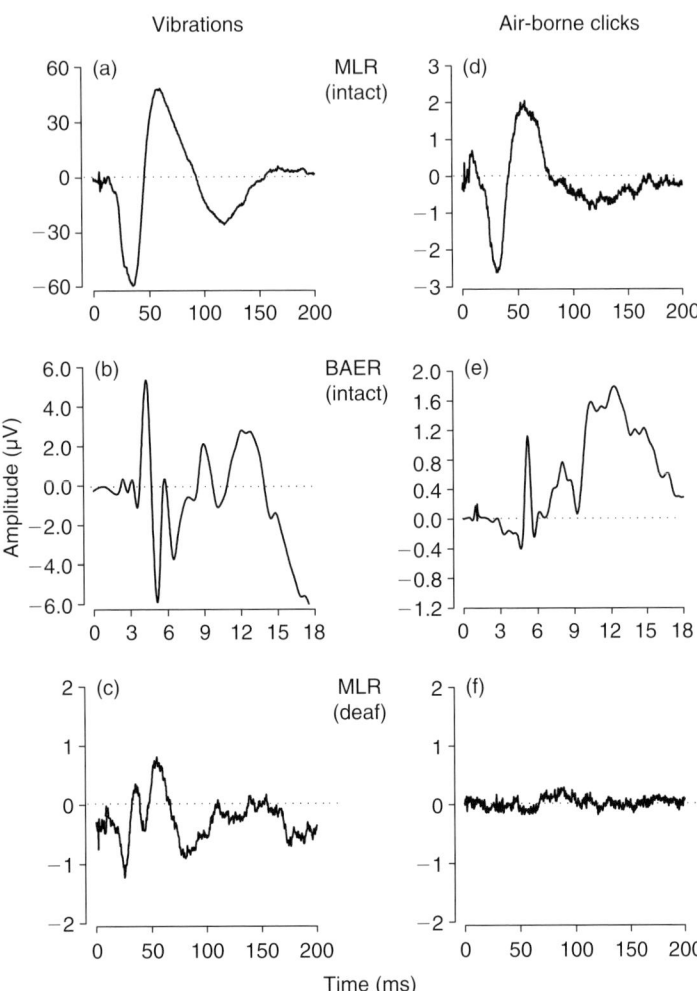

Figure 2.4. Averaged MLR and BAER evoked by repetitive vibratory stimulation (7.8 m/s^2) and by extremely strong air-borne clicks (120 dB SPL, peak intensity) in intact mole rats (a, b and d, e, respectively) and in binaurally deafened animals (c, f). Stimulation rate: 0.5 stimuli/s. Note the similarity in shape and time course of the MLRs evoked by the vibrations and the air-borne clicks but the marked difference in the amplitude (note the differences in the amplitude scales). The residual response to a vibratory stimulus in the deaf animal (c) has a shorter latency as compared with the intact one (a) and the deaf animal (f) lacks a response to the air-borne click (after Rado et al., 1998).

2.6 Behavioral tests

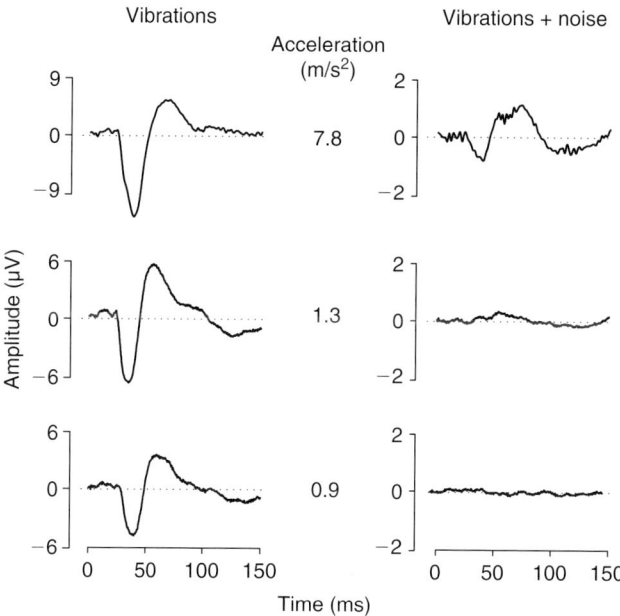

Figure 2.5. Effect of vibratory stimuli at different accelerations on MLR with and without concurrent exposure to high-intensity (120 dB SPL) broadband (0.02–20 kHz) masking noise. Note that in the absence of masking noise (left panel) an acceleration as low as 0.9 m/s² was very effective in eliciting a prominent response, while with noise there was only a residual response at very high accelerations (after Rado et al., 1998).

whether deafening the animals had any effect on their head drumming behavior. This was accomplished by introducing pairs of mole rats, a test animal (control or deaf) and a naive (intact) partner, into a Plexiglas tube with a barrier separating them. The parameters examined were the number of test animals that performed head drumming after a partner was introduced, and the number of head drumming bursts within the first 5 min of the two animals being in the tube. Since mole rats occasionally perform spontaneous head drumming, we considered only animals that performed at least three head drumming bursts within these first 5 min.

Control and deafened animals were tested with the same partners 1 week before deafening and 2, 4 and 8–10 weeks after deafening. The results are summarized in Fig. 2.6. The number of deafened mole rats that had previously dialogued with their control partners gradually declined with time until complete cessation occurred 8–10 weeks after deafening (Fig. 2.6a). The number of head drumming bursts from deaf animals in the 5-min test period also declined

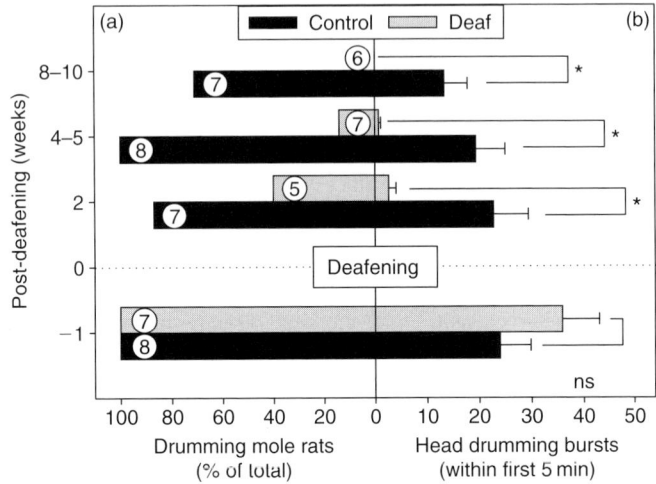

Figure 2.6. Effect of binaural deafening on head drumming behavior of the blind mole rat elicited by confrontation with a naive intact animal at various times pre- and post-deafening. The test animal (deaf or control) and a naive one were introduced into a Plexiglas tube, separated by a barrier to avoid physical contact. (a) Percent test animals that performed head drumming (numbers in the bars are numbers of tested animals both for (a) and (b). (b) Numbers of head drumming bursts produced by the test animal within the first 5 min in the tube. Values represent means ± SEM. Note that control animals reduced the number of bursts only moderately over the test period (weeks) while head drumming dropped significantly in the deaf animals (one sided t-test, $P < 0.05$) (after Rado et al., 1998).

significantly with time, as compared to the control animals (Fig. 2.6b). It seems that the animals were frustrated not to perceive the vibrations. The slight and gradual decrease in the number of head drumming bursts from control animals most probably reflects behavioral habituation. The results of these behavioral experiments further corroborate our assumption that jaw hearing is the major means for detecting the vibratory signals used by blind mole rats for long-distance communication, suggesting a bone conduction pathway for transmitting the signals to the inner ear.

2.7 Morphology

Hearing via the lower jaw is not unique to blind mole rats and has been demonstrated in several other animals, aquatic and terrestrial. The dwarf sperm whale, *Kogia sima*, (Goold and Clarke, 2000) and some limbless burrowing squamates of the order *Amphisbenia* (Gans and Wever, 1972) are just a few examples. This mode of detecting vibratory signals and the transmission of the vibrations to the inner ear via bone conduction asks for morphologic adaptations that may enhance bone conduction. Mason (2001) suggested that the extraordinarily hypertrophied auditory ossicles of some genera of golden moles are adaptations for inertial bone conduction used for the detection of substrate vibrations by the ear. Based on unique structural features of the blind mole rat's middle ear and its morphologic connection with the lower jaw, we proposed (see also Burda *et al.*, 1989; Rado *et al.*, 1989) a different mechanism that not only enhances bone conduction but also explains the low sensitivity of the blind mole rat to air-borne sounds.

Briefly, a low area ratio of the typical tympanic membrane and the relatively large stapes footplate, together with a rather low arm–lever ratio of the middle-ear ossicles explain most of the low efficiency of the blind mole rat's middle ear in transferring air-borne sounds resulting in poor behavioral and electrophysiologic sensitivity to air-borne sounds throughout the entire hearing range (Bruns *et al.*, 1988; Bronchti *et al.*, 1989; Heffner and Heffner, 1992). Bone conduction is enhanced by a peculiar incudo-periotic joint and a unique articulation between the lower jaw and the skull. The middle-ear ossicular chain is of the parallel "freely mobile" type (Fleischer, 1973; Lay, 1973) with the manubrium of the malleus and the long process of the incus running approximately in parallel (Fig. 2.7a). The suspended malleus-incus complex is anchored by means of two ligaments, one connecting the anterior process of the malleus with the anterior limb of the tympanic annulus, and the other the short process of the incus with the periotic bone. The thin bony periotic lamina separating the epitympanic cavity from the antrum, becomes more bulky at its edge forming a cup-shaped protrusion with its concave surface facing the fossa incudis. The short process of the incus fits perfectly into this concavity and is attached firmly to the cup by means of a short and stout ligament and a thin layer of cartilage (Fig. 2.7b and c). This unique structure, which to our knowledge has not been described in any other mammalian species, creates a solid connection between the auditory ossicles and the wall of the tympanic bulla.

Regarding the distinctive articulation between the lower jaw and the skull, it has been shown (Topachevskii, 1976) that caudal of the typical caudo-rostrally oriented glenoid fossa of most rodents, which is partially concealed by the bulla tympanum and formed by its wall and the petromastoideum, there is a widening of the articulation area (Fig. 2.8a) defined as the "fossa pseudoglenoidae" (false articulating fossa). In addition, and unlike most other rodents, the condyles of the lower jaw protrude prominently towards the midline (Fig. 2.8b). When the jaws are not active in digging or mastication, for example, during "jaw hearing",

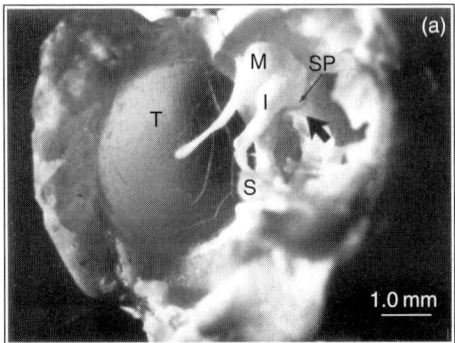

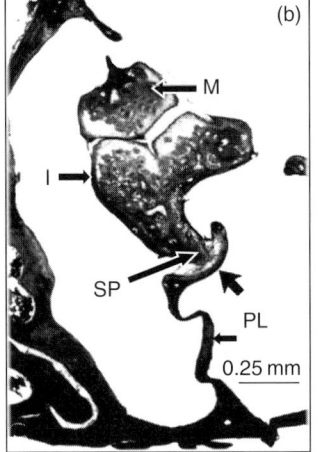
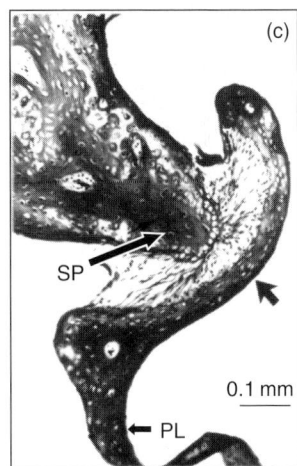

Figure 2.7. Middle ear of the blind mole rat. (a) A view on an open tympanic bulla with the cochlea completely removed. (b) A sagittal section through the bulla, showing the thin periotic bony lamina (PL, which was somewhat deformed during preparation), its unique cup (heavy filled arrow) and the joint between the short process (SP) of the incus (I) and the cup. (c) Enlargement of the peculiar joint between the process of the I and the cup-shaped structure. M: malleus; S: stapes; T: tympanic membrane (after Rado et al., 1998).

the condyle fits into the "pseudo fossa". Thus, physical contact is made between the lower jaw and the bulla. Hence, the firm joint of the auditory ossicles to the wall of the epitympanic cavity form, as explained, a solid continuity between the lower jaw and the middle-ear ossicles. This implies that the jaw hearing apparatus enables a direct and efficient transmission of substrate vibrations from the ground to the stapes and oval window by means of bone conduction, bypassing the eardrum and the malleus.

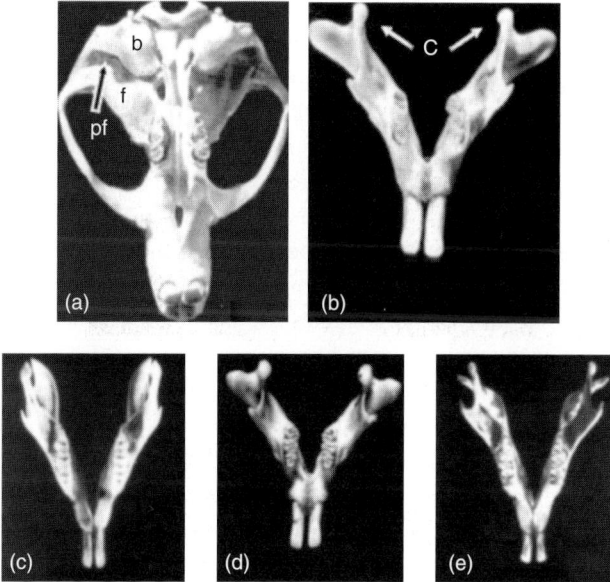

Figure 2.8. Articulation of the mole rat's lower jaw with the skull. (a) A ventral view of the skull. Note the regular caudo-dorsal oriented fossa "f" and the pseudoglenoid fossa "pf" which is partially hidden under the bulla "b". (b) A dorsal view of the lower jaw. Note the markedly caudo-medial protruded condyles "c". (c–e) A comparison between the lower jaw of the blind mole rat (d) with those of two other rodents, *Rattus rattus* (c) and *Meriones sacramenti* (e). Note the different size and orientation of the "c" in the three rodents and the peculiarity of the mole rat in this regard. Scales: 5 mm.

2.8 Discussion

The isolation imposed on the blind mole rat by its subterranean environment, exerted an evolutionary pressure resulting in the development of two parallel communication systems, one for short distance based on air-borne vocalizations (Heth *et al.*, 1986, 1988; Nevo *et al.*, 1987), the other for long distance based on substrate-borne (seismic) vibrations (Heth *et al.*, 1987; Rado *et al.*, 1987). The former is used for communication between mother and young, and between adults encountering each other in the same tunnel system either incidentally or during mating. The latter enables communication between conspecifics inhabiting separate tunnel systems. Seismic vibrations as a means for long-distance intraspecific communication by subterranean rodents are not unique to the blind mole rat. At least one more subterranean mole rat, the Cape mole rat (*Georychus capensis*), also uses vibratory signals for this purpose, this time produced by drumming its hind legs on the burrow floor (Narins *et al.*, 1992).

Theoretically, substrate-borne vibrations can be detected and processed by the somatosensory system, the auditory system or both. The high resemblance in latency, shape and duration of the responses evoked by the vibratory stimuli and by high-intensity air-borne clicks, the similar BAER and MLR evoked by the two kinds of stimuli, the almost complete elimination of these responses by masking noise or by deafening the animal, all provide solid evidence for the primary role of the auditory system in the processing of the vibratory signals.

Judging by the unique morphologic features of the middle-ear ossicles and the peculiar articulation between the lower jaw and the tympanic bulla, it is also apparent that the vibratory signals are transmitted to the inner ear by means of bone conduction. The jaw hearing apparatus enables transmission of seismic signals to the cochlea with minimal loss of energy, thereby compensating for the low efficiency of the middle ear in transferring air-borne sounds to the inner ear (Bronchti et al., 1989; Rado et al., 1989, 1998; Heffner and Heffner, 1992). Despite the poor sensitivity of the blind mole rat to air-borne sounds, the use of seismic signaling provides an alternative means of long-distance communication between individuals inhabiting separate and relatively remote tunnel systems (Heth et al., 1987; Rado et al., 1987).

Applying cyto- and myeloarchitectural procedures and using the metabolic marker 2-deoxyglucose and the projection tracer horseradish peroxidase (HRP) we have shown that in the blind mole rat, the dorsal lateral geniculate nucleus (dLGN) and extensive areas of the occipital cortex (two primary visual areas) can be activated by auditory stimuli (Bronchti et al., 1989; Heil et al., 1991; Bronchti et al., 2002). Auditory information processing in cortical areas normally occupied by the visual system has also been shown by recording event related responses and single unit activity from blind mole rat occipital cortex (Heil et al., 1991; Bronchti et al., 2002; Sadka and Wollberg, 2004). The major source of the auditory input to these "historical" visual areas has been shown to be the central nucleus of the inferior colliculus (IC), abbreviated as ICC, the major midbrain auditory nucleus, which in addition to all typical auditory targets, also projects to the dLGN (Doron and Wollberg, 1994). To the best of our knowledge, this is the first congenitally blind mammal in which auditory compensatory projections have been described and traced. Intriguingly, we found and described similar IC–dLGN projections in neonatally enucleated hamsters but not in normal ones (Izraeli et al., 2002).

Characteristic frequencies (CFs) of single cells in the auditory-activated occipital cortex fell into two distinct subgroups, one around 100 Hz matching the most intensive spectral components of the vibratory signals used by the mole rat for long-distance communication, the other ranging between 2500 and 4400 Hz (Sadka and Wollberg, 2004) corresponding to the main spectral components

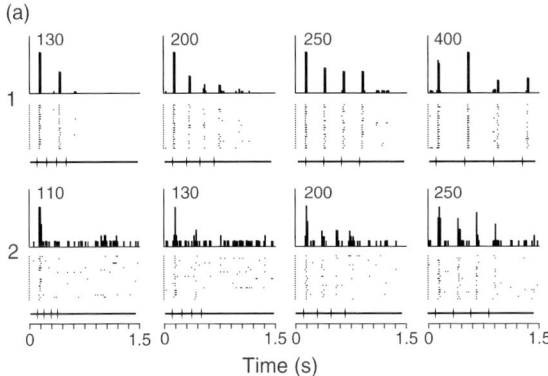

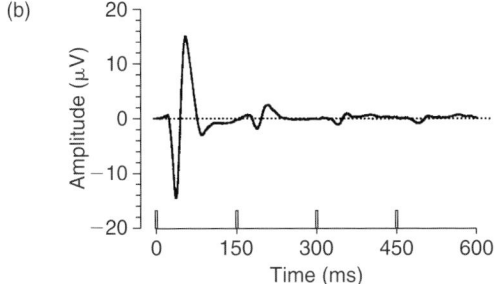

Figure 2.9. (a) Dot rasters (20 repetitions, 2 s apart) and corresponding peristimulus time histograms (PSTHs, 5 ms bin width) illustrating responses of two cells (1, 2) to sequential single cycles of an amplitude modulated 100 Hz tone with various inter-cycle time intervals (values in millisecond on top-left of each PSTH). Timing of stimuli is indicated below the dot rasters. Note the different ability of the two cells to follow the sequential stimuli. (b) Averaged MLR evoked by four sequential vibrations, 150 ms apart. Timing of stimuli is indicated by short rectangles on the timescale. Note that a maximum response was elicited only by the first vibration in each burst, whereas response to all the others faded gradually.

of its air-borne vocalizations. Is this an indication of some functional organization of the auditory-activated visual cortex? One very tempting possibility, though not yet proven, is that the occipital cortex recruited during evolution by the auditory system is dedicated to enhance the processing of the two types of acoustic/seismic communication signals. This possibility is supported by the fact that the most efficient auditory stimuli of those we used were clicks and sinusoidal amplitude modulated (SAM) tones (100 Hz carrier frequency), shaped by a two-quadrant multiplier (responding to the positive voltages and chopping the negative components of the symmetrical sinusoidal modulating signal; Sadka and Wollberg, 2004.) that simulated the vibratory communication signals of the mole rat. As expected, the ability of single cells to follow sequential single cycles of a SAM tone, modulated by a 32 Hz sinusoidal wave, is limited by the inter-cycle time interval. The minimum interval below which cells responded to only the first stimulus ranged between 130 and about 300 ms. Intervals longer than these evoked responses that recovered gradually as the interval increased until complete recovery (Fig. 2.9a; Sadka and Wollberg, 2004). This was in accordance with MLRs elicited by vibratory signals and recorded from the scalp of anesthetized mole rats (Fig. 2.9b; Rado *et al.*, 1998).

2.9 Conclusion

Based on our findings, it is apparent that the vibratory (seismic) signals employed by the blind mole rat for long-distance intraspecific communication is perceived by the animal's auditory system. A unique morphology of the middle ear and of the articulation between the lower jaw and the skull, as well as a peculiar "jaw listening behavior", enable transmission of the seismic vibrations to the inner ear, mainly by bone conduction. We have also shown that the dLGN and extensive areas of the occipital cortex, two typical visual areas, can be activated by various auditory stimuli. Single unit recordings from the occipital cortex revealed that the most efficient of the auditory stimuli that we used were clicks and SAM tones that simulated the vibratory communication signals of the blind mole rat. Answering the question of whether these auditory-recruited areas are indeed involved in the processing of these communication signals, and whether they have any behavioral function, requires further behavioral and electrophysiologic investigation.

2.10 Abbreviations

BAER: brainstem auditory evoked response
CF: characteristic frequency
dLGN: dorsal lateral geniculate nucleus
HRP: horseradish peroxidase
IC: inferior colliculus
ICC: central nucleus of the inferior colliculus
MLR: middle latency response
SAM: sinusoidal amplitude modulation

Acknowledgments

We wish to thank Naomi Paz for help with manuscript preparation, to Miriam Wollberg for doing the histology, Dan Gonen for technical assistance and Amikam Shub for the photography. Our studies described in this review were supported by the German-Israeli Foundation for Scientific Research and Development (G.I.F) grant 1-81-075.1/88; the Jewish National Foundation grant 188-90/91, and the Israel Science Foundation grant 434/91.

REFERENCES

Arieli R, Ar A (1981) Heart rate responses of the mole rat (*Spalax ehrenbergi*) in hypercapnic, hypoxic and cold conditions *Physiol Zool* 54: 14–21.

Arieli R, Heth G, Nevo E, Zamir Y, Neutra O (1986) Adaptive heart and breathing frequencies in 4 ecologically differentiating chromosomal species of mole rats in Israel. *Experientia* 142: 131–133.

Bronchti G, Heil P, Scheich H, Wollberg Z (1989) Auditory pathway and auditory activation of primary visual targets in the blind mole rat (*Spalax ehrenbergi*). I. A 2-deoxyglucose study of subcortical centers. *J Comp Neurol* 284: 253–274.

Bronchti G, Rado R, Terkel J, Wollberg, Z (1991) Retinal projections in the blind mole rat: a WGA-HRP tracing study of natural degeneration. *Dev Brain Res* 58: 159–170.

Bronchti G, Heil P, Sadka R, Hess A, Scheich H, Wollberg Z (2002) Auditory activation of "visual" cortical areas in the blind mole rat (*Spalax ehrenbergi*). *Eur J Neurosci* 16: 311–329.

Bruns V, Muller M, Hofer W, Heth G, Nevo E (1988) Inner ear structure and electrophysiological audiograms of the subterranean mole rat, *Spalax ehrenbergi*. *Hear Res* 33: 1–10.

Burda H, Bruns V, Nevo E (1989) Middle ear and cochlear receptors in the subterranean mole rat, *Spalax ehrenbergi*. *Hear Res* 39: 225–230.

Cei G (1946) Ortogenesi parallela e degenerazione degli organi dello visto negli *Spalacidi*. *Mon Zool Ital* 55: 69–88.

Cooper HM, Herbin M, Nevo E (1993) Ocular regression conceals adaptive progression of the visual system in a blind subterranean mammal. *Nature* 361: 156–159.

Doron N, Wollberg Z (1994) Cross-modal neuroplasticity in the blind mole rat *Spalax ehrenbergi*: a WGA-HRP tracing study. *NeuroReport* 5: 2697–2701.

Fleischer G (1973) Studien am Skelett des Gehörorganes der Säugetiere, einschliesslich des Menschen. *Säugetierkundl Mitt* 21: 131–239.

Francescoli G (2000) Sensory capabilities and communication. In: Lacey E, Patton J, Guy Y, Cameron N (eds), *Life Underground: The Biology of Subterranean Rodents*, pp. 111–144. University of Chicago Press, Chicago.

Francescoli G, Altuna CA (1998) Vibrational communication in subterranean rodents: the possible origin of different strategies. *Evol Commun* 2: 217–231.

Gans C, Wever EG (1972) The ear and hearing in *Amphisbaenia* (reptilia). *J Exp Zool* 179: 17–34.

Gogala M (1985) Vibrational communication in insects (biophysical and behavioral aspects). In: Kalmring K, Elsner N (eds), *Acoustic and Vibrational Communication in Insects*, pp. 117–134. Parey, Berlin.

Goold JC, Clarke MR (2000) Sound velocity in the head of the Dwarf Sperm Whale, (*Kogia sima*), with anatomical and functional discussion. *J Marine Biol Assoc UK* 80: 535–542.

Heffner RS, Heffner HE (1992) Hearing and sound localization in blind mole rats (*Spalax ehrenbergi*). *Hear Res* 62: 206–216.

Heth G, Frankenberg E, Nevo E (1986) Adaptive optimal sound for vocal communication in tunnels of a subterranean mammal (*Spalax ehrenbergi*). *Experientia* 42: 1287–1289.

Heth G, Frankenberg E, Raz A, Nevo E (1987) Vibrational communication in subterranean mole rats (*Spalax ehrenbergi*). *Behav Ecol Sociobiol* 21: 31–33.

Heth G, Frankenberg E, Nevo E (1988) "Courtship" calls of subterranean mole rats (*Spalax ehrenbergi*): physical analysis. *J Mammal* 69: 121–125.

Heil P, Bronchti G, Wollberg Z, Scheich H (1991) Invasion of visual cortex by the auditory system in the naturally blind mole rat. *NeuroReport* 2: 735–738.

Hill PSM, Shadley JR (2001) Talking back: Sending soil vibration signals to lekking prairie mole cricket males. *Amer Zool* 41: 1200–1214.

Izraeli R, Gimseong K, Lamish M, Heicklen AJ, Heffner HE, Heffner RS, Wollberg Z (2002) Cross-modal neuroplasticity in neonatally enucleated hamsters: structure, electrophysiology and behaviour. *Eur J Neurosci* 15: 693–712.

Kimchi T, Terkel J (2003a) Seeing and not seeing. *Curr Opin Neurobiol* 12: 728–734.

Kimchi T, Terkel J (2003b) Mole rats (*Spalax ehrenbergi*) select bypass burrowing strategies in accordance with obstacle size. *Naturwissenschaften* 90: 36–39.

Lay DM (1973) The anatomy, physiology, functional significance and evolution of specialized hearing organs of Gerbilline rodents. *J Morphol* 138: 41–120.

Lewis ER, Narins PM (1985) Do frogs communicate with seismic signals? *Science* 227: 187–189.

Mason MJ (2001) Middle ear structures in fossorial mammals: a comparison with non-fossorial species. *J Zool Lond* 255: 467–486.

Mason MJ, Narins PM (2001) Seismic signal used by fossorial mammals. *Amer Zool* 41: 1171–1184.

Mason MJ, Narins, PM (2002) Seismic sensitivity in the desert golden mole (*Eremitalpa granti*): A review. *J Comp Psychol* 116: 158–163.

Michelsen A, Fink F, Gogala M (1982) Plants as transmission channels for insect vibrational songs. *Behav Ecol Sociobiol* 11: 269–281.

Narins PM (1990) Seismic communication in anuran amphibians. *Bioscience* 40: 268–274.

Narins PM (2001) Vibration communication in vertebrates. In: Barth F, Schmidt A (eds), *Ecology of Sensing*, pp. 127–148. Springer-Verlag, Berlin.

Narins PM, Reichman OJ, Jarvis JUM, Lewis ER (1992) Seismic signal transmission between burrows of Cape mole-rat, *Georychus capensis*. *J Comp Physiol A* 170: 13–21.

Narins, PM, Lewis ER, Jarvis JUM, O'Riain J (1997) The use of seismic signals by fossorial southern African mammals: a neuroethological gold mine. *Brain Res Bull* 44: 641–646.

Nevo E (1961) Observation on Israeli populations of the mole-rat *Spalax ehrenbergi* Nehring 1898. *Mammalia* 25: 127–144.

Nevo E, Heth G, Beiles A, Frankenberg E (1987) Geographic dialects in blind mole rats: Role of vocal communication in active speciation. *Proc Natl Acad Sci USA* 84: 3312–3315.

Nevo E, Heth G, Pratt H (1991) Seismic communication in a blind subterranean mammal: a major somatosensory mechanism in adaptive evolution underground. *Proc Natl Acad Sci USA* 88: 1256–1260.

O'Connell-Rodwell CE, Arnason BT, Hart LA (2000) Seismic properties of Asian elephant (*Elephas maximus*) vocalizations and locomotion. *J Acoust Soc Am* 108: 3066–3072.

O'Connell-Rodwell CE, Hart LA, Arnason BT (2001) Exploring the potential use of seismic waves as a communication channel by elephants and other large mammals. *Amer Zool* 41: 1157–1170.

Rado R, Levy N, Hauser H, Witcher J, Adler N, Intrator N, Wollberg Z, Terkel J (1987) Seismic signalling as a mean of communication in a subterranean mammal. *Anim Behav* 35: 1249–1251.

Rado R, Himelfarb M, Arensburg B, Terkel J, Wollberg Z (1989) Are seismic communication signals transmitted by bone conduction in the blind mole rat? *Hear Res* 41: 23–30.

Rado R, Wollberg Z, Terkel J (1991) The ontogeny of seismic communication during dispersal in the blind mole rat. *Anim Behav* 42: 15–21.

Rado R, Wollberg Z, Terkel J (1992a) Dispersal of young mole rats (*Spalax ehrenbergi*) from the natal burrow. *J Mammal* 73: 885–890.

Rado R, Bronchti G, Wollberg Z, Terkel J (1992b) Sensitivity to light of the blind mole rat. Behavioral and neuroanatomical study. *Israel J Zool* 38: 323–331.

Rado R, Terkel J, Wollberg Z (1998) Seismic communication signals in the blind mole-rat (*Spalax ehrenbergi*): electrophysiological and behavioral evidence for their processing by the auditory system. *J Comp Physiol A* 183: 503–511

Randall JA (1997) Species specific footdrumming in kangaroo rats: *Dipodomys ingens, D. deserti, D. spectabilis. Anim Behav* 54: 1167–1175.

Randall JA (2000) Why do desert rodents drum their feet? *Amer Zool* 40: 1182–1183.

Randall JA (2001) Evolution and function of drumming as communication in mammals. *Amer Zool* 41: 1143–1156.

Randall JA, Lewis ER (1997) Seismic communication between the burrows of kangaroo rats, *Dipodomys spectabilis. J Comp Physiol A* 181: 525–531.

Randall JA, Rogovin KA, Shier, DM (2000) Antipredator behavior of a social desert rodent: footdrumming and alarm calling in the great gerbil, *Rhombomys opiums. Behav Ecol Sociobiol* 48: 110–118.

Sadka RS, Wollberg Z (2004) Response properties of auditory activated cells in the occipital cortex of the blind mole rat: an electrophysiological study. *J Comp Physiol A* 190: 403–413.

Sanyal S, Jansen HG, De Grip WJ, Nevo E, de Jong WW (1990) The eye of the blind mole rat, *Spalax ehrenbergi*: Rudiment with hidden function. *Invest Ophthalmol Vis Sci* 31: 1398–1404.

Schmitt A, Schuster M, Barth FG (1994) Vibratory communication in a wandering spider, *Cupiennius getazi*: female and male preferences for features of the conspecific male's releaser. *Anim Behav* 48: 1155–1171.

Schuch W, Barth FG (1985) Temporal patterns in the vibratory courtship signals of the wandering spider *Cupiennius salei* Keys. *Behav Ecol Sociobiol* 16: 263–271.

Storier D, Wollberg Z, Ar A (1981) Low and non rhythmic heart rate of the mole rat (*Spalax ehrenbergi*): control by autonomic nervous system. *J Comp Physiol* 142: 533–538.

Sturzl W, Kempter R, van Hemmen JL (2000) Theory of arachnid prey localization. *Phys Rev Lett* 84: 5668–5671.

Topachevskii VA (1976) *Fauna of the SSSR: Mammals, Mole Rats, Spalacidae*, p. 308. Amernind Publishing, New-Delhi.

Zuri I, Terkel J (1996) Locomotor patterns, territory, and tunnel utilization in the mole rat *Spalax ehrenbergi*. *J Zool* 240: 123–140.

3
Audiovocal Communication and Social Behavior in Mustached Bats

*Matthew J. Clement, Punita Gupta,
Nicole Dietz and Jagmeet S. Kanwal*

3.1 Introduction

Many species of bats live in large, communal colonies with little or no light. These conditions are expected to select for an extensive and sophisticated system for acoustic social communication. Although auditory communication in bats is gaining importance for neuroethologic studies, their social vocalizations have garnered less attention than their echolocation vocalizations (e.g., Fenton, 1984). Nocturnal activity, inaccessible roosts, and the ability to fly, all have hindered the study of bats' social habits, but there is evidence from several studies that some bat species have extensive vocal repertoires (Nelson, 1964; Bradbury and Emmons, 1974; Brown, 1976; Porter, 1979; Brown et al., 1983; Kanwal et al., 1994; Davidson and Wilkinson, 2002); see Fenton (1985) for review. Several researchers have also described the social functions of some bat calls. For example, mothers and pups use cries to locate each other (Pearson et al., 1952; Gould, 1971; Barclay et al., 1979; Matsumura, 1981; Esser and Schmidt, 1989). Males of some bat species use vocal displays to attract females to their territory (Wickler and Seibt, 1976; Bradbury, 1977; Davidson and Wilkinson, 2004). Vocalizations play a role in other social behaviors as well (Nelson, 1964; Porter, 1979; Thomas et al., 1979). For example, calls are used to defend territory (Barlow and Jones, 1997) and coordinate foraging activities (Wilkinson and Boughman, 1998).

Captive mustached bats, *Pteronotus parnellii*, use a repertoire of at least 19 simple syllables (Kanwal et al., 1994). Fourteen of these are shown in Fig. 3.1. These syllables may be emitted either as simple syllables, or as composites (a sequence of two or more simple syllables without any intervening silent interval), or as a sequence (or train) of similar syllables with short intersyllable intervals. In mustached bats, with a few exceptions, a syllable is from 3 to approximately 100 ms long. *Subsyllables* are short sound segments which are not emitted

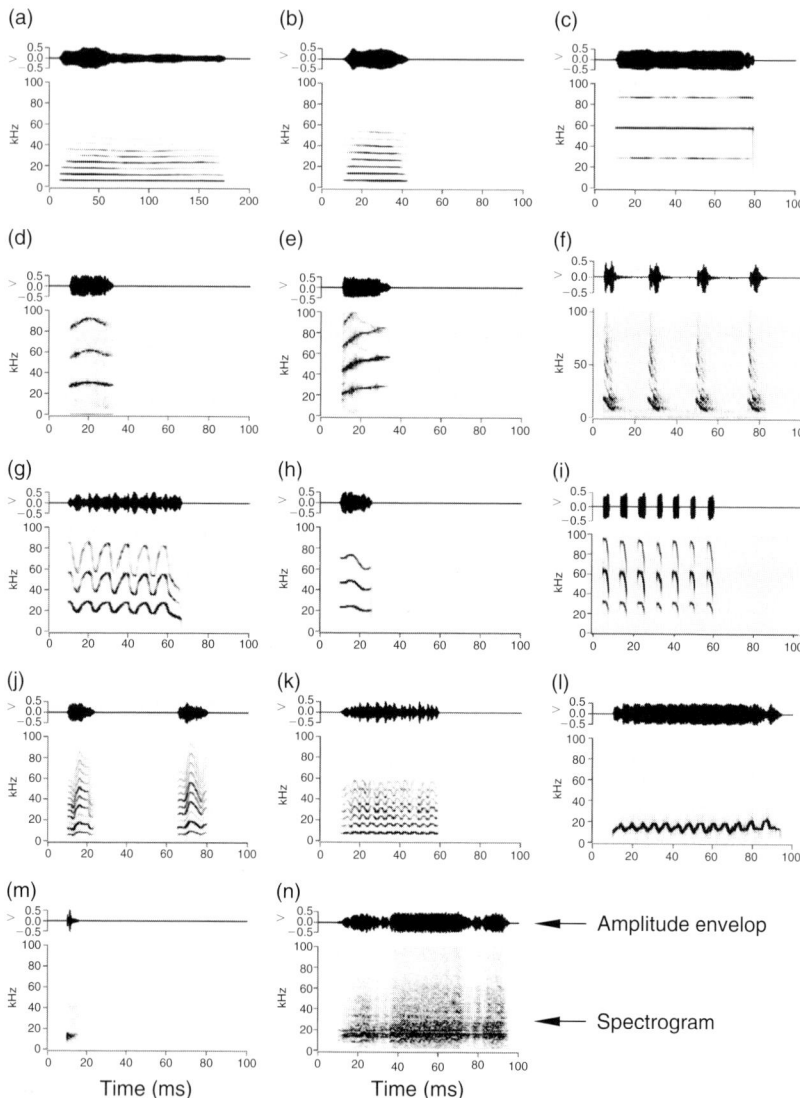

Figure 3.1. Amplitude envelopes (above) and spectrograms (below) of fourteen simple syllabic calls emitted by mustached bats (*P. p. rubigenosus* and/or *P. p. parnellii*). The stretched RFM call (sRFM) is a variant of the dRFM call (a) Long, quasi CF (QCFl); (b) short, quasi CF (QCFs); (c) short, true CF (TCFs); (d) single arched FM (sAFM); (e) bent upward FM (bUFM); (f) bent downward FM (bDFM); (g) stretched rippled FM (sRFM); (h) single humped FM (sHFM); (i) checked downward FM (cDFM); (j) short, wrinkled FM (WFMs); (k) fixed sinusoidal FM (fSFM); (l) fixed rippled FM (fRFM); (m) short, narrowband NB (NNBs); (n) rectangular broadband NB (rBNB).

independently but occur always in combination with other components in a composite. A subsyllable can be potentially reclassified as a simple syllable, if there is evidence that it is emitted independently. *Simple syllables* can be constant frequency (CF), frequency modulated (FM), or noise burst (NB) types. A *composite* consists of a sequence of two or more types of distinct components, each representing a simple CF, NB or FM pattern, combined together without any intervening silent interval. Each component in a composite corresponds to either a simple syllable or a subsyllable. Syllables emitted singly constitute a monosyllabic call. By this definition, a bat's syllable is equivalent to a "note" in a bird song (Kroodsma, 1977) and is acoustically analogous to the syllable in a speech sound in that it consists of either a vowel-like, CF or FM component, or a consonant-like, NB component, or a combination of these elements (Picket, 1980; Lieberman and Blumstein, 1988). This terminology allows one to equate, at least at the acoustic level, mammalian vocalizations with human speech sounds. The "phonetic-like" structural syntax in composites and trains of syllables in mustached bats has been studied at both the acoustic and neurophysiologic levels (Kanwal *et al.*, 1994; Ohlemiller *et al.*, 1996; Esser *et al.*, 1997).

The social contexts in which non-echolocation calls are emitted in mustached bats have not been described in detail previously (Gupta *et al.*, 1998). In nature, *P. parnellii* roost in large groups of hundreds to thousands of individuals inside large caves and feed on the wing for 5–7 h a night. They have been reported to segregate in separate roosts by sex except for the January mating season, when the sex ratio is even (Wilson, 1973). This species usually lives deep inside a cave and individuals are relatively shy so that their social behavior is difficult to study in captivity (Gupta *et al.*, 1998). Only recently, Vater *et al.* (2003) reported vocalizations of *P. parnellii* pups in the wild. This species is being increasingly used for studying neural mechanisms for processing and encoding communication sounds at different levels of the auditory system (Ohlemiller *et al.*, 1994, 1996; Esser *et al.*, 1997; Kanwal, 1999; Kanwal *et al.*, 2000; Portfors and Wenstrup, 2002; Portfors, 2004; Medvedev and Kanwal, 2004; Kanwal, Chapter 7, this volume). Here we describe several body postures and social behaviors and the associated communication sounds from observations in both free-flying and caged in *P. parnellii*. We have annually collected, 30–60 juvenile and adult mustached bats, *P. parnellii*, from either the Windsor and Douglas Caves in Jamaica (*P. p. parnellii*), 11 individuals of (*P. p. puertoricensis*) from Puerto Rico, and about 40 from a cave near Chaguanas, Trinidad (*P. p. rubigenosus*) and describe our behavioral observations gathered over the last several years. We discuss our data and its likely functional significance together with what is known about social vocal behavior in other mammals, including other bat species.

3.2 Social behaviors in captive free-flying bats

3.2.1 Maintenance and recording procedures

Captive bats were maintained at 28–30°C and 60–70% humidity under Biosafety Level II conditions with a 6:18 h light–dark cycle using dimmed fluorescent light (average light level of about 2 lux). The colony was kept in a 4.0 m × 2.5 m × 2.5 m flight room where they could fly at will and roost in two upside-down pots fixed on the ceiling. The inside surface of the pots was coated with a 1:1 cement and plaster-of-Paris mixture to provide a rough surface for hanging that is comparable to what is present inside limestone caves where they frequently reside in the wild. The bats were provided mealworms and vitamin fortified water *ad libitum*.

For a small group of bats, individuals were marked either with a distinctive collar or by a distinctive bare skin pattern created by applying depilatory cream on the head. During this phase, we made audio–video recordings of the bats with a Sony TRV310 digital HI8 video camera with an attached optimus unidirectional condenser microphone (flat, within a 5 dB range, up to 20 kHz; sampling rate of 44 kHz). These recordings were supplemented with ultrasonic recordings made with a bat detector (model U30; Ultrasound Advice), bandpass filtered (between 4 and 100 kHz) with a Krohn-Hite filter (model 3550). The broadband spectrum sounds recorded via the bat detector were digitized with CB-Disk software (version 1.0; Engineering Design) at a sampling rate of 250 kHz (flat within 5–100 kHz). The narrowband recordings were aligned to high-resolution broadband recordings to confirm call identity. The digitized sounds were analyzed with signal software (version 3.0; Engineering Design) using a 512-point Fourier fast transformation (FFT) and a Hanning window to produce spectrograms. We used a Lorex VQ-2120 infrared light so that we could record video in total darkness. Digital video was processed with Macintosh iMovie software. Since the bats spent almost all of their time inside the artificial roosts, we placed our camera 1.5 m below the roost and directed it upward to focus on this small area. Although we lost sight of some bats for short periods of time, the setup allowed us to make detailed behavioral observations on the roosting bats. Individual video recordings, typically lasted for 15–25 min at different times of the day.

3.2.2 Scoring behaviors and calls

We used an event-based system for recording behavior. Each behavior (described below) was a discrete act, with a clear beginning and end, and lasted for a few seconds or less. As we reviewed the video, we recorded each behavior, its context, the bats involved, its start time, and duration. We then used these data for individual recognition (using the markings placed earlier) and for assignment of calls to specific bats. The latter was accomplished off-line by repeated playback and close observations of video frames of interest at paused or slowed speed until correlations between vocalizations and social behaviors and the actors involved were firmly established.

We assigned a call to a particular bat by matching its mouth, head, and body movements to the recorded sounds. This technique was effective because the behavior and vocalization onsets were almost simultaneous. For example, during the most common behavior, vocalization began within 0.5 s of behavior onset 74% of the time (mean = 0.13 s; SD = 0.67 s). In a minority of cases, it was not possible to determine the caller. This difficulty was more frequent with short, soft calls that were not associated with any actions. We considered a call to be directed to a particular bat if the vocalizing bat turned its head towards the receiver when calling. In a minority of cases, the call was not directed toward any particular bat. We classified each syllable based on the spectral criteria described in Kanwal *et al.* (1994). We performed a frame-by-frame analysis of ~5 h (302 min) out of a total of 35.4 h of video recorded for *P. p. rubigenosus* to clearly document those discrete social behavior patterns that either promoted or were the result of interactions between conspecifics and distinguished them from those behaviors, such as autogrooming, that did not. If a bat did not exhibit any discrete social behavior, it was generally either resting or echolocating within the roost (Fig. 3.2c). The echolocation pulses of mustached bats may also have communicative value considering the low level of variability relative to social calls (Barclay, 1982; Fenton, 1994) and as reported recently for African large eared, free-tailed bats (Fenton *et al.*, 2004). However, we did not closely examine the communicative function of echolocation pulses here. We also did not observe any copulation, pregnancies, or births during this study. Our April–October observations did not include the January mating period.

For recordings of tagged bats in the roost, the occurrence or non-occurrence of a call or behavior was scored as a 1 or 0, respectively. Each incidence of call–behavior pair could then be scored as both (1,1), call only (1,0), or behavior only (0,1). Absence of a call or any distinctive behavior was labeled as (0,0). To score a (1,1), the two had to overlap in time. We used a Chi-squared contingency test to evaluate the association of calls and behaviors. To evaluate the effect of a disturbance on the rate of inspection, we treated them as Poisson processes and used confidence intervals from Dowdy and Wearden (1991). We also used a two-sample *t*-test to test for changes in the rate of marking and crouching.

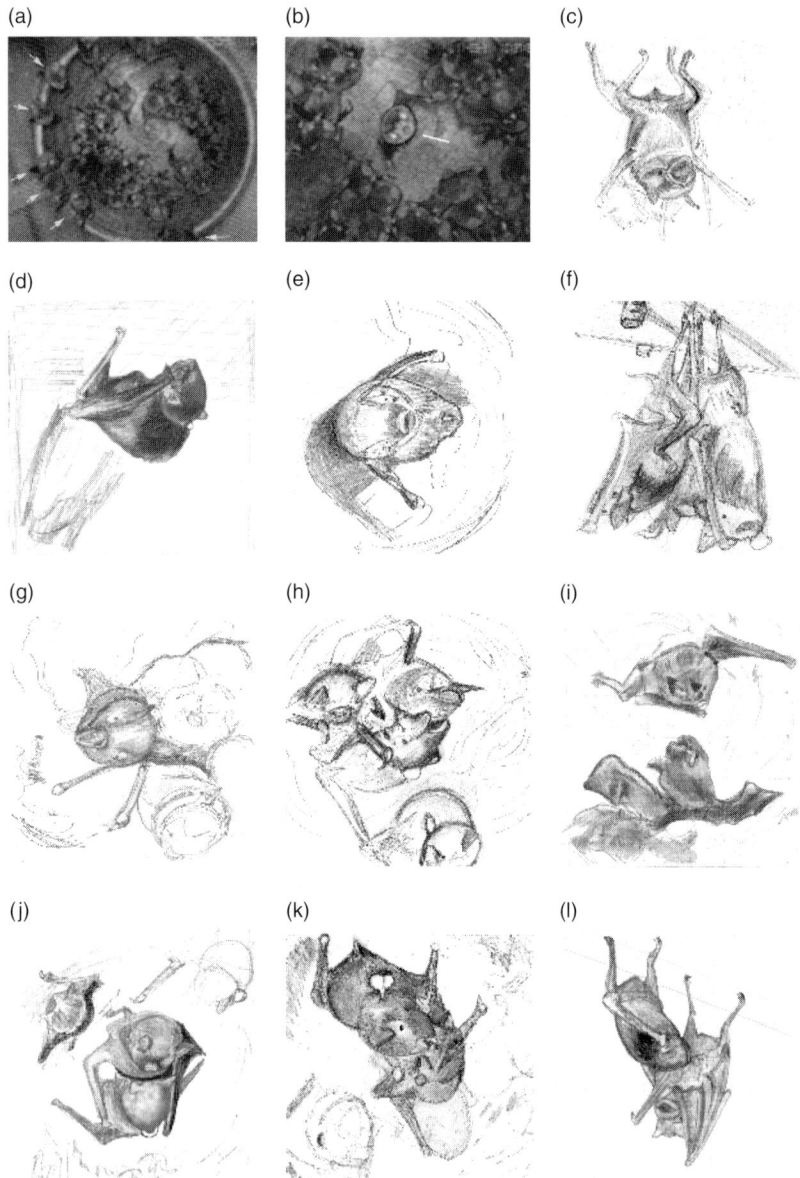

Figure 3.2. (a) Photomicrograph of a colony of bats roosting inside an upside-down clay pot hanging from the ceiling in the flight room. Arrows point to the guardian bats hanging along the rim of the pot. (b) Photomicrograph of bats hanging inside a clay pot with the alpha-male positioned in the center. The mouth of the alpha-male is open and it is emitting echolocation pulses. Both photographs are digitized video frames with the camera placed directly below the roost. (c) through (l) are drawings from digitized video frames. (c) Two bats hanging back-to-back

3.3 Results and discussion

3.3.1 Roosting structure, activity patterns, and social interactions

In captivity, the bats roosted in a tight, mixed-sex cluster inside an artificial roost. In the flight room, mustached bats exhibited a harem-like social structure with guardian bats hanging on the edge of the inverted clay pots. They were the first ones to fly out each time the colony was alerted by the entrance of an intruder/observer into the flight room (Fig. 3.2a). These singly hanging bats were both males and females and were easily alarmed. Typically, after an investigatory flight consisting of circling two to three times around the room, these bats returned to their hanging position along the rim of the pot; that is, at the periphery of the colony. In the presence of a large number of females in one shipment made in the month of November, mustached bats exhibited a harem-like social structure with an alpha-male positioned at the center of the colony of males and females (Fig. 3.2b). Although quantitative data were not recorded, the alpha-male was typically larger, had an approximately 20% higher body weight (normal body weight in *P. p. parnellii* is about 18 g), and exhibited enlarged sexual organs. At least one incidence of parental/altruistic type of behavior was recorded on video, when one individual of an unknown sex brought back a mealworm for feeding another individual roosting in the colony.

Due to the size of the room and the immobility of their food, the captive bats spent relatively little time in flight. In general, we did not observe any spontaneous copulation or pregnancies in captivity. Live births have been observed in our colony multiple times and generally resulted from the shipment of pregnant bats. In two instances, live births were observed after female and male bats were kept housed together in captivity for >1 year. Bat pups did not survive in captivity beyond a 3-month period. Most of the roosting bats tended to congregate in one tight group, although two roosts (pots) were available in the flight room. The roosting bats oriented themselves so that their ventral side

Caption for fig. 3.2. (cont.)
and echolocating. (d–f) Drawings of typical postures corresponding to those behavioral patterns in *P. parnellii* that are not typically associated with any vocalizations. (d) Two bats engaged in autogrooming. (e) One of two bats, hanging back-to-back crouching to touch its nose to the substrate. (f) A male moves his hips forward as he rubs the substrate with his anogenital region. (g) A bat is yawning. This behavior is occasionally accompanied with spitting. (h) One bat nipping another in the roost. (i) Two bats flicking their wings at each other. (j) Sketch showing the posture in which two bats wrestle during a fight inside the roost. (k) Two males turn their heads and "kiss." (l) A male bending to inspect the genital region of a female.

was typically facing the outer edge of the roost. Thus, most of the bats made dorsal-to-ventral bodily contact, although bats in the center of the roost had a tendency to make dorsal-to-dorsal contact. A ventral-to-ventral orientation between two conspecifics was only seen during agonistic behavior.

3.3.2 Behavioral postures and associated calls

The consistently structured relationship between vocalizations and behaviors is a strong indicator of the important role of calling during social interactions in *P. parnellii*. Based on our detailed analysis of both calls and behavioral patterns, we propose below the most likely function for several of these calls. Ten discrete types of behavior patterns together with the call types that were associated with each behavior are itemized below (Clement and Kanwal, submitted). The postures associated with each social behavior are shown in Fig. 3.2d–k.

3.3.2.1 Crouching behavior

During crouching, a hanging male or female slowly bent upwards and touched its nose to the substrate without vocalizing (Fig. 3.2d). This behavior was observed both in a roost as well as when bats were housed in a small cage, either solitary or with the other sex. Male bats, however, tended to crouch more frequently than females.

3.3.2.2 Marking behavior

A hanging bat moved its hips forward and briefly rubbed its anogenital region against the substrate without emitting any vocalizations (Fig. 3.2e). Only males engaged in marking, sometimes alternating with crouching. We found that when we held a male in a cage for a few days and then returned it to the colony, it significantly increased its crouching and marking from 0.27 ± 0.09 to 3.28 ± 0.72 times/min in the first 10 min after rejoining the colony ($t = 4.26$, $P = 0.048$, $n = 6$).

3.3.2.3 Grooming, licking, and yawning behaviors

This group of self-directed behaviors was exhibited spontaneously in a resting state. Grooming consisted of two major parts. The bat hung from one foot and used its other foot to comb its fur and wing membranes. It then inserted the foot into its mouth to clean it before repeating the combing (Fig. 3.2f). Bats could reach all parts of their bodies with their extremely flexible and dexterous legs. In the second component of grooming, a bat opened its wing or tail membrane and cleaned it with its tongue, covering every part of the surface. We only observed autogrooming, never allogrooming.

During yawning, a bat opened its mouth with its lower jaw at an obtuse angle (Fig. 3.2g). Fourteen percent of yawns were accompanied by the short narrow NB (NNBs) syllable, while 88% of NNBs appeared to arise spontaneously. This sound represented just 2% of all syllables recorded. We could also elicit the NNBs sound by putting water drops on a bat's nose, which it would spit out. In repeating this test with five of the bats, we recorded 113 NNBs sounds and only four other vocalizations.

3.3.2.4 Nipping behavior

A bat nipped at another bat with its mouth (Fig. 3.2h). This behavior was associated with rapid withdrawal and was distinct from a biting action, where a bat opened its mouth wide with the teeth showing and clamped its jaws with a purposeful grip. This behavior had less intensity than fighting. Twenty-eight percent of the times that a bat was nipped, it emitted a rectangular broadband NB (rBNB) and 6% of the times it emitted both an rBNB and a fixed sinusoidal FM (fSFM).

3.3.2.5 Wing-flicking behavior

A bat flicked a slightly open wing at another bat (Fig. 3.2i). This behavior was frequently seen when one bat approached other bats. This behavior was often accompanied by the descending rippled FM (dRFM) syllable pairs in case of *P. p. parnellii* or a monosyllabic, stretched rippled FM (sRFM) in case of *P. p. rubiginosus*.

3.3.2.6 Boxing and poking behavior

Bats employed a combination of these gestures that could lead to or, in more intense situations, be a part of fighting that included wrestling and biting. This behavior consisted of a rapid movement towards another individual followed by a rapid withdrawal.

Noisy, broadband "screech" sounds (rBNB and fSFM) were commonly associated with this behavior. In total we recorded 373 syllables of rBNB on 118 occasions, and 33 syllables of fSFM on 23 occasions. Of 62 fights, 45 included rBNB sounds, 5 had fSFM sounds, 10 had both, and only 2 had neither. rBNB were recorded 26 other times and fSFM one other time, either as spontaneous vocalizations or during more common behaviors, such as a bat moving or shifting. Outside the flight room, the bats emitted these calls when being handled. Poking a bat with a blunt probe elicited a few rBNB and fSFM syllables, but more often the bat sat still without vocalizing. In one instance, a poked bat responded with 33 single humped FM (sHFM) and 17 short-wrinkled FM (WFMs) over 30 s even though these calls were very rare in the flight room.

3.3.2.7 Wrestling and biting behavior

Bats clinging to each other for brief periods of time with either their forearms and wings (wrestling) or their jaws clamped on to the other with a purposeful grip (Fig. 3.2j) showed agonistic behavior patterns which seemed to constitute (within a short time frame) a true "fight" between two or three bats. This type of fighting behavior was associated with high-frequency tonal sounds, such as short, true CF (TCFs), fSFM, bent upward FM (bUFM) and long, wrinkled FM (WFMl) sounds. Altogether, we recorded 35 bUFM syllables and 17 TCFs syllables, of which 16 bUFM and 9 TCFs syllables occurred during fights.

When a satellite male intruded into the roost, the nearest resident male would approach and often attack him. During this intrusion behavior, the satellite male often emitted long trains of long, quasi CF (QCFl) syllables. We observed this 23 times.

3.3.2.8 Arching back and "kissing" behavior

A male bat rapidly and repeatedly arched its back to contact the head of another male (Fig. 3.2k). The second male often licked the face of the first male. This "kissing" event was recorded a total of 199 times in 5 h of analyzed video. Eighty-six percent ($n = 199$) of the times, a QCFl call was emitted by the male arching backwards. A total of 536 syllables were present within the 177 calls giving an average of 3 syllables per call.

3.3.2.9 Inspection behavior

During inspections, a male bat bent towards a female bat and brought his nose close to her genital region, presumably to smell her (Fig. 3.2l). The female frequently aided his inspection by turning her hips towards him and slightly spreading her wings. We observed 177 of these inspections and the female emitted a simultaneous QCFl call in 174 (98%) of them, for a total of 2154 syllables. During inspections, vocalization began within 0.5 s of behavior onset 74% of the time (mean = 0.13 s; SD = 0.67 s). The multisyllabic QCFl call was emitted in 174 (98%) of the instances. In a minority of cases, it was not possible to determine the caller. This difficulty was more frequent with short, soft calls that were not associated with any actions. In a more rare event, we observed a male inspect another male 12 times, with the inspected male calling 6 times (50%) – a total of 37 QCFl syllables. We did not see females inspect other male or female bats.

Inspection behavior was triggered when a researcher stepped briefly (for <1 min) into the flight room and then left ($n = 20$) to create a slight, nonspecific disturbance. We recorded the response of the bats for 5 min prior to and 10 min during and after the intrusion. The disturbance increased the rate of inspections. Immediately after the intrusion, the bats were agitated, echolocating constantly, moving about the roost, and sometimes flying away. In the second minute, as the bats' agitation wore off, inspections increased to a peak of 1.80/min. In the first 10 min after the disturbance, we recorded an average of 0.91 inspections per minute. Based on a Poisson distribution of inspection events, this peak rate was significantly higher (at a 95% confidence level) than the average rate under undisturbed conditions (Dowdy and Wearden, 1991).

3.3.2.10 Fly-by behavior

Mustached bats echolocate through the mouth both in flight and when roosting. During the fly-by behavior, a bat flew into, out of, or past the opening of the roost. Nearby bats emitted dRFM calls 158 times (223 syllables) and sRFM calls 91 times (184 syllables) at the time of this behavior. The sRFM is a new call, not reported for *P. p. parnellii*, but apparently common to *P. p. rubigenosus*. It is spectrally similar to a brief echolocation pulse with an extra up-down FM component at the end. The sRFM call can be considered to be structurally similar to the terminal "buzz" (echolocation pulses emitted in rapid succession), but with an upward FM connecting the intervening silence intervals. Of the 293 times that a bat flew either in, out, or past the opening of the roost (a fly-by), the flying bat emitted these calls 177 times (60%). We also recorded these calls in 29 of 62 fights. Re-introducing a bat into the flight room increased the use of these calls. Normally, these calls were used in just 12 of 43 fights (28%) and in 16 of 52 flights (31%). On 9 occasions when we introduced a new bat, these syllables were recorded during 17 of 19 fights (89%) and 161 of 241 flights (67%).

Occasionally, the bats shook their legs in a shivering motion, shifted their bodies, or walked a short distance to explore their environment. Sometimes, the bats emitted calls during this type of exploration behavior, but these movements were too subtle and graded, and their onset and offset too ambiguous to be scored in a consistent manner.

3.3.3 Behavioral context of simple syllabic calls

Altogether, we noted 1053 behavioral events (crouches: 88; marks: 56; yawns: 22; pecks: 79; wing flicks: 65; fights: 62; head turns and kisses: 199; inspections: 189; and fly-bys: 293) and 801 social calls that frequently accompanied some of these behaviors. Calling behavior included the following simple syllables (QCFl: 4349; rBNB: 373; fSFM: 33; bUFM: 35; sTCF: 17; NNBs: 51; fixed RFM (fRFM): 479; ndRFM: 502; WFMs: 2; NNBl: 25). All of these behaviors took up less than 10% of the bats' total time. On average and without regard to gender, the roosting bats divided their time between either resting or surveying their surroundings using echolocation while roosting. Altogether,

Table 3.1 A prediction success table generated from logistic regression of vocal activity (communication calls) against behavior. Numbers in the body of the table provide the classificatory power of the model and show how observations from each level of the dependent variable (call types) are allocated to predicted outcomes. Test results in the lower section of the table provide an indication of the strong association of the different call types (columns) with specific behaviors (rows). The likelihood ratios are highly significant and the specificity index ranges from 0.82 to 1.00. "Fight" includes wrestling and biting (Section 3.3.2.7). "Inspect-1" refers to spontaneously generated inspections and "Inspect-2" refers to inspections triggered by a disturbance due to an intruder (a person entering the flight room).

	Call groups							
	CF calls		FM calls				NB calls	
Behavior	TCFs	QCF1	bUFM	d/sRFM	pRFM	fSFM	rBNB	NNBs
Crouch	0.0	0.0	0.0	0.0	0.0	0.0	0.0	0.0
Mark	0.0	0.0	0.0	0.0	0.0	0.0	0.0	0.0
Yawn	0.0	0.0	0.0	0.0	0.0	0.0	0.0	2.0
Nip	4.5	0.0	4.5*	0.2	0.0	63.9**	11.1**	0.0
Flick	0.0	0.0	1.0	1.7*	2.4*	0.0	1.6	0.3
Fight	34.6**	0.0	42.7**	0.7	1.3	313.9**	156.0**	0.0
Kiss	0.0	5.1**	0.0	0.1	0.1	0.0	0.0	0.1
Inspect-1	0.0	18.4**	0.0	0.1	0.0	0.0	0.2	0.0
Inspect-2	0.0	20.3*	0.0	0.0	0.0	0.0	0.0	0.0
Fly	0.9	0.0	0.5	2.4**	3.4**	1.7	0.6	0.1
Likelihood ratio	58.6**	1084.0**	97.2**	293.0**	250.0**	111.0**	395.0**	67.5**
ρ^2	0.31	0.52	0.36	0.20	0.22	0.46	0.48	0.17
Specificity	0.99	0.82	0.99	0.85	0.90	0.99	0.96	0.97
Pearson coeff.	1.00	0.82	1.00	0.99	1.00	1.00	1.00	1.00

Note:
pFRM: paraboloid RFM.
$*P < 0.05$; $**P < 0.001$.

roughly 66.7% of the time was spent in either of these states, 20% of the time was spent grooming and 13.3% interacting with conspecifics.

A logisitic regression analysis of our data (see Table 3.1) shows a highly specific association between each call type and the behavior that is being concurrently expressed. Every call, except the NNBs syllable, is associated with one or more behaviors, and every behavior, except yawning, marking, and crouching, is associated with at least one call type. The call-to-behavior association is robust as indicated by the high likelihood ratios, and the ρ^2 values, which range from high to very high, with the exception of the NNBs sound. Each call type exhibits a high value for the specificity measure and a Pearson Chi-square of near unity. Some calls are produced almost every time the behavior occurs, whereas a few others are produced less consistently, perhaps because additional factors or contexts determine the reliability with which they can be triggered. Crouching and marking never involved any vocalization; nearly all (95%) of inspections produced QCFl syllables and 89% of fights produced BNB syllables. On average, all of these behaviors accounted for <10% of the daily, 24-h period. During the remainder of the time, when none of these behaviors were expressed, the bats spent <2% of their time emitting social calls. Free-living *P. parnellii* presumably budget their time differently (e.g. foraging is reduced and mating is virtually absent).

3.3.4 Inspection and appeasement

The inspection behavior in mustached bats appears to be a type of greeting behavior, in which the female emits the QCFl call while the males sniff the genitals of females. Most terrestrial mammals use scent as an identifying feature (Bradbury and Vehrencamp, 1998), so male mustached bats may be using scent from sebaceous glands to identify the females. Many other mammals incorporate genital inspection into their greeting behaviors (Bradbury and Vehrencamp, 1998) and at least three bat species have been reported to show this type of behavior (McWilliam, 1982).

The QCFl syllable may be a type of appeasement call. Its tonal structure is suited for appeasement (Morton, 1977) and it is common for mammals to incorporate appeasement gestures into greetings. In addition, the female's genitals are exposed to the male, so an appeasement call would be an appropriate way to protect herself from any aggression. The fact that males also direct this call to each other lead us to propose that this is an affiliative call, used to maintain peaceful relationships. However, since males use QCFl calls repeatedly and at a higher amplitude in male–male compared to male–female interactions, we suggest that there may be an element of competition in the call as well.

Captive mustached bats do not mate or produce offspring, so that a similar study on wild populations of *P. parnellii* would almost certainly uncover new behaviors and vocalizations. Although studies of captive animals may be less conclusive than studies in the natural environment, our results are a significant first step towards understanding audiovocal communication in a scientifically valuable species. Our observations benefited from the fact that the bats were housed as a group in an environment mimicking resting places in a cave and were largely undisturbed by the video recordings. They generally appeared to have a full and active social life and we could not discern any of the obsessive, repetitive, purposeless behaviors that indicate low animal welfare and high enclosure effects of caged animals (Berkson, 1968).

3.3.5 Territoriality and social dominance

Mustached bats live in such close proximity to one another, usually in physical contact, that one does not expect to find evidence of territoriality in the colony. However (careful observations of marked individuals indicate that males maintain their positions relative to each other and their roosting areas overlap very little (Clement and Kanwal, submitted). We propose that these territories are maintained by scent marks laid down during the marking behavior. Crouching may be used to monitor the scent boundaries. The most compelling evidence for this is that reintroduced bats mark their territory at 10 times their normal rate. During this initial phase, they are repeatedly attacked by other males. Within 24 h, the marking behavior and the attacks are greatly reduced. Only the males mark and they maintain more strict territories than females. Scent marking of territories with various exudates has been reported in several bat species (Nelson, 1965; Holler and Schmidt, 1993; Balasingh et al., 1995; Defanis and Jones, 1995), including marking with the anogenital region for at least four species (McWilliam, 1982; French and Lollar, 1998). It would be interesting to investigate if the satellite bats that were excluded from roosting within colonies was a result of scent-marking competition.

During free-ranging observations of the natural roosting behavior of bats, it was observed that some males always stayed outside the roost (pot). When one of these satellite males entered the roost, other males attacked him until he left and found a spot to hang on the ceiling in the flight room. To better understand the social structure in the colony of captive bats, we perturbed the composition of a colony of 20 bats consisting of seven dominant and three satellite males by removing the dominant male bats for 1 week from a colony. We kept dominant males visually and acoustically isolated from the rest of the bats by placing them in a sound attenuated chamber. We successively returned each individual to the flight room over a period of 2 weeks in the month of May. When we removed the seven dominating males from the roost, the three satellite males entered the roost and joined the females. When we reintroduced all of the seven dominating males, the satellite males were once again ejected from the roost and the dominating males regained their previous status in the colony. We also found that any male that was held out of the colony for several days was attacked when he was returned to the flight room. Based on forearm length, the satellite bats were not significantly smaller than the dominating bats ($t = 0.77, P > 0.5$) and based on weight, they were only insignificantly larger ($t = 0.16, P > 0.5$).

The dominating males significantly increased their crouching and marking on reintroduction into the colony. In 6 of 9 cases, a reintroduced male did not immediately join the colony. Instead, he flew around the room, and if he did land in the roost, he was attacked and driven away by other males. Under these circumstances, we did not observe the marking and crouching behaviors. However, on three occasions, a male quickly entered the colony and was not driven away. These reintroduced males marked or crouched an average of 3.28 (SE = 0.72) times per minute for the first 10 min in the roost. On the following day, the same males marked or crouched at a significantly lower rate of 0.27 (SE = 0.09) times per minute ($t = 4.26$, $P = 0.048$; two-sample t-test).

3.4 Social interactions in caged bats

3.4.1 Recording procedures

Placing bats in a small cage (12 in. × 12 in. × 4 in.) with an arched ceiling, a wooden base and sides of nylon mesh and plexiglass facilitated detailed observations of social interactions and acquisition of clean sound recordings (Fig. 3.3). Three to four bats were tagged with either small triangles, squares or circles that were cut from a 3 mm wide strip of white paper and pasted with a

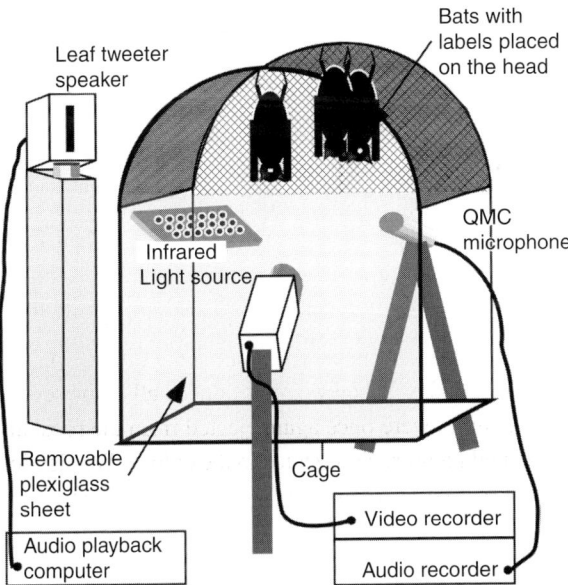

Figure 3.3. Scheme showing the specially constructed cage and set-up for simultaneous audio and video recordings of captive *Pteronotus*. A video recorder, oscilloscope, audio recorder, speaker and video monitor were used to monitor and record behaviors and sounds.

minimal amount of glue (Loctite 411) onto the hair on a bat's head. The encaged bats together with the recording set up was placed inside a 2.1 m by 1.2 m soundproof room maintained at an optimal temperature of nearly 31°C. Recordings (up to 2 h sessions) were restricted to one per day for any bat and conducted after a 2–3 month post-acclimation period after a new shipment of bats.

To record the sounds emitted, a broadband QMC S-200 microphone connected to a RACAL ST0705 tape recorder was placed about 3 ft from the nylon mesh cage. In addition, bat vocalizations were monitored by connecting the microphone to a bat detector with loudspeaker for heterodyning high frequencies emitted by the bats to relatively low frequencies audible to the human ear. Excess noise below 5 kHz and above 100 kHz was filtered using a Krohn-Hite filter (model 3550) with a 24-dB/octave slope. The frequency setting of the bandpass filter matched the frequency range of communication calls from 7 kHz to 100 kHz (120 kHz for biosonar). A Hewlett Packard 465A amplifier amplified (20 dB) the oscilloscope trace and audibility of the bandpassed frequencies. A two-channel Tektronix 2211 digital storage oscilloscope was used to compare the quality of the original and the recorded sounds. A mini-speaker connected to the output of the amplifier was also used at times to monitor the bat vocalizations. For a contextual analysis of vocal behaviors, observations were made at various times in the late afternoon and early to late evening because this species is most active around 7:00 p.m.; that is close to the time at which they emerge from their cave for feeding (Kanwal, unpublished observation). To facilitate video recordings, either one or two custom-made infrared light sources were placed near the cage and aimed at the bats. Recordings were obtained with a near infrared sensitive video camera (Fig. 3.3). Bats have a tendency to hang from the highest elevation; that is, in the arched area of the cage. This allowed us to keep the bats within the field of view and the focal depth of the camera. After the observation cage was placed within the flight room, the bats were given several minutes to settle down before recordings started. Audio and video recorders were turned on simultaneously and synchronization was facilitated by periodically recording the reading from the digital counters on the videocassette recorder (VCR) and the tape recorder.

For call analysis, sounds recorded on the RACAL tape recorder were played back at one-half to one-sixty-fourth the speed at which they were recorded, and filtered and amplified before being analyzed and digitized by the SIGNAL™ computer program (Engineering Design, Boston, MA). With another computer software (RTS™), spectrograms of the calls were generated and visualized on-line as a rolling display. By "freezing" the digitized segment of interest as a spectrogram on SIGNAL™ or RTS™, frequency contours and harmonics were identified and classified as done before (Kanwal *et al.*, 1994). The custom-modified

VCR (Sony ß-L-750) allowed the visual recordings to be analyzed at various slow motion speeds or as single frames. This capability allowed us to analyze actions that were completed in the subsecond range.

3.5 Results and discussion

3.5.1 Natural aggression

Calls were most frequently emitted by both sexes in the context of agonistic interactions. Figure 3.4a shows postures and calls of two males ("m_1" and "m_2") and one female ("f") over an approximately 1 min sequence. The first frame shows m_1 emitting echolocating pulses and exhibiting a threatening posture with the wings folded and spread out. In the second frame, the same bat (m_1) exhibits crouching behavior as he prepares to join with m_2/f. As m_1 and m_2 come in contact with each other in frame 3, the female begins to crouch. Finally, the group rotates with the female "upset" at being displaced from her roosting position where she is in contact with m_1 (frame 4). A vocal exchange ensues in the next 2 s (frames 4–7) with the female emitting a sequence of noisy fSFM type of syllables. The male, m_1, responds with several WFMs syllables. This observation, however, was partially

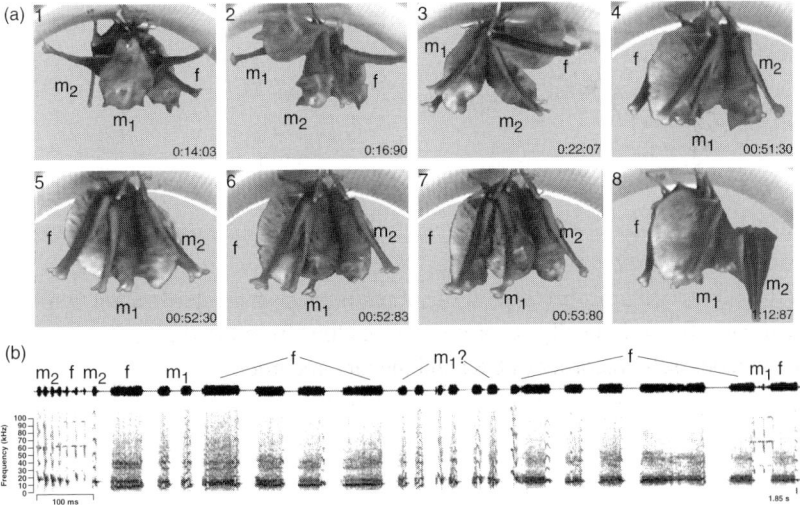

Figure 3.4. (a) Sequence of video frames showing agonistic interactions between two males and one female. (b) Spectrograms (bottom) and amplitude envelopes (top) of a brief episode of vocal exchange between the three bats during the shown behavioral sequence. "m_1" and "m_2" indicate the two males and "f" the female. See text for details. Photographs were retouched to generate a uniformly gray background.

obscured by the limbs of both m_1 and f resting at the side of their mouths (frame 5). The female emitted six more fSFM syllables (frame 7) and a few seconds later, the group reached a "compromise" as suggested by preening of wings by m_2 (frame 8). A spectrographic analysis of the sounds revealed that fSFM, WFMs, and bUFM were the predominant syllables emitted during this interaction (Fig. 3.4b).

Combinations of several different types of syllables may be associated with aggression; the predominant ones include the fSFM and rBNB, as well as noisy fSFM that is intermediate in its acoustic characteristics between the other two syllables. These signals may convey threat, anger, and/or pain. Composites of rBNB and fSFM syllables indicate that both may reflect very similar emotions of the calling animal. The level of noisiness in the fSFM call may correspond to the intensity of anger, but this remains to be tested. The use of broadband sounds in agonistic encounters is consistent with findings for numerous other bat species (Porter, 1979; Fenton, 1985; Pfalzer and Kusch, 2003).

Broadband vocalizations have been observed when one bat is hanging from another, and when one bat is wrestling or biting another. Male *Carollia perspicillata*, emit a similar sound when they are defending female bats (Bradbury and Emmons, 1974). One interpretation of this observation is that loud broadband sounds of relatively low frequency may be related to a warning, or a threatening signal. As observed in free-flying bats, the sTCF and bUFM calls were also produced during fights. A similar use of these two syllables is also consistent with the presence of composites that combine bUFM and sTCF syllables. Their appearance at the beginning of vocal exchanges suggests that they may signal a threat or trigger a fight.

3.5.2 Experimentally elicited aggression

Intraspecific aggression was elicited artificially by (i) introducing a conspecific, usually a male, to a group of three bats isolated from a colony; (ii) playing back a previously recorded communication call; and (iii) probing a single bat roosting among a group of bats. These manipulations were primarily carried out in *P. p. parnellii*. Figure 3.5a shows spectrograms of a sequence of vocalizations containing several "fSFM" type of syllables that were emitted when a male was introduced into a cage containing a male and two females from the same colony. These sounds were punctuated by a series of several bUFMs, sHFM and both single and double arched FMs (sAFM and dAFM). The most commonly emitted sound by males was a noisy fSFM with duration ranging from 30 ms to 100 ms and a fundamental frequency of about 9 kHz. These fSFMs accompanied by boxing and biting were effective in keeping the new male away from the two females roosting with the first male. The variety of syllables emitted by the new male may

3.5 Results and discussion

indicate a desire to be included in the group, although the details of this "language" cannot be deciphered any further at this stage.

The changes in vocalizations resulting from a second example in which an unfamiliar male (obtained from a different colony and kept isolated for nearly one year) was introduced to a group of bats are shown in bar graphs in Fig. 3.5b.

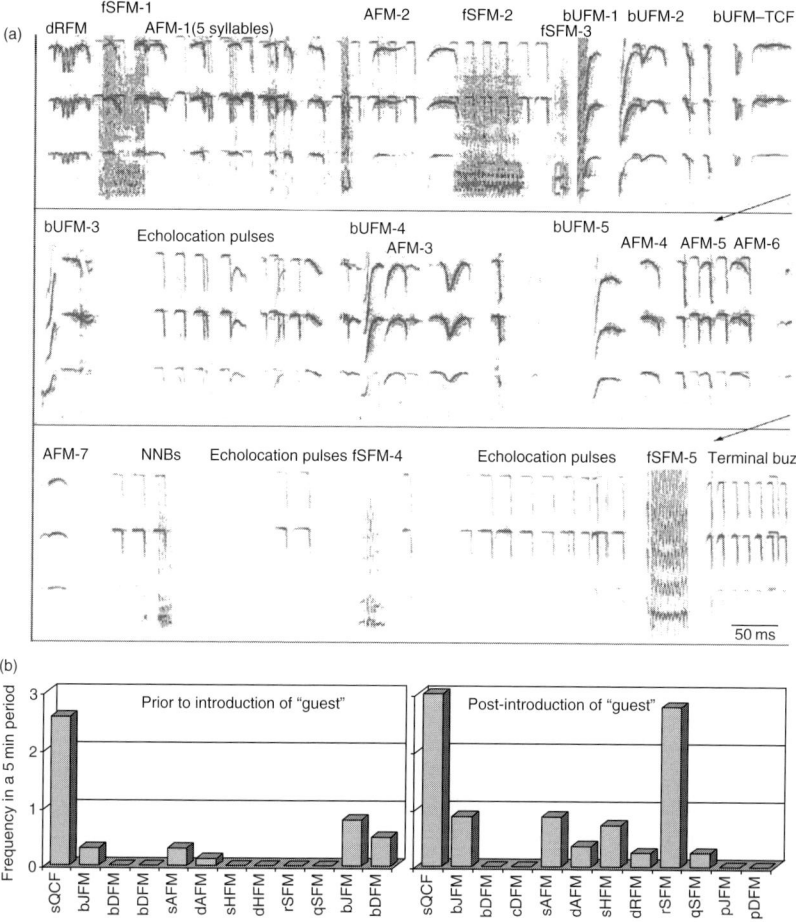

Figure 3.5. (a) A continuous sequence (shown in three panels) of spectrograms of vocalizations consisting of several different syllable types emitted during agonistic interactions that included biting and boxing among three bats (two males and one female). (b) Bar graphs showing the frequency of sounds emitted over a 5-min period immediately before and after introducing a new male "guest" bat (three males and one female).

There was a shift in the type of syllables produced most frequently in a 5 min recording period before and after the introduction of the unfamiliar male. Before introducing the "guest", paraboloid upward FM (pUFM) and paraboloid downward FM (pDFM) subsyllabic elements were present within calls. After the introduction of the guest, these sounds were no longer emitted. fSFM was commonly associated with rBNB as a composite prior to introduction of the guest, and afterwards was emitted almost exclusively as a simple syllable. New syllables were also recorded after introduction of the guest and included fSFM and other FM types of syllables, such as bUFM, sAFM, and sHFM. The total number of syllables produced increased and stereotypic behaviors associated with aggressive interactions, for example, wing flicking and boxing, were commonly observed. Although QCFs was frequently emitted, its frequency did not change much with the introduction of the unfamiliar male. Whereas it is difficult to make valid conclusions from one instance, these data conform to other observations both in free-flying and encaged bats. Altogether, the data indicate that the rBNB, fSFM and bUFM call types are associated with agonistic interactions; the exact role or meaning of each call, however, remains unclear.

Call playback was generally unsuccessful in eliciting vocal responses from the bats. No clear stereotypic behavior patterns were observed in response to a random presentation of the 14 simple syllables shown in Fig. 3.1 at sound intensities ranging from 60 to 100 dB SPL. A group of five or more bats showed an overall reflex flexion of their legs and head movements to some of the noisy syllables presented at intensities of about 90 dB SPL. For other syllables, no consistent vocal or behavioral response was observed. Of the eight sessions of playback experiments, each using at least four different digitized calls presented randomly at two or three intensity levels, there was only one example in which a clear response was observed. In this case, three of the four bats present in the cage clustered together within seconds, leaving the fourth bat to hang on its own. As the excluded bat approached the cluster, the rBNB syllable was played back at ~90 dB SPL. Upon playback of this call, the excluded bat quickly turned away in a direction opposite to its approach. This sequence of video frames is depicted in Fig. 3.6a, where the body circumference of each bat is represented as a circle when viewed from the bottom of the cage. Other calls failed to elicit a similar response turning away reaction in response to vocalizations of broadband sounds was also observed during spontaneous male-to-male approach behavior in free-flying bats within a colony.

The third and final manipulation used to elicit communication sounds was prodding or probing the bats with a blunt instrument. Two sounds elicited during probing were noisy fSFM and rBNB and were accompanied by wing-flicking

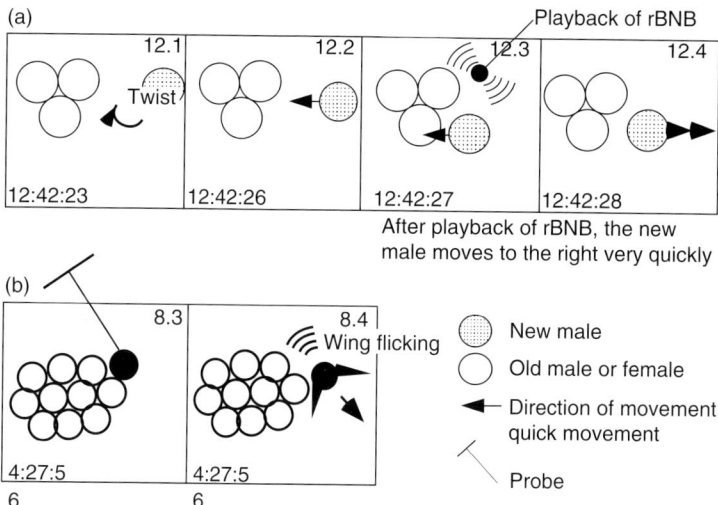

Figure 3.6. (a) A cluster of bats schematized as open circles and an approaching bat (dotted circle), both in a bottom-up view. The sketch, taken from a sequence of video frames, shows the retreat of the approaching bat when rBNB was played back. No vocalizations were emitted in response to the playback. (b) A cluster of bats (unfilled circles) viewed from below. A bat (solid filled circle) was poked by a blunt probe, which triggered the wing-flicking behavior.

as shown in Fig. 3.6b. As a comparison, in non-probing situations, a variety of sounds were emitted but did not include any noisy/pure fSFM or rBNB types of syllables. This suggests, once again, that fSFM and rBNB are not emitted "randomly" and may be related most commonly to aggression.

3.6 Call types and acoustic signal design

Data available from a few species of bats indicate that bat vocalizations generally follow Morton's (1977) predictions of acoustic signal design derived from vocalizations of a number of mammalian species (Fenton, 1985; August and Anderson, 1987). According to Pfalzer and Kusch (2003), harsh broadband calls are widely used during aggression, although buzzes and trills also fill this role. They also report that tonal calls are used between mothers and pups, whereas more complex calls are used during mate attraction behavior. Both sexes of *Megaderma lyra*, an Old World microchiropteran species, use a low, multi-harmonic "grumble" as an aggressive call, and males use a mix of tonal CF and FM calls in a display for females (Leippert, 1994). Male white-lined bats

(*Saccopteryx bilineata*) use harsh, broadband calls to threaten other males and direct tonal calls towards females (Davidson and Wilkinson, in press).

Similar to the stretched RFM (sRFM) in mustached bats, *Noctilio leporinus, Myotis volans,* and *Myotis lucifugus* produce a "honk" by adding a downward frequency sweep to their echolocation pulse (Suthers, 1965; Barclay et al., 1979; Fenton and Bell, 1979). *Pteropus poliocephalus* and *C. perspicillata* both use a "screech" to avoid collisions (Nelson, 1964; Porter, 1979). An inverted V-shaped call has been commonly observed in several species of bats. We did not record any syllable with a classic inverted-V-shaped call structure from mustached bats. If a conspecific of this species is aroused by a stimulus, they investigate it by emitting echolocation pulses directed at the source. Based on published reports, *Antrozous pallidus, C. perspicillata, M. lucifugus,* and *Noctilio albiventris* also lack an inverted-V call (Brown, 1976; Barclay et al., 1979; Porter, 1979; Brown et al., 1983). The best case for an inverted-V call comes from species that are active in well-lit conditions. Inverted-V calls appear to be associated with different functions. Mothers of *P. poliocephalus* emit an inverted-V type call upon locating their pups (Nelson, 1964). Bradbury and Emmons reported that *S. bilineata* males emit an inverted-V during a hover display to females. Davidson and Wilkinson (2004) found in the same species that a similar call, the screech-inverted-V, generally had no contextual association, and described the call to be a neutral notification "bark" for advertising territorial claims. Thus, variants of the RFM syllable, such as the paraboloid RFM (pRFM) and sRFM, may be used to maintain spacing among individuals in a large colony of bats.

Earlier analyses of the acoustic structure of calls revealed that mustached bats produce a rich and complex repertoire of communication calls (Kanwal et al., 1994). The question regarding the social significance of each call type remained unanswered. Here, at least 10 different types of behavioral contexts and body postures were identified in which different types of syllables were emitted. Mustached bats used the harsh, broadband rBNB and fSFM during aggression and the tonal QCFl for appeasement. Tonal TCFs and bUFM were also emitted during agonistic interactions. These may indicated fear leading to appeasement on the part of the receiver in an effort to please the aggressor. One major difference between these syllables and QCFl is that their fundamental frequency either averages or is modulate upwards to about 30 kHz, whereas QCFl is between 6 and 8 kHz depending on the subspecies. In a behavioral context, therefore, the acoustically distinct categories of call types may be sequenced from the noisy, broadband calls to the more tonal calls modulated upward to high frequencies and correspond to four motivationally distinct states (Fig. 3.7; also, see Ehret, Chapter 4, this volume). These states generally

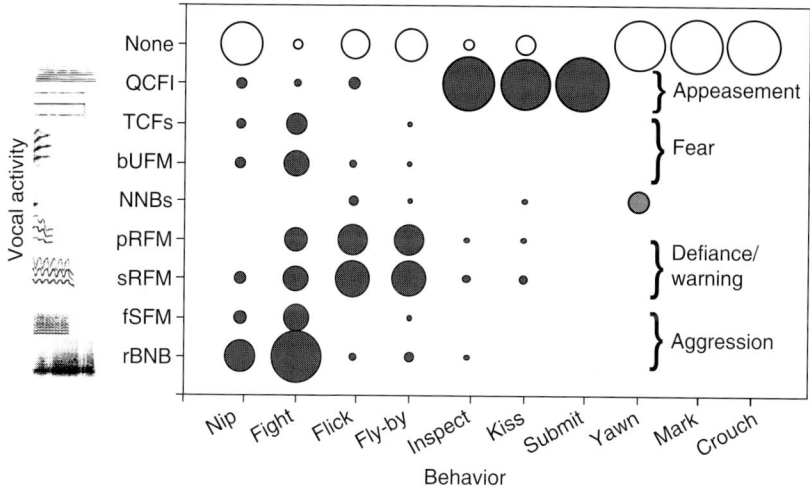

Figure 3.7. A bubble-plot (filled dark gray circles) showing the association of different calls with different behavioral states or emotions that influence group cohesion. An act or state of *appeasement* impacts positively on the affiliation between two individuals and may be triggered either spontaneously or because of an impending uncertainty; a state of *fear* implies increased awareness to external stimuli in preparation for fight or flight and preempts a sensation of pain or stress; a state of *defiance* or to *warn* implies readiness to combat when confronted by the aggressor and may be triggered by either fear or insecurity; a state of *aggression* is an expression of dominance and one possible outcome of a state of defiance and warning. The size of the bubble is proportional to the normalized (percentage of total events) frequency of occurrence of a call type. Unfilled circles are proportional to the percentage of the total number of events when a call did not accompany a particular behavior. Filled light gray circle indicates an "organic" (spitting) sound that is sometimes associated with yawning, but is not a vocalization. No calls were emitted during yawning, marking, and crouching. Fighting, which can be a complex and long-lasting behavior, produced the largest variety of call types.

follow Morton's (1977) Motivation–Structure hypothesis. However, one may also take this classification scheme to represent only the extreme ends of a continuum of motivational and behavioral states. Thus, in general, our data are consistent with the Motivation-Structure hypothesis which appears to provide a useful though a possible over-simplification of acoustic design in the animal world.

Recent neurophysiologic studies reveal that mustached bats have the cognitive machinery to identify not only the various call types, but also the syntax between simple syllables (Ohlemiller *et al.*, 1996; Esser *et al.*, 1997). In addition, there is a good chance that bats can detect subtle variations in call types (Esser and Lud, 1997) and use this information to label the identity and/or the mood of the emitter as is the case in some birds, such as the crested cockatoo

(*Calyptorhyncus funereus latirostris*), where 12 different calls, each announcing a different mood, have been identified (Saunders, 1983). In primates too, the loudness and spectral form of syllables can vary with the mood of the emitter. Variants of a single syllable, however, can also have a different meaning when used in different contexts (Petersen, 1982; Snowdon, 1982). Clearly, bats are capable of extracting large amounts of information from small variations in echolocation signals, which suggests that they, at least, have the potential to extract information from communication signals as well. Thus, the number of perceptual categories due to cognitive processing may be much larger given the variety of cortical areas and neuronal responses in the brain of mustached bats (see Kanwal, Chapter 7, this volume). Furthermore, the meaning of call variants may be influenced by the history/memory and expectation of an animal as well as the context in which they are perceived. Some calls may also be used to deceive or transmit information with little or no emotional content, such as about the availability, variety and location of food (Karakashian *et al.*, 1988).

Long-term studies in mustached bats may show that variants of social calls can also be learnt during maturation and growth. Studies in the Mexican free-tailed bat, *Tadarida brasilensis*, show that pups can recognize or distinguish their mother's calls from those of another female (Balcombe, 1990), and the echolocation calls of *Rhinolophus ferrumequinum* vary seasonally and over a life time in a predictable manner (Jones and Ransome, 1993). Recent studies in adult spear-nosed bats, *Phyllostomus discolor*, show that pups can learn to mimic calls of their mother and that conspecifics of *Phyllostomus hastatus* can modify their calls according to their auditory environment (Esser and Schmidt, 1989; Boughman, 1998). These studies suggest that infants and adults at least of some bat species are capable of learning and modifying their calls.

Except for composites that include the rBNB syllable, we are still unable to assign a context or meaning to composites and complex sequences of calls that are less commonly observed. Since captive mustached bats do not mate and produce offspring, field studies on wild populations of *P. parnellii* will almost certainly uncover new behaviors and vocalizations. Although studies of captive animals may be less conclusive than studies in the natural environment, our results are a significant first step towards understanding auditory processing and perception in a species that emits a complex echolocation pulse and employs a rich repertoire of calls for audiovocal communication. Clearly, more work is needed to understand the role of long sequences of vocalizations compared to the association of discrete syllables with behavioral postures before it will be possible to unravel the full potential of audiovocal communication in this bat species.

3.7 Conclusions

Mustached bats, *P. parnellii*, are both social and highly vocal, and emit an acoustically rich repertoire of calls. These calls consist of at least 19 simple syllables emitted either singly or in a train as well as complex combinations of multiple syllables within composites and call bouts. From audio–video recordings made over a number of years of both, captive free-flying and encaged animals, we conclude that in addition to having an extensive call repertoire, this species exhibits a variety of body postures and behavior patterns, such as agonistic displays, scent marking, and fighting. A statistical evaluation of the data on freely behaving bats established a tight coupling between many simple syllabic calls and specific behavior patterns. We found that two echolocation-like calls (sRFM and fRFM) are used to maintain spacing in the colony, three broadband calls (rBNB, fSFM, and bUFM) are frequently emitted during agonistic interactions and one tonal appeasement call, QCFl, is produced in a variety of social situations. More complex vocal exchanges were also elicited and reliably recorded with encaged bats. According to the Motivation–Structure hypothesis of acoustic signal design, usage of a call type is governed or prompted by the animal's motivational state. Our data on the association of calls with behaviors is generally consistent with this hypothesis and provides one valid framework within which to further study audiovocal communication behavior. Overall, our results indicate that even though bats are specialized to survive in a unique ecologic niche, their calls and behavior patterns are similar to those observed in many other social mammals. Thus, studies of audiovocal communication in mustached bats are not only interesting from a behavioral perspective, but a knowledge of the usage and meaning of calls will be critical in guiding future neuroethologic and neurophysiologic research on auditory processing in this species.

Acknowledgments

We thank Mr. F. Muradali for collecting the bats and the Department for Wildlife in Trinidad for permission to export bats. Drs. G. Ehret and D. Solick provided many helpful suggestions to improve this manuscript. Some of the behavioral work and acoustic analysis was initiated in the laboratory of N. Suga and in collaboration with S. Matsumura, who contributed preliminary observations that stimulated a systematic analysis of the social behavior as presented in this study. This material is based in part upon work supported by the National Science Foundation under Grant No. 0080822 to J.K. Additional support was provided by grant Nos. DC02054 to J.K. and DC04733 to C. Portfors from the National Institutes of Health.

REFERENCES

August PV, Anderson JGT (1987) Mammal sounds and motivation-structural rules: a test to the hypothesis. *J Mammal* 68: 1–9.

Balasingh J, Koilraj J, Kunz TH (1995) Tent construction by the short-nosed fruit bat *Cynopterus sphinx* (Chiroptera, Pteropodidae) in southern India. *Ethology* 100: 210–229.

Balcombe JP (1990) Vocal recognition of pups by mother Mexican free-tailed bats: do pups recognize their mothers? *Anim Behav* 39: 980–986.

Barclay RM (1982) Interindividual use of echolocation calls eavesdropping by bats. *Behav Ecol Sociobiol* 10: 271–275.

Barclay RM, Fenton MB, Thomas DW (1979) Social behavior of the little brown bat, *Myotis lucifugus* II vocal communication. *Behav Ecol Sociobiol* 6: 137–146.

Barlow KE, Jones G (1997) Function of pipistrelle social calls: field data and a playback experiment. *Anim Behav* 53: 991–999.

Berkson G (1968) Development of abnormal stereotyped behaviors. *Develop Psychobiol* 1: 118–132.

Boughman JW (1998) Vocal learning by greater spear-nosed bats. *Proc R Soc Lond B* 265: 227–233.

Bradbury JW (1977) Lek mating behavior in the hammer-headed bat. *Z Tierpsychol* 45: 225–255.

Bradbury JW, Emmons LH (1974) Social organization in some Trinidad bats: I. Emballonuridae. *Z Tierpsychol* 36: 137–183.

Bradbury JW, Vehrencamp SL (1998) *Principles of Animal Communication*. Sinauer Associates, Sunderland, MA.

Brown P (1976) Vocal communication in the pallid bat, *Antrozous pallidus*. *Z Tierpsychol* 141: 34–54.

Brown PE, Brown TW, Grinnell AD (1983) Echolocation, development, and vocal communication in the lesser bulldog bat, *Noctilio albiventris*. *Behav Ecol Sociobiol* 13: 287–298.

Davidson SM, Wilkinson GS (2002) Geographic and individual variation in vocalizations by male *Saccopteryx bilineata* (Chiroptera: Emballonuridae). *J Mammal* 83: 526–535.

Davidson SM, Wilkinson GS (2004) Function of male song in the greater white-lined bat, *Saccopteryx bilineata*. *Anim Behav* 67: 883–891.

Defanis E, Jones G (1995) The role of odor in the discrimination of conspecifics by pipistrelle bats. *Anim Behav* 49: 835–839.

Dowdy S, Wearden S (1991) *Statistics for Research*. Wiley and Sons, New York.

Esser KH, Lud B (1997) Discrimination of sinusoidally frequency-modulated sound signals mimicking species-specific communication calls in the FM bat *Phyllostomus discolor*. *J Comp Phys* 80: 513–522.

Esser KH, Schmidt U (1989) Mother-infant communication in the lesser spear-nosed bat, *Phyllostomus discolor* – evidence for acoustic learning. *Ethology* 82: 156–168.

Esser KH, Condon CJ, Suga N, Kanwal JS (1997) Syntax processing by auditory cortical neurons in the FM–FM area of the mustached bat, *Pteronotus parnellii*. *Proc Natl Acad Sci USA* 94: 14019–14024.
Fenton MB (1984) Echolocation: implications for ecology and evolution of bats. *Q Rev Biol* 59: 33–53.
Fenton MB (1985) *Communication in Chiroptera*. Indiana University Press, Bloomington.
Fenton MB (1994) Assessing signal variability and reliability – to thine ownself be true. *Anim Behav* 47: 757–764.
Fenton MB, Bell GP (1979) Echolocation and feeding behavior in four species of *Myotis* (Chiroptera). *Can J Zool* 57: 1271–1277.
Fenton MB, Jacobs DS, Richardson EJ, Taylor PJ, White E (2004) Individual signatures in the frequency-modulated sweep calls of African large-eared, free-tailed bats Otomops martiensseni (Chiroptera: Molossidae). *J Zool* 262: 11–19.
French B, Lollar A (1998) Observations on the reproductive behavior of captive *Tadarida brasiliensis mexicana* (Chiroptera: Molossidae). *Southwest Nat* 43: 484–490.
Gould E (1971) Studies of maternal-infant communication and development of vocalizations in the bats *Myotis* and *Eptesicus*. *Comm Behav Biol* 5: 263–313.
Gupta P, Dietz N, Kanwal JS (1998) Vocal communication and stereotypic social behavior patterns in the mustached bat, *Pteronotus parnellii*. *Assoc Res Otolaryngol* 19: 456.
Holler P, Schmidt U (1993) Olfactory communication in the lesser spear-nosed bat *Phyllostomos discolor* (Chiroptera: Phyllostomidae). *Z Säugetierkunde* 58: 257–265.
Jones GJ, Ransome RD (1993) Echolocation calls of bats are influenced by maternal effects and change over a lifetime. *Proc R Soc Lond B* 252: 125–128.
Kanwal JS (1999) Processing species-specific calls by combination-sensitive neurons in an echolocating bat. In: Hauser MD, Konishi M (eds), *The Design of Animal Communication*, pp. 134–157. MIT press, Cambridge, MA.
Kanwal JS, Matsumura S, Ohlemiller K, Suga N (1994) Analysis of acoustic elements and syntax in communication sounds emitted by mustached bats. *J Acoust Soc Am* 96: 1229–1254.
Kanwal JS, Gordon M, Peng JP, Esser KH (2000) Auditory responses from the frontal cortex in the mustached bat, *Pteronotus parnellii*. *NeuroReport* 11: 367–372.
Karakashian SJ, Gyger M, Marler P. (1988) Audience effects on alarm calling in chickens (*Gallus gallus*). *J Comp Psychol* 102: 129–135.
Kroodsma DE (1977) A re-evaluation of song development in the song sparrow. *Anim Behav* 25: 390–399.
Leippert D (1994) Social behavior on the wing in the false vampire, Megaderma lyra. *Ethology* 98:111–127.
Lieberman P, Blumstein SE (1998) Speech physiology, speech perception, and acoustic–phonetics. Cambridge Univ. Press, New York.
Matsumura S (1981) Mother-infant communication in a horseshoe *bat* (*Rhinolophus ferrumequinum nippon*): vocal communication in three-week-old infants. *J Mammal* 62: 20–28.
McWilliam AN (1982) Adaptive responses to seasonality in four species of Microchiroptera in coastal Kenya. PhD Thesis, University of Aberdeen, Aberdeen, Scotland.
Medvedev AV, Kanwal JS (2004) Local field potentials and spiking activity in the primary auditory cortex in response to social calls. *J Neurophysiol* 92: 52–65.

Morton ES (1977) On the occurrence and significance of motivation-structural rules in some bird and mammal sounds. *Amer Natur* 111: 855–869.
Nelson JE (1964) Vocal communication in Australian flying foxes (Pteropodidae, Megachiroptera). *Z Tierpsychol* 21: 857–870.
Nelson JE (1965) Behaviour of Australian Pteropodidae (Megachiroptera). *Anim Behav* 8: 544–557.
Ohlemiller KK, Kanwal JS, Butman JA, Suga N (1994) Stimulus design for auditory neuroethology: synthesis and manipulation of complex communication sounds. *Audit Neurosci* 1: 19–37.
Ohlemiller KK, Kanwal JS, Suga N (1996) Facilitative responses to species-specific calls in cortical FM–FM neurons of the mustached bat. *NeuroReport* 7: 1749–1755.
Pearson OP, Koford MR, Pearson AK (1952) Reproduction of the lump-nosed bat (*Corynorhinus rafinesquei*) in California. *J Mammal* 33: 273–320.
Petersen MR (1982) The perception of species-specific vocalizations by primates: a conceptual framework. In: Snowdon CT, Brown CH, Petersen MR (eds), *Primate Communication*, pp. 171–211. Cambridge University Press, New York.
Pfalzer G, Kusch J (2003) Structure and variability of bat social calls: implications for specificity and individual recognition. *J Zool* 261: 21–33.
Pickett JM (1980) The sounds of speech communication. University Park Press, Baltimore.
Porter FL (1979) Social behavior in the leaf-nosed bat, *Carollia perspicillata* II social communication. *Z Tierpsychol* 50: 1–8.
Portfors CV (2004) Combination-sensitivity and processing of communication calls in the inferior colliculus of the moustached bat, Pteronotus parnellii. *An Acad Bras Cienc* 76: 253–257.
Portfors CV, Wenstrup JJ (2002) Excitatory and facilitatory frequency response areas in the inferior colliculus of the mustached bat. *Hear Res* 168: 131–138.
Saunders DA (1983) Vocal repertoire and individual vocal recognition in the short-billed white-tailed black cockatoo, *Calyptorhyncus funereus latirostris* Carnaby. *Aust Wildlife Res* 10: 527–536.
Snowdon CT (1982) Linguistic and psycholinguistic approaches to primate communication. In: Snowdon CT, Brown CH, Petersen MR (eds), *Primate Communication*, pp. 212–238. Cambridge University Press, New York.
Suthers RA (1965) Acoustic orientation by fish-catching bats. *J Exp Zool* 158: 319–348.
Thomas DW, Fenton MB, Barclay RMR (1979) Social behavior of the little brown bat, *Myotis lucifugus I.* mating behavior. *Behav Ecol Sociobiol* 6: 129–136.
Vater M, Kössl M, Foeller E, Coro F, Mora E, Russell IJ (2003) Development of echolocation calls in the mustached bat, *Pteronotus parnellii*. *J Neurophysiol* 90: 2274–2290.
Wickler W, Seibt U (1976) Field studies on the African fruit bat, *Epomophorus wahlbergi* (Sundevall), with special reference to male calling. *Z Tierpsychol* 40: 345–376.
Wilkinson GS, Boughman JW (1998) Social calls coordinate foraging in greater spear-nosed bats. *Anim Behav* 55: 337–350.
Wilson DE (1973) Reproduction in neotropical bats. *Period Biol* 75: 215–217.

4
Common Rules of Communication Sound Perception

Günter Ehret

4.1 Introduction

Animal sounds may contain information about the identity of a species or an individual of a given species, about age, sex and rank of a group member, about the biological context and the emotions/motivations the sender is experiencing when producing the sounds, or just about the location in space of another animal. We can certainly build upon the fact that natural selection shaped mechanisms of sound processing and perception in animals in such a way that they can make the potential information in the sounds available for producing an adaptive response behavior. Since most sounds of terrestrial vertebrates are vocal, that is they are generated by vibrations of membranes such as the vocal chords with subsequent filtering, amplifying, and temporal shaping in the vocal tract, these vocalizations share many structural features. This means, from the perceptual point of view, that one can assume common physiological mechanisms of sound analysis and representation in the auditory systems adapted to work out and to differentiate from vocalizations those information-bearing elements and information-bearing parameters (in the sense of Suga, 1988) that serve as the basis of perceptions. Hence, before going into the discussion of the topic of this chapter, I will mention some basic abilities in hearing characterized by psychoacoustical measures and relations among them, which may be useful to detect and discriminate information-bearing elements and parameters. Taken together, these abilities provide some auditory "tools" for communication sound perception. We will see how far these tools are actually used to specify and discriminate auditory percepts, in our case communication sounds.

What is perceived? This question is posed by Hirsh and Watson (1996) in their review on auditory psychophysics and perception. The answers from the literature to this question disagree in the term used to characterize an auditory percept. The percept may be an "event" (Julesz and Hirsh, 1972; Handel, 1989), an "entity" (Hartmann, 1988), an "object" (Moore, 1989; Yost and Sheft, 1993), or an "image" or "scenes of images" (Bregman, 1990; Yost, 1992). At least, all auditory percepts constitute "patterns", which means that they are complex in one or the other acoustical (physical) and auditory (physiological) dimension. I will use the above-mentioned terms when they seem appropriate for a subject without discussing the terminology of perception.

These terms derived from studies on perception in humans miss two aspects important for our discussion on animal communication sound perception:

(i) Vocalizations (of communication sounds) and percepts of these vocalizations have an evolutionary history for every species and, thus, are selected for mutual adaptation. Hence, the percepts are not independent of the vocalizations, they are bound to them *and* to the contexts in which the vocalizations are emitted. In cases in which communication sounds release an instinctive behavioral response without learning of the acoustical patterns, the percepts of the sounds *are* the meaning.

(ii) As human observers of animal behavior, we can only guess what animals experience as percepts from their responses to the sounds presented. For human observers, there is no percept without a behavioral response of the animal who perceives it. Even from an evolutionary point of view discussed on a general level by Owings and Morton (1998), acoustic communication is not selected for "information transfer" but for the consequences of perception (see also discussion in Seyfarth and Cheney, 2003).

By combining these two aspects, we have to consider not only acoustic patterns but also the responses they elicit when we will discuss rules for communication sound perception.

The main aim of this chapter is to derive rules from common properties of hearing and responding for the perception of communication sounds. The rules are not meant to be "principles" but rather hypotheses, abstracted and generalized from available data, for guiding the study of the acoustical communication system of a species. The examples used will mainly be taken from the literature on mammals including humans and will be specified for the house mouse (*Mus musculus*). The rules derived are meant, however, to apply to other vertebrate species. The following six rules, which will be discussed in detail later, make no pretence to be complete:

- Rule 1: Perception is individualized to the level of matched groups.
- Rule 2: Perception of acoustic meanings or objects refers to just meaningful differences (JMDs) between acoustic patterns.
- Rule 3: Perception of acoustically expressed levels of arousal and emotions refers to the variations in acoustic patterns.

- Rule 4: The audible parameter space is partitioned for the perception of three basic meanings.
- Rule 5: The integrated extent of the ranges activated in the audible parameter space is related to the perception of the urgency of the response.
- Rule 6: Perceptions of categories from continuums of acoustic patterns are built upon natural perceptual boundaries.

4.2 The basis: common psychoacoustical measures and relations

Most valuable here is the compilation of psychophysical functions in Fay's (1988) psychophysics databook for hearing in vertebrates. Several properties, at least common to mammals, become obvious.

4.2.1 Audiograms

The relationships between minimum auditory thresholds (MTs) for tones and tone frequency show the general characteristic of a band-pass filter providing a best frequency with a frequency range of low thresholds around, a high-frequency and a low-frequency limit of hearing. The shape of the audiogram is determined by physical properties of the outer, middle, and inner ear (e.g. Dallos, 1973; Henson, 1974; Shaw, 1974; Ehret, 1977). Additional relative maximums of sensitivity at about three times the best frequency may be present as in humans and house mice (Ehret, 1977) or at a special fovea region of the inner ear as in the greater horseshoe bat (Long and Schnitzler, 1975; Bruns, 1976), an extension of the hearing range to high ultrasounds often coupled with a restriction of hearing low frequencies (below about 1 kHz) as in bats and cetaceans (see Fay, 1988), a loss of high-frequency (ultrasonic) hearing as in humans, or an extension to hear low frequencies in small rodents such as the gerbil or the kangaroo rat with specialized middle ears (see Fay, 1988). The shape of the audiogram and the frequency range covered provide the frame not only for the simple audibility of communication sounds in the frequency and the intensity domains but also for the perception of differences in meaning (see rule 4).

4.2.2 Temporal summation

Temporal summation describes the relationship between the MT and sound duration. In general, MTs decrease by about 10 dB or less for an increase of tone duration over a decade, for example, from 1 to 10 ms (Fay, 1988; Ehret, 1989). Lowest thresholds are reached between about 100 ms (high frequencies of hearing of a given species) and 1000 ms (low frequencies, respectively) sound duration. This shows that the auditory system can integrate sound energy over signal duration before the threshold of sound detection is reached. The implication for sound perception is obvious since short-duration sounds require considerably high sound levels to be detected and resolved, especially at low frequencies. To enhance the perception of low-frequency sounds, they should be either long or consist of short-duration pulses that are repeated several times in a second (see rule 4).

4.2.3 Frequency discrimination

The relation between a just noticeable difference (JND) in frequency (Δf) and a given frequency (f) in a sequence of tones ($f, f + \Delta f, f, f + \Delta f, \ldots$) follows a rather linear function in log/log coordinates ($\Delta f/f$ = constant = 0.1–10%; Fay, 1988), except for frequencies lower than about half of the respective best frequency of a given species, where Δf is nearly constant (Fay, 1988; Ehret, 1989). For a given frequency, Δf decreases with increasing sound level up to levels of about 60 dB above MT at that frequency in house mice and humans (Ehret, 1975b; Wier et al., 1977). These relationships can be explained by excitation patterns of the cochlea (Maiwald, 1967; Ehret, 1989) and provide the frame for the perception of frequency changes in communication sounds. Such frequency changes are part of the acoustical variability in sound patterns (see rules 3 and 5).

4.2.4 Intensity discrimination

The relation between a JND in sound level (ΔL) and the sound level above MT (sensation level, SL) in a sequence of tones ($SL, SL + \Delta L, SL, SL + \Delta L, \ldots$) follows a rather linearly decreasing function for all frequencies. Humans can discriminate differences of less than 1 dB, while other mammals have difference limens of about 3–4 dB (Fay, 1988; Ehret, 1989). The relationship between ΔL and SL can be explained by excitation patterns of the cochlea (see above) and provides the frame for the perception of intensity changes in communication sounds. Like frequency differences, intensity differences are part of the variability in sound patterns (see rules 3 and 5).

4.2.5 Duration discrimination

The relation between a JND in duration (Δd) and a tone duration (d) in a sequence of tones ($d, d + \Delta d, d, d + \Delta d, \ldots$) follows a rather linearly increasing function in log/log coordinates ($\Delta d/d$ = constant = 4–100%, is very species- and method-dependent; Fay, 1988). Duration discrimination provides a frame for the perception of duration differences in communication sounds (see rules 3, 5, and 6).

4.2.6 Spectral resolution, spectral summation

Spectral resolution refers to the ability to separate simultaneously occurring frequency components of a complex sound so that they can be detected rather independently of each other. Spectral resolution requires frequency filters. Spectral summation refers to the integration of sound energy within the bandwidth of a frequency filter to provide, for example, the perception of a loudness value for that filter. A psychoacoustical measure of frequency resolution and spectral integration is the so-called critical bandwidth (CBW; Fletcher, 1940; Zwicker and Feldtkeller, 1967; Scharf, 1970). Perceptually important properties of CBWs are: (i) the increase of the CBW with increasing center frequency of the filter except for center frequencies of less than about half of the frequency at MT (CBWs constant) and (ii) the constancy of CBWs with changing sound level. The CBW increase is a little less than one octave for one octave increase of center frequency. The CBWs vary among mammals ranging between about 20% (humans) and about 40% (house mouse) of the center frequency (Fay, 1988; Ehret, 1995). Taking the cochlear basilar membranes of mammals (except some bats such as *Rhinolophus ferrumequinum* and *Pteronotus pernellii* with a special fovea region (Bruns, 1976; Suga and Jen, 1977)) as models derived from a common scale (Greenwood, 1961, 1990), CBWs cover equal distances of 0.7–1 mm of the basilar membrane (Ehret, 1977, 1988). This explains why small mammals with short basilar membranes (mice) have poor frequency resolution. CBWs are encoded in the frequency tuning of auditory nerve fibers and are shaped to their perceptual properties up to or at the level of the auditory midbrain (Ehret and Merzenich, 1985, 1988; Ehret, 1995). CBWs provide the framework for the perception of complex frequency spectra of communication sounds including the level independence (perceptual constancy) of the perception (see rules 4–6).

4.2.7 Pitch perception and discrimination

Complex sounds composed of a fundamental frequency and higher harmonics arise, for example, from smooth vibrations of the vocal chords. Every harmonic, if heard in isolation, creates its own pitch percept according to its frequency. The harmonic complex, however, is perceived with a global pitch corresponding to the fundamental frequency even if the fundamental frequency is physically absent (e.g. Terhardt, 1974, 1978; Moore, 1989). This pitch of a harmonic complex, which can even be a virtual pitch, is heard also as the modulation frequency (periodicity pitch) when a pure tone is amplitude modulated with a low-frequency ($f < \sim 250\,\text{Hz}$) sine wave (e.g. Langner, 1992). Pitch perception (including virtual pitch) has been shown to exist in the common starling (Cynx and Shapiro, 1986), the cat (Heffner and Whitfield, 1976), monkey (Tomlinson and Schwarz, 1988), and even in a bat's ultrasonic hearing range (Preisler and Schmidt, 1995) and, thus, may be a common percept in birds and mammals. Hearing and discriminating pitches provide the frame for separating harmonically from non-harmonically structured sounds, for the perception of regularities of pitch shifts in harmonically structured vocalizations, and for the perception of repetition rates (see rules 3–6).

4.3 The six rules of communication sound perception

4.3.1 Rule 1: Perception is individualized to the level of matched groups

Without the supplement "to the level of matched groups", this rule represents a very general rule that has been discussed in various forms in the literature on the psychology of human perception (e.g. Hentschel *et al.*, 1986; Dennett, 1988; Humphrey, 1992). In most of the experiments on animal perceptual abilities, the individuals observed or tested are matched in as many respects as possible, for example, in age, sex, rank, experience with certain stimuli, behavioral context in the group, and environmental context. It is immediately clear, for example, from the developments of hearing abilities, sexual maturity, and experience with certain sounds and the contexts in which they occur that the perception of a given sound pattern changes with regard to the heard, discriminated and perceived acoustical attributes and its meaning (e.g. Ehret, 1980; Fischer, 2003). Hence, this rule emphasizes that whatever is characterized for a given species as a percept, an acoustic object or a certain class of perceived auditory patterns, this description is valid initially only for the animals of the matched group. As a consequence, comparative studies in other matched groups of a given species and other species are most welcome and numerous such studies have actually been done and more have to be conducted in order to see, in how far findings reflect specializations or whether they can be generalized and common meanings or mechanisms of perception defined.

4.3.2 Rule 2: Perception of acoustic meanings or objects refers to JMDs between acoustic patterns

Nelson (1988), in weighting acoustical properties in the songs of field sparrows for the recognition of the species' own song, found that the limits of hearing and discrimination abilities characterized by JNDs do not predict the discrimination performance when sparrows recognize their own song. In this respect, JMDs between acoustic patterns are important. Miller and Hauser (2004) in discussing the recognition of monkey (cotton-top tamarin, *Sanguinus Oedipus*) combination long calls came to a very similar conclusion. Call recognition in cotton-top tamarins seems to be based on a combination of acoustical properties in the frequency and time domain which all contribute to establish JMDs for the call, not necessarily equal to JNDs for each acoustical property.

Figure 4.1 shows spectrograms of mouse pup ultrasounds and mouse pup wriggling calls (Ehret, 1975a). There is a large variability among exemplars of each call type with regard to duration, frequency bandwidth, distribution of spectral energy, patterns of frequency and intensity changes over time, and the presence of noise components. All of this variability, although being above JNDs and the frequency-resolution capability (Ehret, 1975b, 1976b; Klink and Klump, 2004), is ignored by adult mice perceiving the calls and showing an adaptive response (Smith, 1976; Ehret and Haack, 1982; Ehret, 1992; Ehret and Riecke, 2002; Geissler and Ehret, 2002). Similarly, Japanese macaque monkeys largely ignore differences above JNDs and frequency-resolution capability (Fay, 1988) in duration, harmonic structure, amplitude contour, and noise components when responding to and differentiating species-specific *coo* calls vocalized by different individuals (Green, 1975; May *et al.*, 1988, 1989). Ignoring acoustic differences between calls is also true for pygmy marmosets when they respond to synthesized models of their closed mouth trills (Snowdon and Pola, 1978) and for vervet monkeys classifying their species-specific alarm calls (Seyfarth *et al.*, 1980a, b; Owren, 1990).

These examples indicate a discrepancy between what the auditory system is able to detect, analyze, discriminate, and resolve and what finally becomes significant in producing a difference in perception, that is in response behavior to the acoustic patterns. In perceiving communication sounds as acoustic objects or images in an environment that is comparable to their natural one, animals seem not to exploit all the analytical capacities of their auditory systems but rely on analyses that produce JMDs. This, of course, does not exclude that animals can be trained to discriminate stimuli at their JNDs. In the latter case, however, the experimenter defines both the test paradigm and the information-bearing element and/or information-bearing parameter to be differentially perceived. This is a completely different question to the animal compared to the situation of communication sound perception in which the animal "asks itself" for JMDs and meaning that have been defined by evolution and/or experience in the natural habitat. What the JMDs consist of, has to be clarified for every call type or acoustic continuum of vocalizations and for every matched group who perceives them.

92   4 Common rules of communication sound perception

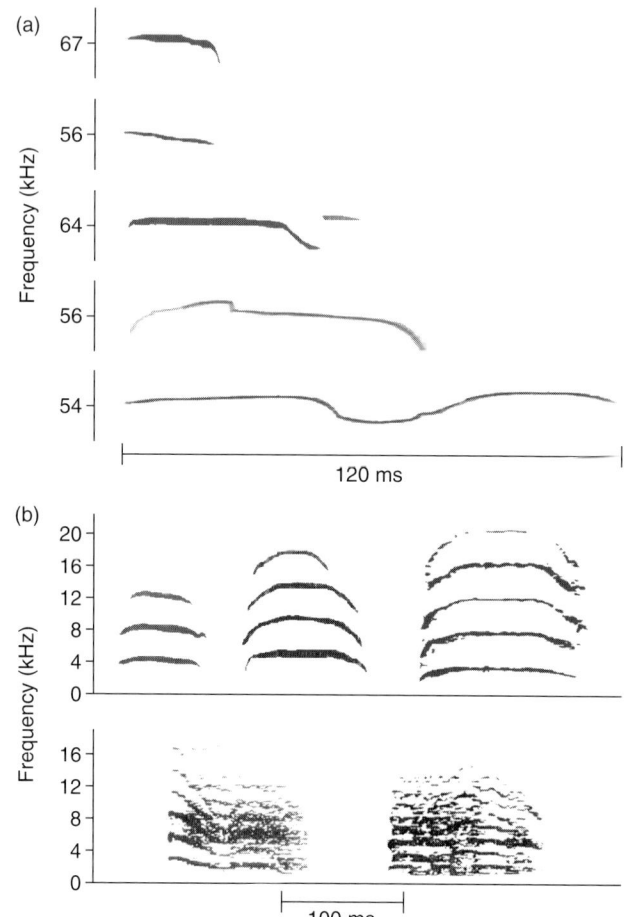

Figure 4.1. (a) Examples of ultrasonic vocalization (USVs) of mouse pups produced in situations of isolation or handling. The USVs are rather pure tones of various durations and frequency contours. (b) Examples of wriggling calls of mouse pups produced in the nest when pushing for the mother's nipples while she is nursing her pups. The vocalizations have either a clear harmonic structure with the fundamental frequency between about 3 and 5 kHz and two or more overtones, or a more noisy structure. The total frequency range rarely exceeds 20 kHz. Most calls with a clear harmonic structure have frequency contours resembling an arch.

4.3.3 Rule 3: Perception of acoustically expressed levels of arousal and emotions refers to the variations in acoustic patterns

Arousal refers to an unspecific background activation of the nervous system (reticular formation). Hence, it is very unlikely – and I am not aware of descriptions for any vertebrate – that a certain arousal level is expressed by an associated type of vocalization. On the other hand, different types of vocalizations are often associated with context-specific emotions/motivations, as demonstrated for primates (e.g. Jürgens, 1979) and developing mammals (e.g. Ehret, 1980). Arousal and a given emotion vary on an intensity scale. I hypothesize that the levels of arousal and of a given emotion are encoded by acoustical parameters of vocalizations such as the repetition rate of calling, the call duration, the duration of elements and inter-element intervals within calls, the sound pressure level and the rate of modulations of frequency, amplitude, and pitch (e.g. for human speech, Scherer, 1989; for baboon grunt variants, Rendall, 2003; for infant rodent ultrasounds, Ehret, 2005), all of which can be varied continuously and, thus, can represent continuous variations of levels and intensities. In this context, the analytical capabilities of the auditory system of the perceiving animal may be fully exploited. By hearing small differences and subtle variations in the vocalizations of a sender, the levels of the arousal and emotions in the perceiving animal and, thus, its response can accurately be adjusted to be adequate to the sender's level of arousal and emotions.

According to the mutual adaptation of vocalizations and perceptions during the evolution of intra-species communication, given levels of arousal and emotions encoded by values of the above-mentioned and further parameters are expected to induce equivalent levels of arousal and emotions in the perceiver. The hypothesis that vocalizations are arousal-produced and arousal-producing was explicitly formulated for the call type of ultrasounds emitted by rodent pups in situations of social isolation (Bell, 1974). The transmission of emotions in acoustical communication is emphasized by Dawkins and Krebs (1978) with the notion that advertisement calls are not there to inform or to misinform but to persuade.

In any case, the level of arousal as a property of communication sounds can be assumed to be processed automatically (without percept) to contribute to the adjustment of the perceiver's own level of arousal. The processing of the level of emotions may be part of the processing for the perception of the basic meanings of vocalizations (rule 4) or may lead to an additional percept (rule 5).

4.3.4 Rule 4: The audible parameter space is partitioned for the perception of three basic meanings

Audible parameter space relates to the audible ranges of acoustical properties in vocalizations. The properties of pure tones are simple ones such as frequency, sensation level, duration, location in space, direction and speed of change of frequency, and level and location in space. Spectrally and temporally more complex sounds comprise additional dimensions such as frequency bandwidth, spectral shape and density, pitch, bandwidth of sensation level, modulation rates if sounds are sinusoidally amplitude- or frequency-modulated, and repetition rates of single elements.

This rule refers to Morton's (1977) motivation-structural rules which describe and explain from an evolutionary point of view the relationship between the physical structure of a sound and the motivational/emotional background of sound production expressed by the acoustical properties. The rules were derived from the sounds of 28 species of birds and mammals each. In summary, aggressive or hostile motivations/emotions are expressed by harsh, noisy, low-frequency sounds, while friendly, appeasing or fearful tendencies are expressed by rather tonal, high-frequency sounds. These rules were tested by August and Anderson (1987) using sounds from 50 species of mammals. They confirmed Morton's rule for sounds produced by aggressive animals but proposed to make distinctions between sounds from friendly and sound from fearful animals by using additional acoustic parameters. Low-frequency sounds, even if noisy, can express friendly motivations/emotions especially when they are of a low sensation level and of repetitions in the noisy elements producing a low pitch like purrs of carnivores such as cats (Peters, 1984).

On the basis of Morton's (1977) rules with the modifications of August and Anderson (1987) and considering a coevolution of sound production and perception we ask here, what kind of rules may be derived from the message side to characterize the perception. According to rule 2, we have to deal with JMDs and meaning.

I hypothesize that sounds are perceived to have three basic meanings: *attraction*, *cohesion*, and *aversion*. Primarily, these meanings refer to contexts of instinctive behavior. These meanings may be taken as the immediate perceptual counterparts of the three basic motivations/emotions expressed in animal vocalizations (Morton, 1977; August and Anderson, 1987). Appeasing or fearful emotions/motivations may be perceived as attraction, friendliness as cohesion, and aggression as aversion. These perceived meanings do not necessarily predict the perceiver's response, because the response depends on the result of the processing of the meaning with the perceivers internal state of arousal and motivation. Since each of the three basic emotions/motivations appear to be expressed vocally by different acoustical properties (Morton, 1977; August and Anderson, 1987; see above), I hypothesize that the three associated meanings are related to perceptions from different parts of the audible parameter space.

In the following I show for the mouse the association of a given meaning with a stimulation of a certain part of the audible parameter space, which is a very simple relationship in this case, because only one parameter is of main concern,

4.3 The six rules of communication sound perception

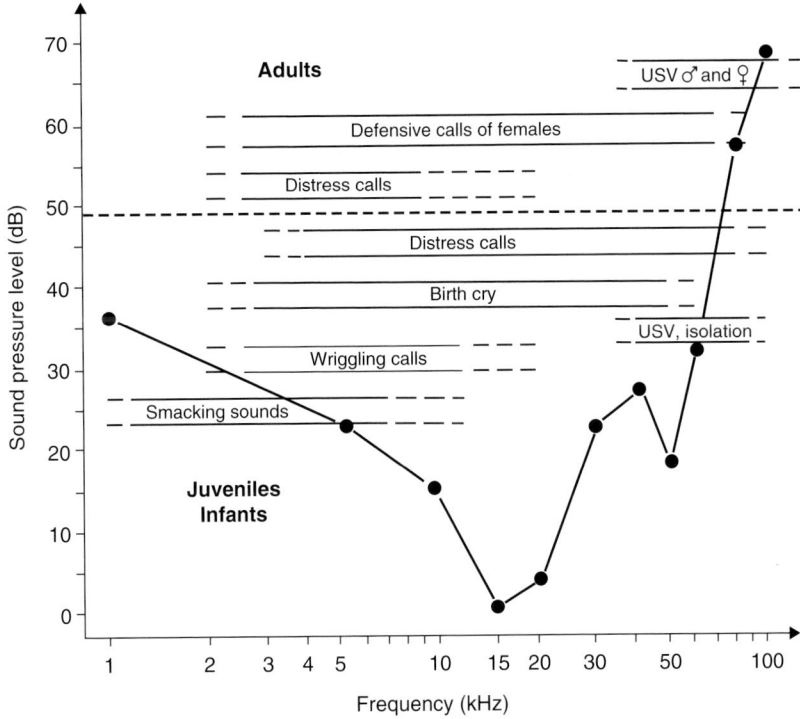

Figure 4.2. The frequency range of hearing and auditory sensitivity of the adult mouse illustrated by the behavioral audiogram (heavy line; Ehret, 1974). The horizontal lines show the main frequency ranges covered by the indicated vocalizations of the mouse (Ehret, 1975a; Haack et al., 1983; Whitney and Nyby, 1983). Calls of adults (upper part) are separated by a dashed line from calls of infants and juveniles (lower part).

namely the frequency range. Figure 4.2 shows the audiogram of the house mouse (Ehret, 1974) representing the lower border of the audible sound pressure level as a function of frequency, and the frequency ranges of the nine vocalizations known so far for this species (Ehret, 1975a; Haack et al., 1983; Whitney and Nyby, 1983). The "high-frequency" group consists of single-frequency (tonal) ultrasounds emitted either by mouse pups in isolation (ultrasonic vocalization (USV) pup), or by males investigating an unfamiliar female (USV male) or by a "confused" female (USV female). All energy of these sounds is far above the best hearing range (about 15–20 kHz). The "broadband" group consists of broadband sounds composed of harmonics and noise and covering most of the audible frequency range (about 2–80 kHz). They are emitted as the first cry after birth (birth cry), by mouse pups when injured or roughly handled (distress infant), or by non-receptive females rejecting a sexually interested male (defense female). The "low-frequency" group consists of broadband sounds composed of harmonics and/or noise covering only a range of low frequencies (about 2–20 kHz). The main energy is always at frequencies well below the best hearing range. The sounds are emitted by pups when wriggling in the nest while an adult is in a nursing position (wriggling calls), or by pups when huddling up in the nest or being warmed (smacking sounds), or by adults when injured or by a male when repeatedly chased by another male (distress adults). The sound structures follow the modified Morton's rules (Morton, 1977; August and Anderson, 1987).

Observations and behavioral tests done on some of the sounds and synthesized models of them have shown the following main effects on the behavior of a perceiving adult (although emitting different types of vocalizations, the young infants are still deaf; Ehret, 1976a). For two vocalizations, JMDs also have been measured.

4.3.4.1 High-frequency group

USVs of pups attract adults (perceived meaning is attraction). The JMD for a discrimination of pup USVs against sounds of the other two groups is that sound energy has to be concentrated within a single CBW in the ultrasonic range with at least 20 dB less sound level in neighboring CBWs (Ehret and Haack, 1982). A JMD for the discrimination of pup USVs against any other sounds such as environmental noises is that the USVs should have a duration between about 25 and 270 ms (Ehret and Haack, 1982; Ehret, 1987).

USVs of males seem to be courtship vocalizations and should attract females (perceived meaning is attraction; Whitney and Nyby, 1983).

4.3.4.2 Broadband group

A *birth cry* stops a mother from cleaning a pup after birth for some moments (perceived meaning is aversion). If a pup does not cry and does not move, the probability is high that it is eaten (Ehret, 1975a).

Distress calls of pups emitted when they are picked up by an adult make the adult drop the pup (Ehret, 1975a). The perceived meaning is aversion.

Defensive calls of females stop the male from investigating her for some moments (Ehret, 1975a). The perceived meaning is once again aversion.

4.3.4.3 Low-frequency group

Wriggling calls of pups release pup-caring behavior in mothers (Ehret and Bernecker, 1986), that is the mother remains in the nest, shows cohesion with the pups, and acts according to her maternal motivation. The JMDs for a discrimination of wriggling calls against vocalizations of the other two groups and against other sounds are: they should consist of three resolved (by CBWs) harmonics or formants in the low-frequency range, have a minimum duration of about 100 ms,

4.3 The six rules of communication sound perception

and a repetition of single calls at a rate between about 1.5 and 3.5/s (Ehret and Riecke, 2002; Geissler and Ehret, 2002; Gaub and Ehret, unpublished).

The available data on the responses to mouse vocalizations show that vocalizations of the high-frequency group release *attraction*, those of the broadband group *aversion*, and those of the low-frequency group (as far as tested) *cohesion* in the responding animal. The available data suggest that the calls representing these three basic meanings are well separated by JMDs. The JMDs partly reflect, at least emphasize, differences in the emitted vocalizations, that is differences in the covered frequency range and in the spectral and temporal structure. A further discrimination of mouse vocalizations of a given basic meaning on the basis of acoustic cues may be possible (see Liu *et al.*, 2003 for an acoustic separation of pup USVs from male USVs), it is not necessary, however, for the release of different responses (Ehret, 1989). Animals hearing different types of vocalizations of a perceived common basic meaning can adjust their response according to their own motivation and the perception of other than acoustical cues from the behavioral context. For example, females hearing USVs of a male in a sexual encounter perceive other non-acoustical stimuli and can be assumed to be differently motivated compared to a female caring for her pups and listening to pup USVs. Hence, the release of an adaptive behavioral response by one of the different types of mouse vocalizations can be interpreted as the result of the perception of one of the three basic meanings processed in the context of the perceiver's own arousal and motivation.

In the mouse, the meanings are mainly represented by the frequency ranges perceived. This must not necessarily be so in other species. However, certain general assumptions follow from the modified Morton's rules (Morton, 1977; August and Anderson, 1987):

(i) The meaning of *attraction* may be associated with high-frequency, rather tonal sounds. Examples are ultrasounds of infant rodents (e.g. Noirot, 1972; Smith and Sales, 1980; Ehret, 2005).
(ii) The meaning of *aversion* may be associated with sounds of a rapidly varying frequency spectrum covering a rather broad frequency range and having noisy components and/or harmonics that are not resolved by CBWs (roughness percept). Examples are "panic calls" of suricates (Manser, 2001; Manser *et al.*, 2001).
(iii) The meaning of *cohesion* may be associated with low-pitched or rhythmic noises or sounds of resolved frequency components that both are rather constant in their properties over time. Examples are grunts associated with group movements in baboons (Cheney *et al.*, 1995; Rendall *et al.*, 1999) and infrasonic contact calls of African elephants (McComb *et al.*, 2000).

This rule with its specifications now needs to be tested for the perception of meanings and specified for the parameter spaces associated with the meanings in other species including those listed by Morton (1977) and August and Anderson (1987). As mentioned before, these described meanings and the associated acoustical patterns refer primarily to contexts of instinctive behavior. When cognitive functions modify instinctive behavior and exert an increased influence on perception, a significant differentiation of and expansion from the three basic meanings and their associated acoustic patterns can be expected and has actually been demonstrated. Examples are the acoustical perception of predator types in vervet monkeys (Seyfarth *et al.*, 1980a, b), Diana monkeys (Zuberbühler, 2000), and suricates (Manser *et al.*, 2001), the perception of individuals and/or kin in vervet monkeys (e.g. Cheney and Seyfarth, 1982), rhesus monkeys (Rendall *et al.*, 1996), spotted hyenas (Holekamp *et al.*, 1999), elephants (McComb *et al.*, 2000), and pigs (Illmann *et al.*, 2002).

Finally, an experiment on perceptual strategies of rhesus monkeys, a species of well-developed cognitive abilities, shows that rule 4 applies to the perception of species-typical vocalizations when their meaning seems not to be learned (Hauser, 1998; Gifford *et al.*, 2003). Rhesus monkeys produce two acoustically distinct types of vocalizations, the harmonic arch and the warble, to indicate the presence of high-quality or rare food items, and produce series of grunts when signaling low-quality or common food items or when being in non-food related contexts such as grooming each other or moving together (Hauser and Marler, 1993; Hauser, 1998). Harmonic arches and warbles are long, tonal vocalizations with high-frequency harmonics so that, applying rule 4, they are perceived as having the meaning *attraction*. Grunts are short, rhythmically repeated, noisy vocalizations. Applying rule 4, they are perceived as having the meaning *cohesion*. Hauser (1998) and Gifford *et al.* (2003) habituated the animals with either harmonic arches or warbles and found no dishabituation response to the other vocalization type. Although both vocalizations provide different auditory patterns, they are of the same meaning and, thus, are not discriminated. This could be predicted from rules 2 and 4. A habituation to grunts, however, leads to a dishabituation response both to harmonic arches and warbles. The interpretation is that a behavioral change occurred in response to a perceived change of meaning. Interestingly, a habituation to either harmonic arches or warbles did not lead to a dishabituation to grunts, although a change in meaning must have been perceived. In this case, the change of meaning indicated, however, a shift from the perception of *attraction* to *group cohesion*, the latter being something rather general and unimportant compared to the meaning of attraction to a high-quality or rare food. Hence, the monkey may not be motivated to show a response to grunts after listening to warbles or harmonic arches.

Altogether, these data are in accordance with rules 2 and 4 and suggest that the animals experienced perceptions of basic meanings. Obviously, their responses did not reflect just hearing differences in acoustic patterns. In addition, not all changes in the perceived meaning are expressed in the responses observed by the human experimenter so that, without dishabituation to grunts after having heard harmonic arches or warbles, we can only suggest a change in the perceived meaning which has to be demonstrated, however, in other behavioral tests.

4.3.5 Rule 5: The integrated extent of the ranges activated in the audible parameter space is related to the perception of the urgency of the response

I have argued before that the level of arousal expressed in sounds (rule 3) is processed to adjust the perceiver's level of arousal without leading to a percept. I hypothesize here that the level of a given emotion in the perceiving animal may lead to the percept of the *urgency of the response*, while the response itself represents the meaning as detailed under rule 4. This urgency of the response constitutes the perceptual counterpart of the level of emotion/motivation in the vocalizing animal. It varies on an intensity scale, and may be measured in the latency of the response or, in case the response consists of several behaviors, in the temporal patterning of the behavioral sequence. I further hypothesize that increasing levels of urgency of the response are perceived via increasing ranges of the audible parameter space being activated by the vocalizations.

To specify this rule for the mouse, we have to look at Fig. 4.2 and consider the responses to the vocalization types (described under rule 4). Vocalizations of the broadband group cover large ranges of the parameter "frequency bandwidth". They also cover a large range in the space of the parameter "bandwidth of sensation level" because they stimulate far above the absolute auditory threshold in the best hearing range and less in other frequency ranges of the audiogram. Further, the spectral shapes vary over about 120 ms call duration. Thus, these calls cover considerable ranges of several parameters of the audible space and, therefore, should be perceived as requiring a high urgency of a response. In fact, the responses to these vocalizations occur at a short latency (less than 3 s). The low-frequency group wriggling calls also cover considerable bandwidths of frequency and sensation level, less however than the vocalizations of the broadband group. Spectral shape varies over about 120 ms call duration. In comparison to the calls of the broadband group, wriggling calls should be perceived of somewhat reduced urgency. This is actually the case, because a high-urgency response within 3 s after the vocalizations occurs only to about 60% of the vocalizations (Ehret and Riecke, 2002). Vocalizations of the high-frequency group cover only a small range of the parameter bandwidths of frequency and sensation level. In addition, they are shorter than the broadband calls, therefore covering less range in duration (Ehret, 1975a). Hence, these calls are perceived as of low urgency of a response. In fact, responses to USVs of mouse pups often occur only after many seconds or even minutes depending on the ongoing behavior (motivational background) of the perceiving animal (Ehret and Haack, 1984).

The repetition rate of calls and/or elements in calls may be of potential importance for the perception of the urgency of a response. Systematic tests in mice for this parameter have, however, not been done yet.

Examples for the perception of the urgency of the response are given by alarm calls of suricates (Manser, 2001; Manser et al., 2001). The call to be expected to have the highest urgency of the response in perception (panic call) is noisy, and has the most amplitude variations of all alarm calls while covering a comparably large frequency range (Manser, 2001). Thus, parameter ranges for frequency, spectral density and rate of amplitude modulation are large predicting a high urgency of the response. In fact, most (75%) of the perceiving animals immediately run for shelter which they don't do at that high incidence when they perceive less urgency in other types of alarm calls (Manser et al., 2001).

4.3.6 Rule 6: Perceptions of categories from continuums of acoustic patterns are built upon natural perceptual boundaries

In the discussion of the rules 2 and 4, the term "meaning" is used to specifically label an acoustical pattern that is discriminated from other patterns by the response it elicits. The labeling and discrimination of acoustical patterns and, thus, the formation of perceptual classes or meanings may be categorical in the sense used in human speech perception (e.g. Liberman et al., 1957; Studdert-Kennedy et al., 1970). Applying the criteria for categorical perception of human speech on animal vocalizations means that we have to take one audible parameter of a given vocalization after the other and test each for the occurrence of JMDs or perceptual boundaries between potential categories of meaning. Categorical perception can be stated if both labeling and discrimination functions indicate a perceptual boundary at the same values of the tested parameter. This means that stimuli from the two sides of a perceptual boundary are labeled differently and are discriminated well, while stimuli from the same side are labeled the same and are not discriminated (see Ehret, 1987). The formation of perceptual classes or meanings may be categorical also in the sense of key-stimulus identification, which is to identify the audible parameter space an acoustical stimulus has to cover in order to release a certain response behavior (see discussion

4.3 The six rules of communication sound perception

in Ehret, 1987). In this case, several acoustical properties of a vocalization may add together (with equal or different weights) in order to form a certain category of meaning. Since it is highly challenging and time consuming to test the perceptual boundaries in many audible parameter dimensions in order to define the key-stimulus configuration (the acoustical Gestalt), this has not been done (to the best of my knowledge) for any vertebrate's vocalization. What has been done for a number of vocalizations are tests for one or few perceptual boundaries on one or few dimensions.

From a general point of view and in search of rules for communication sound perception, we have to ask for the nature of perceptual boundaries, whether they separate meanings categorically (speech perception criteria applied) or not. Boundary formation and categorical perception depends very much on the knowledge of and experience with the acoustical patterns to which attention has to be directed when they are perceived (Spiegel and Watson, 1981). Categorization is most obvious in untrained animals as in our experiments in mice (Ehret and Haack, 1981, 1982; Ehret 1987, 1992) or under high-uncertainty conditions in monkeys (May et al., 1989). If human listeners are extensively trained to pay the same attention to all differences between speech phonemes, categorical perception in human speech disappears, although it is present under normal listening conditions (Samuel, 1977). These results stress the fact that categorization of acoustic patterns, at least in mammals (including humans) and birds, is related to and reflects JMDs, not necessarily JNDs. In other words, sounds that are discriminated in tests exploring the sensory limits of a species (JNDs) may be non-discriminated, that is perceived as belonging to the same category, in tests in which the animals respond to sounds according to their meanings.

Considering vocalizations produced and perceived in instinctive contexts as discussed under rules 2 and 4, I hypothesize that the discrimination and perception of meanings are based on natural perceptual boundaries that are established by properties of sound processing in the auditory system. If perceptual boundaries occur as a result of learning to perceive differences in acoustic patterns important for directing the perceiver's response, I hypothesize that natural perceptual boundaries on continuums of given audible parameters serve as prototypes for generating modified and/or additional boundaries on the same continuums.

The idea of categorization of sounds at natural perceptual boundaries is not new. It was introduced by Kuhl and Miller (1975, 1978) when they found that chinchillas had the same category boundaries on the voice-onset time (VOT) continuum as humans for labeling phonemes (/da/ versus /ta/, /ba/ versus /pa/, /ga/ versus /ka/) of human speech. The question, however, still is: how are natural perceptual boundaries generated? In other words, can we identify mechanisms of the auditory system and/or perceptual mechanisms from which natural perceptual boundaries and further learned boundaries can be derived? In the following, I will consider boundaries in the three audible parameter dimensions: (i) spectral structure, (ii) direction of frequency sweeps, and (iii) duration of information-bearing elements and parameters. Boundaries in these parameter dimensions are most important for the discrimination of phonemes in human speech, namely the formant structure of vowels, formant transitions, and the VOT (e.g. Liberman et al., 1956, 1957; Liberman and Pisoni, 1977).

4.3.6.1 Spectral (formant) structure

Certain frequency ranges of vocalizations are amplified by resonances of the vocal tract. As in vowels of human speech, the spectral energy in these frequency ranges represents the formants of the vocalizations. Thus, a formant may be expressed by a single harmonic of a vocalization if its fundamental frequency is rather high and the vocalization is devoid of noise and rapid amplitude and/or frequency modulations (e.g. ultrasounds of bats and rodents), or a group of harmonics or a band of noise if its fundamental frequency is low (i.e. the harmonics are close together) and the sound contains noise and/or rapid amplitude/frequency modulations. Each vowel in human speech has a characteristic formant composition and can be identified and discriminated from other vowels by the position of its first formant on the frequency scale and the frequency ratios between the first two or three formants, largely independent of the pitch and intensity of the individual voice (Peterson and Barney, 1952; Flanagan, 1972; Summerfield and Haggard, 1977). It is important to notice that the absolute frequencies or frequency ranges of formants of a given vowel (except for the first formant) have little importance for vowel perception compared to the frequency ratios between the spectral peaks. Thus, the perception of meaning, that is the identification of a certain vowel, includes a process of generalization across the variations of individual voices. Under many test conditions, vowels are categorized both by adults and children (e.g. Flanagan, 1972; Kuhl, 1979; Grieser and Kuhl, 1989).

Similar to vowels of human speech, many animal vocalizations including those of the mouse (Figs. 4.1 and 4.2, except smacking sounds) consist of formants. Using vowel perception by humans as the basis, I hypothesize that animal vocalizations with formant structure are perceived by the number and the specific frequency ratios of their spectral peaks and the position of the lowest-frequency peak.

In tests of the perception of mouse pup wriggling calls by adults (Ehret and Riecke, 2002) we have shown the following importance: (i) the number of formants, (ii) the presence of the lowest-frequency peak, and (iii) the frequency distances between the peaks. Like in human vowel perception (Darwin, 1984, 1992; ter Keurs *et al.*, 1992), only resolved spectral peaks (formants) can be grouped together for the perception of a vowel equivalent wriggling call. In human vowel and in mouse wriggling-call perception and in the perception of vocalizations having a formant structure, in general, a "harmonic sieve" acting on the resolved spectral peaks (Darwin, 1992; Cohen *et al.*, 1995) is necessary for normalizing vocalizations from different individuals so that the formant reference frame for perception can emerge.

What are the physiological mechanisms of frequency resolution, spectral peak detection, the harmonic sieve, spectral envelope integration, and formant reference-frame setting? Frequency resolution and spectral peak detection refers to the critical band analysis explained before. Neuronal coding of critical bands with their perceptual properties (e.g. adequate bandwidths, intensity independence) has been found in the auditory midbrain inferior colliculus (Ehret and Merzenich, 1985, 1988; Schreiner and Langner, 1997). The harmonic sieve may function as explained for critical-band based perception of mouse ultrasounds (Ehret, 1983, 1987). Integration over spectral profiles has been found in neurons of the inferior colliculus (Escabí and Schreiner, 2002; Escabí *et al.*, 2003) and auditory cortex (Schreiner and Calhoun, 1994; Miller *et al.*, 2002). Spectral envelope integration together with the function of selectivity (harmonic sieve) is also present in combination-sensitive neurons of the inferior colliculus (e.g. Ehret and Merzenich, 1988; Mittman and Wenstrup, 1995; Leroy and Wenstrup, 2000; Egorova *et al.*, 2001) and the auditory cortex (e.g. Suga, 1988; Fitzpatrick *et al.*, 1993; Fishman *et al.*, 2000) which show a facilitated response to simultaneously occurring spectral peaks at certain frequencies often of distances comparable to critical bands (Fishman *et al.*, 2000). The formant reference frames and, thus, the natural perceptual boundaries may be set by the local selectivity of combination-sensitive neurons at the level of the auditory cortex as found for the echolocation calls of the mustached bat (Suga *et al.*, 1979, 1983; Suga and Tsuzuki, 1985; see Suga, 1988).

In conclusion, I hypothesize that perceptual boundaries between vocalizations differing in spectral (formant) structure are established by neurons in the auditory midbrain responding in a critical-band related way and by auditory cortical neurons that are sensitive to combinations of certain spectral peaks especially those having frequency distances of a critical band. Since spectral receptive fields of auditory cortical neurons change with learning (e.g. Weinberger, 1995; Ohl and Scheich, 1996, 1997), I suggest that combination sensitivity and therewith perceptual boundaries of neurons can change in the course of learning new formant patterns.

4.3.6.2 Direction of frequency sweeps (formant transitions)

Many animal vocalizations such as mouse pup wriggling calls and USVs (Fig. 4.1) are frequency modulated. One form of frequency modulation is a frequency increase in the first part of the call and a decrease in the last part as demonstrated in some of the USVs and wriggling calls (Fig. 4.1). In the mouse vocalizations, frequency changes are very variable and do not carry any specific meaning (Ehret and Haack, 1982; Ehret and Riecke, 2002) except possibly contributing to the perception of the urgency of the response (rule 5). In Japanese macaque monkeys, however, contact calls ("coo" calls) are discriminated categorically on the basis of the direction of frequency change in the calls (May et al., 1989). So-called early-high calls reaching the frequency maximum early in the call are typical of isolated animals, while late-high calls (the frequency maximum is late in the call) are emitted by a subordinate to a dominant animal (Green, 1975). The perceptual boundary between early- and late-high calls is the position of the frequency peak. If the peak is in about the first third of the call duration the meaning is early-high, if it is in the last third the meaning is late-high (May et al., 1988, 1989). The Japanese monkeys easily learn to selectively pay attention to the position of the frequency peak while for other monkey species and humans this is more difficult (Zoloth et al., 1979; Hopp et al., 1992). In human speech perception, the direction of frequency sweeps at the beginning of the second formant (formant transition) is decisive for the categorization of the phonemes /ba/ (upward sweep), /da/ (no sweep), and /ga/ (downward sweep) (Liberman et al., 1957). In the mustached bat, the perception of decreasing frequency sweeps at the end of the formants of echolocation calls is decisive for the perception of the echo delay which is equivalent to the meaning of a distance between the bat and its pursued prey.

These examples show that the directions (upward and downward) of frequency sweeps can be perceived as a special meaning which changes at the perceptual boundary, that is between upward and downward sweep, or no sweep.

The natural perceptual boundaries for the perception of sweep direction have their neural correlates in populations of neurons in the inferior colliculus and primary auditory cortex. Studies in the mouse inferior colliculus (Hage and Ehret, 2003) and the cat auditory cortex (Mendelson et al., 1993) have shown that most neurons preferring upward sweeps are spatially separated from neurons preferring downward sweeps. Therefore, I hypothesize that the natural perceptual boundaries for meanings perceived via sweep direction are established by the spatial separation of direction-selective neurons from the auditory midbrain upwards in the auditory system. It is conceivable that learning to pay attention to the activation of local clusters of combination-sensitive neurons with one main characteristic of sweep-direction selectivity is easy, if

4.3 The six rules of communication sound perception

the activation by a given vocalization is associated consistently with a certain behavioral and/or perceptual context as is the case with the coo calls of Japanese monkeys and the phonemes of human speech.

4.3.6.3 Duration of information-bearing elements or parameters

Ultrasounds of mouse pups are perceived by their mothers categorically in the time domain with a perceptual boundary near 25 ms duration (Ehret, 1992). A perceptual boundary near 22 ms is seen also in the categorical perception of species-specific communication calls of the bat, *Phyllostomus discolor* (Zimmer et al., 1998). Thus, call duration is an information-bearing element (Suga, 1988) that can be perceived as different meanings depending from which side of the perceptual boundary on a duration continuum the actual duration is taken.

Temporal gaps between sounds can be information-bearing parameters (Suga, 1988) if the gaps contribute to perceptual differences. In the mustached bat auditory cortex (frequency modulation–frequency modulation area), temporal gaps of about 1–18 ms between the onset of the downward frequency sweep at the end of an echolocation call and the onset of the sweep in the echo are spatially represented in a systematic way by response preferences of the neurons (Suga and O'Neill, 1979; Suga et al., 1983; see Suga, 1988). Thus, the auditory cortex of this bat categorizes with its neural map temporal gaps of less than about 18 ms as the meaning "pursuable object". Since longer gap widths seem not to be represented, such gaps will not release a pursuing response and, hence, would have the meaning "non-pursuable object".

VOT is an information-bearing parameter with perceptual boundaries separating the meanings of phonemes of human speech. The shortest VOT boundary is that separating /ba/ from /pa/ at about 25 ms in humans (Abramson and Lisker, 1970; Pisoni and Lazarus, 1974), and 23 ms in chinchillas (Kuhl and Miller, 1978). Of the same duration (20–25 ms) are gaps that are just detectable in a noise (Penner, 1975), or gaps between a noisy sound onset and a buzz-like stimulus continuation (Stevens and Klatt, 1974).

The above-mentioned studies indicate perceptual boundaries on the duration continuum which are all close to 20 ms. Duration tuning of neurons most probably starts in the inferior colliculus (Feng et al., 1990; Covey and Casseday, 1999) and is found in the auditory cortex (Galazyuk and Feng, 1997). Compatible with the short echolocation calls, inferior colliculus neurons in bats prefer short durations below about 30 ms (Casseday et al., 1994; Ehrlich et al., 1997; Jen and Feng, 1999; Jen and Zhou, 1999). In the mouse, however, almost 70% of the neurons prefer longer durations (Brand et al., 2000) in accordance with the duration of their communication calls (Ehret, 1975a). In the auditory cortex, duration tuning of neurons is extended to about 100 ms in bats (Galazyuk and Feng, 1997) and 200 ms in cats (He *et al.*, 1997). The selectivity of such cortical neurons is suggested to be responsible for boundaries on the duration continuum separating: (i) the meanings of closed- and open-mouth trills of pygmy marmoset monkeys (boundary near 253 ms; Snowdon and Pola, 1978), (ii) preferred from non-preferred models of wriggling calls of mice (boundary near 90 ms; Geissler and Ehret, 2002), and (iii) preferred from non-preferred USVs of mice (boundary near 280 ms; Ehret and Haack, 1982; Ehret, 1987).

In conclusion, I hypothesize that natural perceptual boundaries along the duration continuum are set by clusters of duration-tuned neurons in the auditory midbrain and cortex. In the auditory cortex of the little brown bat, neurons preferring either long or short durations are spatially separated (Galazyuk and Feng, 1997) so that a perceptual boundary becomes obvious in the spatial separation of the neurons.

4.4 Conclusions

An attempt to establish common rules of communication sound perception in animals including humans may be meaningful only if based on (i) evolutionary considerations of mutual adaptations between acoustical properties of vocalizations and auditory capabilities of a given species, and (ii) considerations about what is called "perception" which, in my view, cannot be defined for animals without knowing their responses to acoustical stimuli in natural habitats or, at least, in rather natural behavioral contexts. Knowledge about auditory sensitivity and acuity for hearing and discriminating sounds as obtained in psychoacoustical tests helps to suggest how members of a given species may perceive their species-specific vocalizations. Such knowledge, however, is not sufficient to predict which sound parameter or which parameter combination are most important to generate differences in perceptions of meaning. JNDs do not necessarily equal JMDs. JMDs seem to be based on natural perceptual boundaries established by basic properties in the auditory system, such as neural coding of spectral resolution and characteristics of combination-sensitive neurons and parameter maps.

The perception of meaning from vocalizations may be very simple in animals who are not trained to pay attention to certain acoustical properties in sounds that discriminate between meanings. I have discussed the perception of the three basic meanings: *attraction*, *cohesion*, and *aversion*. Each of these perceptions are integrated with non-vocal information from the behavioral context of the perception and with information about the perceiver's own motivational/emotional state in order to generate an adaptive response. In addition, the motivational/emotional level of the vocalizing animal may be perceived as *the urgency of the response*.

Acknowledgments

The author thanks D.B. Geissler, J. Kanwal, and H. Wolf for helpful comments on the manuscript. This study benefited much from discussions with colleagues of the Forschergruppe 499 of the Deutsche Forschungsgemeinschaft (DFG). Supported by the DFG, EH 53/19-1.

REFERENCES

Abramson A, Lisker L (1970) Discriminability along the voicing continuum: cross-language tests. In: *Proceedings of the 6th International Congress Phonetic Science, Prague 1967*, pp. 569–573. Academia, Prague.

August PV, Anderson JGT (1987) Mammal sounds and motivation-structural rules: a test to the hypothesis. *J Mammal* 68: 1–9.

Bell RW (1974) Ultrasounds in small rodents: arousal-produced and arousal-producing. *Dev Psychobiol* 7: 39–42.

Brand A, Urban A, Grothe B (2000) Duration tuning in the mouse auditory midbrain. *J Neurophysiol* 84: 1790–1799.

Bregman AS (1990) *Auditory Scene Analysis*. MIT Press, Cambridge, MA.

Bruns V (1976) Peripheral auditory tuning for fine frequency analysis by the Cf-FM bat, *Rhinolophus ferrumequinum*. II. Frequency mapping in the cochlea. *J Comp Physiol* 106: 87–97.

Casseday JH, Ehrlich D, Covey E (1994) Neural tuning for sound duration: role of inhibitory mechanisms in the inferior colliculus. *Science* 264: 847–850.

Cheney DL, Seyfarth RM (1982) Recognition of individuals within and between groups of free-ranging vervet monkeys. *Am Zool* 22: 519–529.

Cheney DL, Seyfarth RM, Silk JB (1995) The role of grunts in reconciling opponents and facilitating interactions among adult female baboons. *Anim Behav* 50: 249–257.

Cohen MA, Grossberg S, Wyse LL (1995) A spectral network model of pitch perception. *J Acoust Soc Am* 98: 862–879.

Covey E, Casseday JH (1999) Timing in the auditory system of the bat. *Annu Rev Physiol* 61: 457–476.

Cynx J, Shapiro M (1986) Perception of missing fundamental by a species of songbird (*Sturnus vulgaris*). *J Comp Psychol* 100: 356–360.

Dallos P (1973) *The Auditory Periphery*. Academic Press, New York.

Darwin CJ (1984) Perceiving vowels in the presence of another sound: constraints on formant perception. *J Acoust Soc Am* 76: 1636–1647.

Darwin CJ (1992) Listening to two things at once. In: Schouten MEH (ed.), *The Auditory Processing of Speech. From Sounds to Words*, pp. 133–147. de Gruyter, Berlin.

Dawkins R, Krebs JR (1978) Animal signals: information or manipulation? In: Krebs JR, Davies NB (eds), *Behavioural Ecology*, pp. 282–309. Sinauer, Sunderland, MA.

Dennett DC (1988) Quining qualia. In: Marcel AJ, Bisiach E (eds), *Consciousness in Contemporary Science*, pp. 42–77. Clarendon Press, Oxford.

Egorova MA, Ehret G, Vartanyan IA, Esser KH (2001) Frequency response areas of neurons in the mouse inferior colliculus I. Threshold and tuning characteristics. *Exp Brain Res* 140: 145–161.

Ehret G (1974) Age-dependent hearing loss in normal hearing mice. *Naturwissenschaften* 61: 506–507.

Ehret G (1975a) Schallsignale der Hausmaus (*Mus musculus*). *Behaviour* 52: 38–56.

Ehret G (1975b) Frequency and intensity difference limens and non-linearities in the ear of the house mouse (*Mus musculus*). *J Comp Physiol A* 102: 321–336.

Ehret G (1976a) Development of absolute auditory thresholds in the house mouse (*Mus musculus*). *J Am Audiol Soc* 1: 179–184.

Ehret G (1976b) Critical bands and filter characteristics in the ear of the house mouse (*Mus musculus*). *Biol Cybernet* 24: 35–42.

Ehret G (1977) Comparative psychoacoustics: perspectives of peripheral sound analysis in mammals. *Naturwissenschaften* 64: 461–470.

Ehret G (1980) Development of sound communication in mammals. *Adv Study Behav* 11: 179–225.

Ehret G (1983) Auditory processing and perception of ultrasound in house mice. In: Ewert JP, Capranica RR, Ingle DJ (eds), *Advances in Vertebrate Neuroethology*, pp. 911–918. Plenum Press, London.

Ehret G (1987) Categorical perception of sound signals: facts and hypotheses from animal studies. In: Harnad S (ed.), *Categorical Perception: The Groundwork of Cognition*, pp. 301–331. Cambridge University Press, Cambridge.

Ehret G (1988) Frequency resolution, spectral filtering and integration on the neuronal level. In: Edelman GM, Gall WE, Cowan WM (eds), *Auditory Function. Neurobiological Bases of Hearing*, pp. 363–384. Wiley, New York.

Ehret G (1989) Hearing in the mouse. In: Dooling RJ, Hulse SH (eds), *The Comparative Psychology of Audition: Perceiving Complex Sounds*, pp. 3–32. Lawrence Erlbaum, Hillsdale.

Ehret G (1992) Categorical perception of mouse-pup ultrasounds in the temporal domain. *Anim Behav* 43: 409–416.

Ehret G (1995) Auditory frequency resolution in mammals: from neuronal representation to perception. In: Manley GA, Klump GM, Köppl C, Fastl H, Oeckinghaus H (eds), *Advances in Hearing Research*, pp. 387–397. World Scientific, Singapore.

Ehret G (2005) Infant rodent ultrasounds – a gate to the understanding of sound communication. *Behav Genet* 35: 19–29.

Ehret G, Bernecker C (1986) Low-frequency sound communication by mouse pups (*Mus musculus*): wriggling calls release maternal behaviour. *Anim Behav* 34: 821–830.

Ehret G, Haack B (1981) Categorical perception of mouse pup ultrasounds by lactating females. *Naturwissenschaften* 68: 208.

Ehret G, Haack B (1982) Ultrasound recognition in house mice: key-stimulus configuration and recognition mechanisms. *J Comp Physiol A* 148: 245–251.

Ehret G, Haack B (1984) Motivation and arousal influence sound-induced maternal pup-retrieving behavior in lactating house mice. *Z Tierpsychol* 65: 25–39.

Ehret G, Merzenich MM (1985) Auditory midbrain responses parallel spectral integration phenomena. *Science* 227: 1245–1247.
Ehret G, Merzenich MM (1988) Complex sound analysis (frequency resolution, filtering and spectral integration) by single units of the inferior colliculus of the cat. *Brain Res Rev* 13: 139–163.
Ehret G, Riecke S (2002) Mice and humans perceive multiharmonic communication sounds in the same way. *Proc Natl Acad Sci USA* 99: 479–482.
Ehrlich D, Casseday JH, Covey E (1997) Neural tuning to sound duration in the inferior colliculus of the big brown bat, *Eptesicus fuscus*. *J Neurophysiol* 77: 2360–2372.
Escabí MA, Schreiner CE (2002) Nonlinear spectrotemporal sound analysis by neurons in the auditory midbrain. *J Neurosci* 22: 4114–4131.
Escabí MA, Miller LM, Read HL, Schreiner CE (2003) Naturalistic auditory contrast improves spectrotemporal coding in the cat inferior colliculus. *J Neurosci* 23: 11489–11504.
Fay RR (1988) *Hearing in Vertebrates: A Psychophysics Databook*. Hill-Fay Association, Winnetka, IL.
Feng AS, Hall JC, Gooler DM (1990) Neural basis of sound pattern recognition in anurans. *Prog Neurobiol* 34: 313–329.
Fischer J (2003) Developmental modifications in the vocal behavior of non-human primates. In: Ghazanfar AA (ed.), *Primate Audition: Ethology and Neurobiology*, pp. 109–125. CRC Press, Boca Raton.
Fishman YI, Reser DH, Arezzo JC, Steinschneider M (2000) Complex tone processing in primary auditory cortex of the awake monkey. II. Pitch versus critical band representation. *J Acoust Soc Am* 108: 247–262.
Fitzpatrick DC, Kanwal JS, Batman JA, Suga N (1993) Combination-sensitive neurons in the primary auditory cortex of the mustached bat. *J Neurosci* 13: 931–940.
Flanagan JL (1972) *Speech Analysis, Synthesis, and Perception*. Springer-Verlag, Berlin.
Fletcher H (1940) Auditory patterns. *Rev Mod Phys* 12: 47–65.
Galazyuk AV, Feng AS (1997) Encoding of sound duration by neurons in the auditory cortex of the little brown bat, *Myotis lucifugus*. *J Comp Physiol Am* 180: 301–311.
Geissler DB, Ehret G (2002) Time-critical integration of formants for perception of communication calls in mice. *Proc Natl Acad Sci USA* 99: 9021–9025.
Gifford W, Hauser MC, Cohen YE (2003) Discrimination of functionally referential calls by laboratory-housed rhesus macaques: implications for neuroethological studies. *Brain Behav Evol* 61: 213–224.
Green S (1975) Variation of vocal pattern with social situation in the Japanese monkey (*Macaca fuscata*): a field study. In: Rosenblum LA (ed.), *Primate Behavior*, Vol. 4, pp. 1–102. Academic Press, New York.
Greenwood DD (1961) Auditory masking and the critical band. *J Acoust Soc Am* 33: 484–502.
Greenwood DD (1990) A cochlear frequency-position function for several species – 29 years later. *J Acoust Soc Am* 87: 2592–2605.
Grieser DA, Kuhl PK (1989) Categorization of speech by infants: support for speech–sound prototypes. *Dev Psychol* 25: 577–588.
Haack, B, Markl H, Ehret G (1983) Sound communication between parents and offspring. In: Willott JF (ed.), *The Auditory Psychobiology of the Mouse*, pp. 57–97. Charles C Thomas, Springfield, IL.

Hage SR, Ehret G (2003) Mapping responses to frequency sweeps and tones in the inferior colliculus of house mice. *Eur J Neurosci* 18: 2301–2312.

Handel S (1989) *Listening: An Introduction to the Perception of Auditory Events*. MIT Press, Cambridge, MA.

Hartmann WM (1988) Pitch perception and the organization and integration of auditory entities. In: Edelman GW, Gall WE, Cowan WM (eds), *Auditory Function: Neurobiological Bases of Hearing*, pp. 623–645. Wiley, New York.

Hauser MD (1998) Functional referents and acoustic similarity: field playback experiments with rhesus monkeys. *Anim Behav* 55: 1647–1658.

Hauser MD, Marler P (1993) Food-associated calls in rhesus macaques (*Macaca mulatta*) I. Socioecological factors influencing call production. *Behav Ecol* 4: 194–205.

He J, Hashikawa T, Ojima H, Kinouchi Y (1997) Temporal integration and duration tuning in the dorsal zone of cat auditory cortex. *J Neurosci* 17: 2615–2625.

Heffner H, Whitfield IC (1976) Perception of the missing fundamental by cats. *J Acoust Soc Am* 59: 915–919.

Henson OW (1974) Comparative anatomy of the middle ear. In: Keidel WD, Neff WD (eds), *Handbook of Sensory Physiology, Vol V/1, Auditory System, Anatomy, Physiology (Ear)*, pp. 39–110. Springer-Verlag, Berlin.

Hentschel U, Smith G, Draguns JG (1986) *The Roots of Perception: Individual Differences in Information Processing Within and Beyond Awareness*. North-Holland, Amsterdam.

Hirsh IJ, Watson CS (1996) Auditory psychophysics and perception. *Annu Rev Psychol* 47: 461–484.

Holekamp KE, Boydston EE, Szykman M, Graham I, Nutt KJ, Birch S, Piskiel A, Singh M (1999) Vocal recognition in the spotted hyaena and its possible implications regarding the evolution of intelligence. *Anim Behav* 58: 383–395.

Hopp SL, Sinnot JM, Owren MJ, Petersen MR (1992) Differential sensitivity of Japanese macaques (*Macaca fuscata*) and humans (*Homo sapiens*) to peak position along a synthetic coo call continuum. *J Comp Psychol* 106: 128–136.

Humphrey N (1992) *A History of the Mind*. Chatto & Windus, London.

Illmann G, Schrader L, Špinka M, Šustr P (2002) Acoustical mother-offspring recognition in pigs (*Sus scrofa domestica*). *Behaviour* 139: 487–505.

Jen PHS, Feng RB (1999) Bicuculline application affects discharge patterns and pulse-duration tuning characteristics of bat inferior colliculus neurons. *J Comp Physiol Am* 184: 185–194.

Jen PHS, Zhou XM (1999) Temporally patterned pulse trains affect duration tuning characteristics of bat inferior colliculus neurons. *J Comp Physiol A* 185: 471–478.

Julesz B, Hirsh IJ (1972) Visual and auditory perception: an essay of comparison. In: David EE, Denes P (eds), *Human Communication: A Unified View*, pp. 283–340. McGraw-Hill, New York.

Jürgens U (1979) Vocalizations as an emotional indicator. A neuroethological study in the squirrel monkey. *Behaviour* 69: 88–117.

Klink KB, Klump GM (2004) Duration discrimination in the mouse (*Mus musculus*). *J Comp Physiol A* 190: 1039–1046.

Kuhl PK (1979) Models and mechanisms in speech perception. Species comparisons provide further contributions. *Brain Behav Evol* 16: 374–408.

Kuhl PK, Miller JD (1975) Speech perception by the chinchilla: voiced-voiceless distinction in alveolar plosive consonants. *Science* 190: 69–72.

Kuhl PK, Miller JD (1978) Speech perception by the chinchilla: identification functions for synthetic VOT stimuli. *J Acoust Soc Am* 63: 905–917.

Langner G (1992) Periodicity coding in the auditory system. *Hear Res* 60: 115–142.

Leroy SA, Wenstrup JJ (2000) Spectral integration in the inferior colliculus of the mustached bat. *J Neurosci* 20: 8533–8541.

Liberman AM, Pisoni DB (1977) Evidence for a special speech-perceiving subsystem in the human. In: Bullock TH (ed.), *Recognition of Complex Acoustic Signals*, pp. 59–76. Abakon Verlagsgesellschaft, Berlin.

Liberman AM, Delattre PC, Gertman LJ, Cooper FS (1956) Tempo of frequency change as a cue for distinguishing classes of speech sounds. *J Exp Psychol* 52: 127–137.

Liberman AM, Harris KS, Hoffman H, Griffith BC (1957) The discrimination of speech sounds within and across phoneme boundaries. *J Exp Psychol* 54: 358–368.

Liu RC, Miller KD, Merzenich MM, Schreiner CE (2003) Acoustic variability and distinguishability among mouse ultrasound vocalizations. *J Acoust Soc Am* 114: 3412–3422.

Long GL, Schnitzler HU (1975) Behavioural audiograms from the bat *Rhinolophus ferrumequinum*. *J Comp Physiol* 100: 211–219.

Maiwald D (1967) Ein Funktionsschema des Gehörs zur Beschreibung der Erkennbarkeit kleiner Frequenz- und Amplitudenänderungen. *Acustica* 18: 81–92.

Manser MB (2001) The acoustic structure of suricates' alarm calls varies with predator type and the level of response urgency. *Proc Roy Soc Lond B* 268: 2315–2324.

Manser MB, Bell, MB, Fletcher LB (2001) The information that receivers extract from alarm calls in suricates. *Proc Roy Soc Lond B* 268: 2485–2491.

May B, Moody DB, Stebbins WC (1988) The significant features of Japanese macaque coo sounds: a psychophysical study. *Anim Behav* 36: 1432–1444.

May B, Moody DB, Stebbins WC (1989) Categorical perception of conspecific communication sounds by Japanese macaques, *Macaca fuscata*. *J Acoust Soc Am* 85: 837–847.

McComb K, Moss C, Sayialel S, Baker L (2000) Unusually extensive networks of vocal recognition in African elephants. *Anim Behav* 59: 1103–1109.

Mendelson JR, Schreiner CE, Sutter ML, Grasse KL (1993) Functional topography of cat primary auditory cortex: responses to frequency-modulated sweeps. *Exp Brain Res* 94: 65–87.

Miller CT, Hauser MD (2004) Multiple acoustic features underlie vocal signal recognition in tamarins: antiphonal calling experiments. *J Comp Physiol A* 190: 7–19.

Miller LM, Escabí MA, Read HL, Schreiner CE (2002) Spectrotemporal receptive fields in the lemniscal auditory thalamus and cortex. *J Neurophysiol* 87: 516–527.

Mittman DH, Wenstrup JJ (1995) Combination-sensitive neurons in the inferior colliculus. *Hear Res* 90: 185–191.

Moore BCJ (1989) *An Introduction to the Psychology of Hearing*, 3rd Edition. Academic Press, London.

Morton ES (1977) On the occurrence and significance of motivation-structural rules in some bird and mammal sounds. *Amer Nat* 111: 855–869.

Nelson DA (1988) Feature weighting in species song recognition by the field sparrow (*Spizella pusilla*). *Behaviour* 106: 158–182.
Noirot E (1972) Ultrasound and maternal behavior in small rodents. *Dev Psychobiol* 5: 371–387.
Ohl FW, Scheich H (1996) Differential frequency conditioning enhances spectral contrast sensitivity of units in auditory cortex (field AI) of the alert Mongolian gerbil. *Eur J Neurosci* 8: 1001–1017.
Ohl FW, Scheich H (1997) Learning-induced dynamic receptive field changes in primary auditory cortex (AI) of the unanesthetized Mongolian gerbil. *J Comp Physiol A* 181: 685–696.
Owings DH, Morton ES (1998) *Animal Vocal Communication: A New Approach*. Cambridge University Press, Cambridge.
Owren MJ (1990) Acoustic classification of alarm calls by vervet monkeys (*Cercopithecus aethiops*) and humans (*Homo sapiens*): I. Natural calls. *J Comp Psychol* 104: 20–28.
Penner MJ (1975) Persistence and integration: two consequences of a sliding integrator. *Percept Psychophys* 18: 114–120.
Peters G (1984) On the structure of friendly, close range vocalizations in terrestrial carnivores (*Mammalia: Carnivora: Fissipedia*). *Z Säugetierk* 49: 157–182.
Peterson GE, Barney HL (1952) Control methods used in a study of the vowels. *J Acoust Soc Am* 24: 175–184.
Pisoni DB, Lazarus JH (1974) Categorical and noncategorical modes of speech perception along a voicing continuum. *J Acoust Soc Am* 55: 328–333.
Preisler A, Schmidt S (1995) Virtual pitch formation in the ultrasonic range. *Naturwissenschaften* 82: 45–47.
Rendall D (2003) Acoustic correlates of caller identity and affect intensity in the vowel-like grunt vocalizations of baboons. *J Acoust Soc Am* 113: 3390–3402.
Rendall D, Rodman PS, Emond RE (1996) Vocal recognition of individuals and kin in free-ranging rhesus monkeys. *Anim Behav* 51: 1007–1015.
Rendall D, Seyfarth RM, Cheney DL, Owren MJ (1999) The meaning and function of grunt variants in baboons. *Anim Behav* 57: 583–592.
Samuel AG (1977) The effect of discrimination training on speech perception: noncategorical perception. *Percept Psychophys* 22: 321–330.
Scharf B (1970) Critical bands. In: Tobias JV (ed.), *Foundations of Modern Auditory Theory*, Vol. 1, pp. 159–202. Academic Press, New York.
Scherer KR (1989) Vocal correlates of emotional arousal and effective disturbance. In: Wagner H, Manstead A (eds), *Handbook of Social Psychophysiology*, pp. 165–197. Wiley, New York.
Schreiner CE, Calhoun MB (1994) Spectral envelope coding in rat primary auditory cortex: properties of ripple transfer functions. *Audit Neurosci* 1: 39–62.
Schreiner CE, Langner G (1997) Laminar fine structure of frequency organization in auditory midbrain. *Nature* 388: 383–386.
Seyfarth RM, Cheney DL (2003) Signalers and receivers in animal communication. *Annu Rev Psychol* 54: 145–173.
Seyfarth RM, Cheney DL, Marler P (1980a) Monkey responses of three different alarm calls: evidence of predator classification and semantic communication. *Science* 210: 801–803.

Seyfarth RM, Cheney DL, Marler P (1980b) Vervet monkey alarm calls: semantic communication in a free-ranging primate. *Anim Behav* 28: 1070–1094.
Shaw AG (1974) The external ear. In: Keidel WD, Neff WD (eds), *Handbook of Sensory Physiology, Vol V/1, Auditory System, Anatomy, Physiology (Ear)*, pp. 455–490. Springer-Verlag, Berlin.
Smith JC (1976) Responses of adult mice to models of infant calls. *J Comp Physiol Psychol* 90: 1105–1115.
Smith JC, Sales GD (1980) Ultrasonic behavior and mother–infant interactions in rodents. In: Bell RW, Smotherman WP (eds), *Maternal Influences and Early Behavior*, pp. 105–133. Spectrum, New York.
Snowdon CT, Pola YV (1978) Interspecific and intraspecific responses to synthesized pygmy marmoset vocalizations. *Anim Behav* 26: 192–206.
Spiegel MF, Watson CS (1981) Factors in the discrimination of tonal patterns. III. Frequency discrimination with components of well-learned patterns. *J Acoust Soc Am* 69: 223–230.
Stevens KN, Klatt DH (1974) Role of formant transitions in the voiced-voiceless distinction for stops. *J Acoust Soc Am* 55: 653–659.
Studdert-Kennedy M, Liberman AM, Harris KS, Cooper FS (1970) Motor theory of speech perception: a reply to Lane's critical review. *Psychol Rev* 77: 234–249.
Suga N (1988) Auditory neuroethology and speech processing: complex-sound processing by combination-sensitive neurons. In: Edelman GM, Gall WE, Cowan WM (eds), *Auditory Function. Neurobiological Bases of Hearing*, pp. 679–720. Wiley, New York.
Suga N, Jen PHS (1977) Further studies on the peripheral auditory system of "CF-FM" bats specialized for fine frequency analysis of Doppler-shifted echoes. *J Exp Biol* 69: 207–232.
Suga N, O'Neill WE (1979) Neural axis representing target range in the auditory cortex of the mustache bat. *Science* 206: 351–353.
Suga N, Tsuzuki K (1985) Inhibition and level-tolerant frequency tuning in the auditory cortex of the mustached bat. *J Neurophysiol* 53: 1109–1145.
Suga N, O'Neill WE, Manabe T (1979) Harmonic-sensitive neurons in the auditory cortex of the mustached bat. *Science* 203: 270–274.
Suga N, O'Neill WE, Kujirai K, Manabe T (1983) Specialization of "combination-sensitive" neurons for processing of complex biosonar signals in the auditory cortex of the mustached bat. *J Neurophysiol* 49: 1573–1626.
Summerfield Q, Haggard M (1977) On the dissociation of spectral and temporal cues to the voicing and distinction in initial stop consonants. *J Acoust Soc Am* 62: 435–448.
Terhardt E (1974) Pitch, consonance and harmony. *J Acoust Soc Am* 55: 1061–1069.
Terhardt E (1978) Psychoacoustic evaluation of musical sounds. *Percept Psychophys* 23: 483–492.
ter Keurs M, Festen JM, Plomp R (1992) Effect of spectral envelope smearing on phoneme identification. In: Schouten MEH (ed.), *The Auditory Processing of Speech. From Sounds to Words*, pp. 282–288. de Gruyter, Berlin.
Tomlinson RWW, Schwarz DWF (1988) Perception of the missing fundamental in non-human primates. *J Acoust Soc Am* 84: 560–565.
Weinberger MM (1995) Dynamic regulation of receptive fields and maps in the adult sensory cortex. *Annu Rev Neurosci* 18: 129–158.

Whitney G, Nyby J (1983) Sound communication among adults. In: Willott JF (ed.), *The Auditory Psychobiology of the Mouse*, pp. 98–129. Charles C Thomas, Springfield, IL.

Wier CC, Jesteadt W, Green DM (1977) Frequency discrimination as a function of frequency and sensation level. *J Acoust Soc Am* 61: 178–184.

Yost WA (1992) Auditory image perception and analysis. *Hear Res* 56: 8–19.

Yost WA, Sheft SS (1993) Auditory perception. In: Yost WA, Popper AN, Fay RR (eds), *Human Psychophysics*, pp. 193–236. Springer-Verlag, New York.

Zimmer U, Ehret G, Esser KH (1998) Categorical perception of sounds mimicking species-specific communication calls in FM-bat *Phyllostomus discolor*. In: Elsner N, Wehner R (eds), *Göttingen Neurobiology Report 1998*, Vol. II, p. 326. Thieme, Stuttgart.

Zoloth SR, Petersen MR, Beecher MD, Green S, Marler P, Moody DB, Stebbins W (1979) Species-specific perceptual processing of vocal sounds by monkeys. *Science* 204: 870–873.

Zuberbühler K (2000) Referential labeling in Diana monkeys. *Anim Behav* 59: 917–927.

Zwicker E, Feldtkeller R (1967) *Das Ohr als Nachrichtenempfänger*. Hirzel, Stuttgart.

Behavioral and Physiologic Adaptations: Summary and Discussion

Günter Ehret and Jagmeet S. Kanwal

Evolutionary forces are expected to have shaped the morphology, anatomy, physiology, and behavior of audiovocal communication systems to an adequate functional level. Thus, for every species, properties of the production and the perception of communication sounds appear mutually adapted in various ways depending on how sounds are physically produced, from which substrate they are perceived, and in which behavioral contexts they occur. The way sounds are produced defines their acoustic properties (frequency–amplitude–time structure), the way sounds are perceived defines the pathways for the transfer of the sound's physical waveform into neural activity patterns, and the diversity of an animal's internal states and external circumstances leading to sound production defines differences in messages emitted to be perceived as differences in meaning. All of these issues are discussed in the first part of the book by drawing upon some marvelous examples of anatomic, physiologic, and behavioral adaptations in sound communication systems and by introducing a general hypothesis about how communication sounds and their potential meanings are perceived.

Anatomic and physiologic adaptations

Different from mammals, birds usually generate sounds in the syrinx, which is a specialized area at the division of the trachea into the two main bronchi. Since Greenewalt's (1968) classical description of the acoustics and physiology of bird song, a deeper understanding of the syringeal functions has evolved, especially with regard to differences in the various taxa of birds and the versatility of structural elements of the syrinx in sound production. Suthers *et al.* (Chapter 1, this volume) describe many of these advances and bring them into the context of sound communication. We emphasize here only one major aspect. In songbirds, the vocal apparatus is doubled in many respects because the left and the right part of the syrinx can be used independently to generate sound trains via motor programs from the left and right half of the brain, respectively (e.g. Nottebohm *et al.*, 1982; Wild *et al.*, 2000). This considerably increases the possible complexity of vocalizations by introducing interferences in the frequency and time domains between the sounds from the two voices. For example, beating amplitude modulations and dissonant sounds can be produced by shifts of left- versus right-side fundamental frequencies, and syllable repetition rates can be doubled by bringing sounds from both sides into one temporal sequence. From the perceptual side, the doubled vocal apparatus with its immense possibilities in producing complexity of vocal expressions is certainly a challenge for the evolution of adaptations in analytical and discrimination power of the perceptual system in order to cope with the heard sound patterns. At least songbirds who are able to adjust their individual song to the song of a tutor during vocal learning are very efficient in using the analytical capabilities of their auditory systems for this adjustment. An enhanced capability of songbirds compared to humans in the discrimination of temporal fine structure in complex harmonic sounds (Dooling *et al.*, 2002; Leek *et al.*, 2004) may be one of these adaptations on the sensory side. Further, a doubling of the vocal apparatus brings about the question of which syrinx, left or right, dominates in song production. Earlier studies reported a left syringeal dominance for song production (e.g. Nottebohm, 1980); recent progress, however, indicates that this is not generally the case (Suthers *et al.*, Chapter 1, this volume). In summary, the advances in the understanding of vocal mechanisms in birds present a picture of specialized adaptations in vocal mechanisms, even among different taxa of birds. These adaptations are complemented by specializations in the perceptual apparatus (e.g. Dooling, 1989) and specialized brain circuits for song learning (e.g. Nottebohm, 1991; Gahr, 2000). Together, these data initiate a new discussion about how far mechanisms of audiovocal behavior and learning in songbirds are representative and can be taken as models for audiovocal behavior and language learning in humans.

The blind mole rat represents another extreme of audiovocal adaptations. Wollberg *et al.* (Chapter 2, this volume) report about seismic communication via the auditory channel in these animals. Mole rats live mostly solitary in a subterranean tunnel system. When they communicate across tunnels, one animal taps with its head against the tunnel walls while the other is listening with its lower jaw which is connected with a unique bony appendix to the tympanic bulla. The lower jaw serves the functions of both the outer and middle ear because it couples the sound via bone conduction directly to the inner ear. In these animals, the usual middle ear pathway of sound transfer is rather insensitive. However, with this exceptional use of the auditory channel for seismic communication, the animals are well adapted to signal their presence to close neighbors in order to direct the digging direction, usually to avoid falling into each other's tunnels. As another adaptation to subterranean life, the auditory pathways in the blind mole rat have invaded parts of the neocortex that are normally reserved for processing visual stimuli (Bronchti *et al.* 1989, 2002; Heil *et al.*, 1991). The invasion of the visual by the auditory systems starts with projections of the auditory midbrain to the dorsal lateral geniculate body, which connects to the occipital (visual) cortex (Doron and Wollberg, 1994). Thus, the auditory system gains new neural substrate for processing communication sounds and seismic vibrations. In how far auditory capabilities of discrimination and resolution are actually improved by this extension of the auditory cortex is not known yet. This case is the reverse of the invasion of the auditory cortex by the visual system which is found in ferrets when they are deprived from auditory input early in development (Pallas *et al.*, 1990; Roe *et al.*, 1990). These two examples show the highly remarkable plasticity of the neocortical circuitry to process signals of several modalities. In other words, different regions of the sensory neocortex are adapted for performing very similar neural operations regardless of the original sensory input.

Behavioral adaptations and the perception of meaning

In a rather unique way, the audiovocal communication system of echolocating bats serves two purposes: self-communication in the behavioral context of echolocation and social communication in contexts, such as sexual behavior, parental behavior, and agonistic interactions in dominance and territory conflicts, that bats share with many other mammals. How do bats separate either the acoustic structure of social and echolocation calls, or the perceptual mechanisms of the calls, or both in order to optimize communication for both purposes? Clement *et al.* (Chapter 3, this volume) analyze social communications calls of two subspecies of mustached bats, (*Pteronotus parnellii parnellii* and *P. p. rubiginosus*) and provide a description of the behavioral context in which most of the simple syllabic calls are emitted in contrast to the earlier acoustic analysis in *P. p. parnellii* (Kanwal *et al.*, 1994). Clearly, social calls differ from echolocation calls in several respects: they are more variable, often much longer in duration, and often cover a lower, broader, or narrower frequency range. The frequency and time structure of social communication calls follow, in general, the motivation–structural rules proposed by Morton (1977) and supplemented by August and Anderson (1987). That is, noisy, broadband, harsh sounding calls occur in aggressive and agonistic encounters, tonal calls with high-frequency components are typical of defeated or fearful animals, and rapidly frequency modulated (rhythmic) broadband calls are emitted in contact seeking or contact avoiding situations. These calls evoke vocalizations or other responses on the part of the receiver only in a relevant behavioral context; that is, if its internal state, its own ongoing behavior, and perceptions are compatible with the perception of a given call. This is an important result because it demonstrates a general relationship between message and meaning (e.g. Seyfarth and Cheney, 2003). An acoustic message can be transformed to a meaningful message only via an adaptive (physiologic and behavioral) response of the perceiving animal. In other words, acoustic communication is not selected for information transfer but for the consequences of perception (for further discussion, see Owings and Morton, 1998). Since the consequences of perception are very different for social communication compared to echolocation, the two systems have certainly evolved in different ways. This is not only reflected by different sound structures and behaviors but also by different pathways of processing in the higher-auditory system. In the mustached bat, the neural pathways and the physiology of representing echolocation calls have been studied in much detail (see reviews in e.g. Suga, 1988; Pollak and Casseday, 1989; Suga et al., 1998), whereas the search for the representation of social communication sounds has just begun (see Kanwal, Chapter 7, this volume).

As indicated above, the acoustic structure of a vocalization may encode a message but not necessarily any meaning or a certain meaning because the meaning is determined by the receiver through a valuation process of the perceived sound properties together with many other receiver-inherent properties. Considering modern evolutionary theory, we have to postulate, however, mutual adaptations between vocalized messages and perceived meanings in order to make a communication system function adequately to the advantage of the communicators' genes. Assuming that vocalizations, by their acoustic properties, express basically three emotional/motivational messages, namely aggressive, fearful, or friendly tendencies (Morton, 1977; August and Anderson, 1987), we can ask, what kind of basic meanings are perceived that fit, in the sense of closing a communication loop, to these messages. Using the audiovocal communication of house mice as a basis, Ehret (Chapter 4, this volume) puts receiver properties found in many mammalian species together to propose six rules for communication sound perception. Rules 2, 4, and 5 deal with the perception of meaning. An important point is that among the multitude of sounds that can potentially be discriminated by the auditory system on the basis of their acoustic structure only few are actually discriminated with respect to the perceived meaning as judged by observations of an animal's behavior. That is, just meaningful differences in sounds do not (necessarily) coincide with just noticeable differences. Thus, there is ample auditory space for grouping sounds of noticeably different acoustic structure into one perceptual category of meaning. It is suggested that the auditory parameter space is perceived as having the basic meaning of either attraction, cohesion, or aversion, and all three with the additional meaning of "the urgency of the response". These meanings arise from the three basic emotional/motivational messages in the vocalizations by the above-mentioned valuation process in the receiver. The result of a general co-evolution of emotional/motivational states of a sender expressed in certain acoustic structures of its sounds, and the adapted responses of a conspecific receiver is visible in communication systems where we observe the following correspondence between the types of message and meaning; that is, aggression causes aversion, fear causes attraction, and friendliness causes cohesion; the latter being expressed in gaining or keeping a certain group structure or doing something together. In audiovocal communication systems as described above, both the delivery of an acoustic message and the perception of a meaning are rooted in the animals' instincts. Since instinctive actions are considered to have some hardwired neural background (e.g. Schildberger and Elsner, 1994), the next great challenge will be to find differences in the pathways and activations in the brain resulting from the processing and perception, for example, of harsh, broadband sounds to aversive behavior or of high-frequency tonal sounds to attraction. Given the evolutionary connection, another great challenge will be to bridge Morton's (1977) basic sound categories (supplemented by August and Anderson, 1987) and their relationship with basic emotions in animals to the dramatically enriched shades of emotions and meanings of sounds perceived by humans.

REFERENCES

August PV, Anderson JGT (1987) Mammal sounds and motivation–structural rules: a test to the hypothesis. *J Mammal* 68: 1–9.

Bronchti G, Heil P, Scheich H, Wollberg Z (1989) Auditory pathway and auditory activation of primary visual targets in the blind mole rat (*Spalax ehrenbergi*). I. A 2-deoxyglucose study of subcortical centers. *J Comp Neurol* 284: 253–274.

Bronchti G, Heil P, Sadka R, Hess A, Scheich H, Wollberg Z (2002) Auditory activation of "visual" cortical areas in the blind mole rat (*Spalax ehrenbergi*). *Eur J Neurosci* 16: 311–329.

Dooling RJ, Leek MR, Gleich O, Dent ML (2002) Auditory temporal resolution in birds: discrimination of harmonic complexes. *J Acoust Soc Am* 112: 748–759.

Dooling RJ (1989) Perception of complex, species-specific vocalizations by birds and humans. In: Dooling RJ, Hulse SH (eds). The Comparative Psychology of Audition. *Perceiving Complex Sounds*, pp. 423–444. Lawrence Erlbaum, Hillsdale, NJ.

Doron N, Wollberg Z (1994) Cross-modal neuroplasticity in the blind mole rat *Spalax ehrenbergi*: a WGA–HRP tracing study. *NeuroReport* 5: 2697–2701.

Gahr M (2000) Neural song control system of hummingbirds: comparison to swifts, vocal learning (songbirds) and nonlearning (suboscines) passerines, and vocal learning (budgerigars) and nonlearning (dove, owl, gull, quail, chicken) non-passerines. *J Comp Neurol* 426: 182–196.

Greenewalt CH (1968) *Bird Song: Acoustics and Physiology*. Smithsonian Institution Press, Washington, DC.

Heil P, Bronchti G, Wollberg Z, Scheich H (1991) Invasion of visual cortex by the auditory system in the naturally blind mole rat. *NeuroReport* 2: 735–738.

Kanwal JS, Matsumura S, Ohlemiller K, Suga N (1994) Analysis of acoustic elements and syntax in communication sounds emitted by mustached bats. *J Acoust Soc Am* 96: 1229–1254.

Leek MR, Dooling RJ, Gleich O, Dent ML (2004) Discrimination of temporal fine structure by birds and mammals. In: Pressnitzer D, de Cheveigné A, McAdams S, Collet L (eds), *Auditory Signal Processing: Physiology, Psychoacoustics, and Models*, pp. 471–477. Springer-Verlag, New York.

Morton ES (1977) On the occurrence and significance of motivation–structural rules in some bird and mammal sounds. *Am Nat* 111: 855–869.

Nottebohm F (1980) Brain pathways for vocal learning in birds: a review of the first 10 years. *Prog Psychobiol Physiol Psychol* 9: 85–124.

Nottebohm F (1991) Reassessing the mechanisms and origins of vocal learning in birds. *TINS* 14: 206–211.

Nottebohm F, Kelley DB, Paton JA (1982) Connections of vocal control nuclei in the canary telencephalon. *J Comp Neurol* 207: 344–357.

Owings DH, Morton ES (1998) *Animal Vocal Communication: A New Approach*. Cambridge University Press, Cambridge.

Pallas SL, Roe AW, Sur M (1990) Visual projections induced into the auditory pathway of ferrets. I. Novel inputs to primary auditory cortex (AI) from the LP/pulvinar complex and the topography of the MGN-AI projection. *J Comp Neurol* 298: 50–68.

Pollak GD, Casseday JH (1989) *The Neural Basis of Echolocation in Bats*. Springer-Verlag, Berlin.

Roe AW, Pallas SL, Hahm JO, Sur M (1990) A map of visual space induced in primary auditory cortex. *Science* 250: 818–820.

Schildberger K, Elsner N (eds) (1994) *Neural Basis of Behavioural Adaptations*. Gustav Fischer Verlag, Stuttgart.

Seyfarth RM, Cheney DL (2003) Signalers and receivers in animal communication. *Annu Rev Psychol* 54: 145–173.

Suga N (1988) Auditory neuroethology and speech processing: complex-sound processing by combination-sensitive neurons. In: Edelman GM, Gall WE, Cowan WM (eds), *Auditory Function. Neurobiological Bases of Hearing*, pp. 679–720. Wiley, New York.

Suga N, Yan J, Zhang Y (1998) The processing of species-specific complex sounds by the ascending and descending auditory systems. In: Poon PWF, Brugge JF (eds), *Central Auditory Processing and Neural Modeling*, pp. 55–70. Plenum Press, New York.

Wild JM, Williams MN, Suthers RA (2000) Neural pathways for bilateral vocal control in songbirds. *J Comp Neurol* 423: 413–426.

PART TWO

Neural adaptations and plasticity

5
Neural Mechanisms of Vocal Communication: Interfacing with Neuroendocrine Mechanisms

Andrew H. Bass

5.1 Introduction

A number of recent reviews have considered advances in studies of the neural and behavioral mechanisms of auditory communication in teleost fish (e.g. Bass and McKibben, 2003; Lu, 2004; Bass *et al.*, 2005). The past 5 years in particular have witnessed a burgeoning of studies addressing auditory mechanisms, ranging from the delineation of central vocal and auditory pathways extending from forebrain to hindbrain levels to the discovery of auditory neurons that encode the temporal properties of vocalizations. This work has been further enriched by studies showing how the neuroendocrine system interfaces with a vocal–acoustic framework (see Goodson and Bass, 2002; Goodson *et al.*, 2003). It is in this context that I will now highlight the most recent advances in showing how steroid hormones modulate the neurophysiologic properties of vocal and auditory neurons.

5.2 Hormonal control of vocalizations

One example of how steroid hormones can influence neural mechanisms of vocal communication comes from recent studies of the effects of steroids on the vocal motor system of the plainfin midshipman fish, *Porichthys notatus*. Midshipman fish migrate seasonally from deep offshore sites along the northwestern coast of North America into the shallow intertidal zone where they reproduce under rocky shelters (Bass, 1996). There, males produce broadband grunts that are used in agonistic contexts and long duration, multiharmonic hums that function as advertisement calls that attract females to the nest (Fig. 5.1a and b; Brantley and Bass, 1994; Bass *et al.*, 1999). Neurophysiologic studies in midshipman and the closely related toadfish (same order and family) delineate a vocal pattern generator in the caudal hindbrain and rostral spinal cord that includes pacemaker-like neurons that set the firing frequency of the nearby

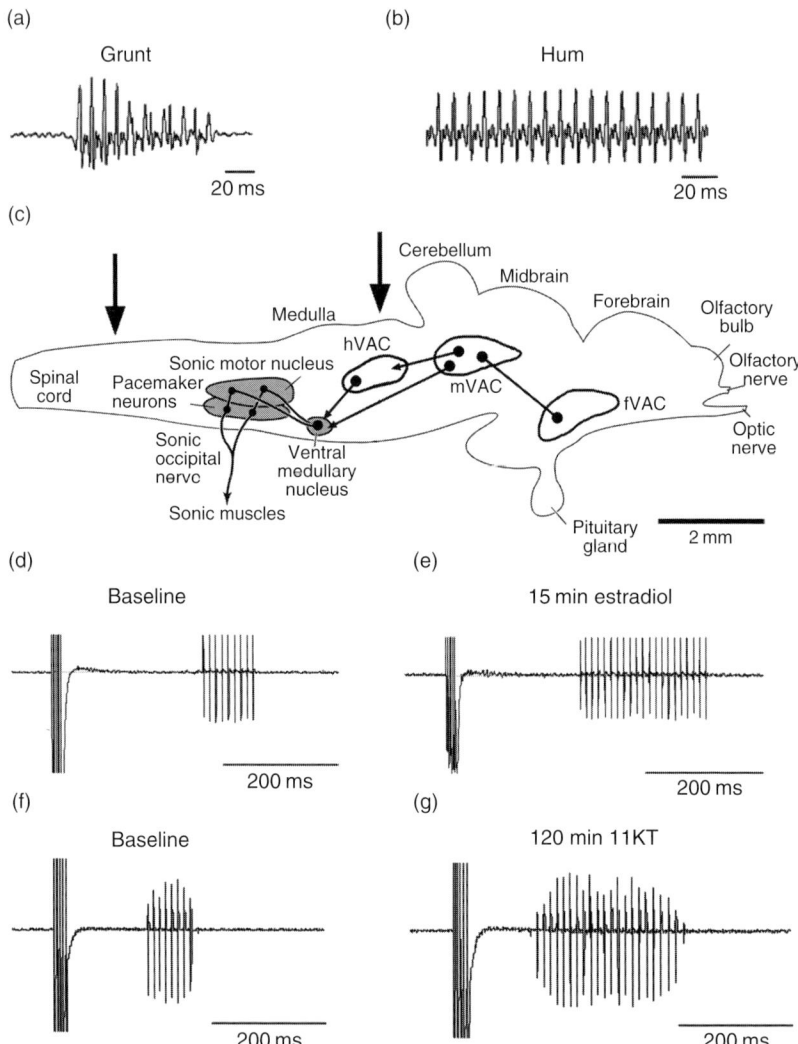

Figure 5.1. Steroid modulation of fictive vocalizations. (a and b) During the breeding season, male midshipman fish produce agonistic grunts and advertisement calls known as hums. (c) Sagittal view of a midshipman brain showing the descending vocal–acoustic control system that includes forebrain (fVAC), midbrain (mVAC) and hindbrain (hVAC) vocal–acoustic integration centers (see Goodson and Bass, 2002). Also shown is the vocal pattern generator that includes a column of pacemaker neurons that innervate sonic motor neurons; a ventral medullary nucleus provides for bilateral coupling of the pacemaker–motor neuron circuit (see Bass and Baker, 1990; Bass et al., 1994). The pattern generator produces a rhythmic vocal motor volley leading to the simultaneous contraction of paired sonic muscles attached to the walls of the swimbladder. The rapid contraction of these muscles produces vocalizations such as grunts and hums. (d–g) Oscillogram

motor neurons that innervate paired sonic muscles attached to the walls of the swimbladder (Fig. 5.1c; Bass and Baker, 1990). The vocal pattern generator can be activated by electric stimulation in midbrain and forebrain vocal nuclei that together form a descending vocal control system (Fig. 5.1c; Goodson and Bass, 2002). This output, referred to as a fictive vocalization, sets the fundamental frequency and duration of natural vocalizations and is easily monitored using intracranial recordings from occipital nerve roots that give rise to the sonic nerve that innervates each of the swimbladder muscles (Bass and Baker, 1990). This preparation has been used to study the influence of both neuropeptides (Goodson and Bass, 2000a,b) and steroid hormones (Remage-Healey and Bass, 2004) on fictive vocalizations that are predictive of the characteristic features of natural vocalizations. Remage-Healey and Bass (2004) just reported that both adrenal- and gonad-derived steroids have rapid affects on vocal patterning in midshipman fish. Thus, within 5 min of an intramuscular injection of either 17β-estradiol, cortisol or 11-ketotestosterone which is a teleost-specific androgen that is non-aromatizable, that is, will not be converted to estradiol by the aromatase enzyme, there is an increase in the duration of the fictive vocalizaiton (Fig. 5.1d–g). Superimposed upon this general steroid effect on duration are steroid-specific effects on the sustainability of the increased duration. Thus, the 17β-estradiol induced increase in duration only lasts for about 30 min, while the cortisol effect disappears after 60 min and the 11-ketotestosterone effect continues for at least 120 min (Fig. 5.1d–g). Remage-Healey and Bass (2004) were able to show by surgical isolation of the region that contains the vocal pattern generator (large vertical arrows, Fig. 5.1c) that the short-lived cortisol effect and a similar, short-term component of the 11-ketotestosterone effect ($\leq$30 min) appear to be localized to the influence of steroids on the hindbrain–spinal pattern generator. Steroid effects beyond 30 min are dependent on steroid action upon more upstream vocal centers in midbrain and/or forebrain vocal centers that provide inputs to midbrain vocal neurons (see Goodson and Bass, 2002). The results imply a hierarchical pattern of descending control of vocal patterning.

How do these results relate to natural behaviors? Perhaps the most relevant set of experiments are recent field studies of the Gulf toadfish, *Opsanus beta*, that also

Caption for fig. 5.1. (cont.)
records of fictive vocalizations evoked by midbrain electric stimulation in male midshipman fish; each fictive vocalization represents the vocal motor volley as recorded from one side of the brain (both sides fire in synchrony; see Bass and Baker, 1990). The stimulus artifact appears on the left side of each trace. Baseline recordings prior to steroid treatment are shown to the left (d and f). To the right are shown fictive vocalizations at 15 min following an intramuscular injection of 17-β-estradiol (e) and 120 min following androgen (11-ketotestosterone, 11KT) injection (g). From Remage-Healey and Bass (2004).

has a steroid-sensitive vocal motor circuit (Remage-Healey and Bass, 2003). Playbacks of tone stimuli that mimic the fundamental frequency and duration of advertisement calls (known as boatwhistles for toadfish; see Bass and McKibben, 2003) can evoke rapid increases ($\leqslant 20$ min) in both the rate and duration of boatwhistle calling by males that are targets of the playbacks (Remage-Healey and Bass, 2003; 2005). Moreover, these behavioral changes are concurrent with rapid changes ($\leqslant 20$ min) in plasma levels of 11-ketotestosterone. While comparable experiments have not been completed for midshipman fish, they exhibit elevated 11-ketotestosterone levels during advertisement calling (Knapp et al., 2001). Egg-guarding male midshipman fish also show changes in the duration of individual agonistic calls as well as their repetitiveness (referred to as grunt trains; M. Marchaterre and A. Bass, unpublished observations; also see Bass et al., 1999) that are consistent with the magnitude of steroid-induced changes in fictive vocalizations. We expect that steroids act synergistically with neuropeptides of the nine amino acid family of arginine vasotocin and isotocin (fish homologs of respectively mammalian arginine vasopressin and oxytocin) that can also influence the duration of fictive vocalizations (Goodson and Bass, 2000a,b).

5.3 Vocal-auditory coupling

A second example of how steroids can influence auditory mechanisms also comes from studies in the midshipman fish. All of the available evidence indicates that males produce advertisement hums only during the breeding season and that the hum is both necessary and sufficient to elicit positive phonotaxis by females (Bass et al., 1999; Bass and McKibben, 2003). Females apparently release all of their eggs in a single evening within the nest of one male and soon leave the nest after spawning, to return to offshore habitats. By contrast, males remain to guard fertilized eggs and newly hatched embryos, while also humming on subsequent nights to attract more females and enhance their reproductive success (Brantley and Bass, 1994). Sisneros and Bass (2003) reported for female midshipman fish that there was a seasonal change in the degree of temporal encoding of the upper harmonics of the hum by primary afferents of the eighth nerve that innervate the saccular epithelium, the main auditory end organ of the inner ear in this and many other groups of teleosts (Popper and Fay, 1999; Bass and McKibben, 2003). Temporal encoding is shown by the degree of phase locking, measured as the vector strength of synchronization or VS (see Goldberg and Brown, 1969; Fay, 1978), of the afferent spike train to auditory tone stimuli that span the range of frequencies that are prominent in midshipman calls ($\leqslant 500$ Hz). During the non-reproductive months of the fall and winter seasons,

5.3 Vocal-auditory coupling

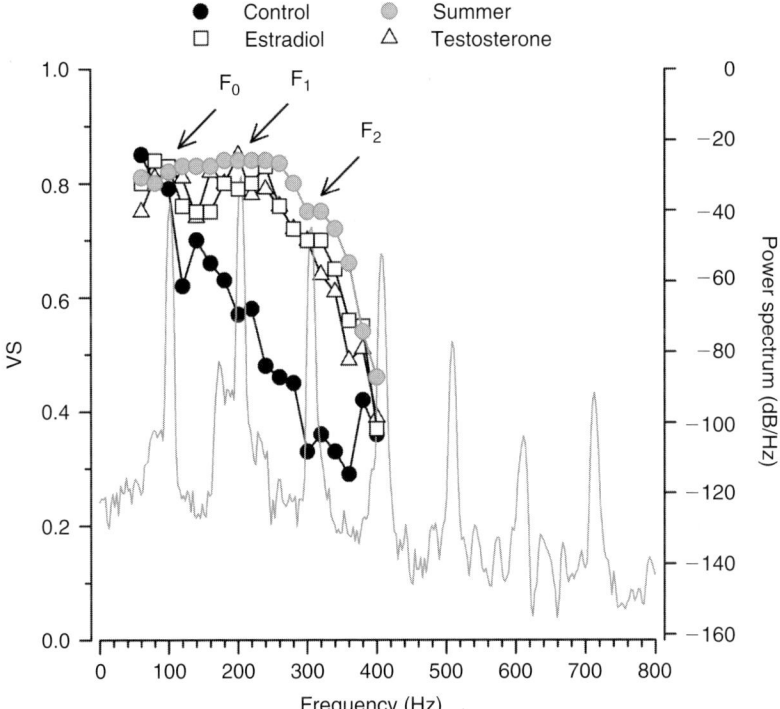

Figure 5.2. Vocal–auditory coupling in midshipman fish. VS is used as a measure of the degree of phase locking of spike trains (Y-axis to the left) to an acoustic stimulus such as tones. Also shown is a power spectrum of the advertisement call hum (also see Fig. 5.1b); relative dB values are shown on the Y-axis to the right. Frequency is plotted along the X-axis for both measures. Shown here are median VS values obtained from single unit recordings from the afferents that innervate the sacculus (main auditory end organ of the inner ear) of non-reproductive females that are untreated controls (dark-shaded circles), treated with testosterone (triangles) and treated with 17β-estradiol (squares). Non-reproductive females show high VS values for frequencies close to the fundamental (F_0), while steroid-treated females show a wider range of similarly robust encoding that extends up to the second and third harmonics (F_1, F_2) of the hum advertisement call. The encoding observed for steroid-treated females overlaps that observed for wild-caught reproductive females from the breeding season (light-shaded circles). Adopted from Sisneros et al. (2004a).

saccular afferents show the largest VS values for frequencies ≤100 Hz (Fig. 5.2). However, during the reproductive months of late spring and summer, saccular afferents show large increases in VS values for frequencies >100 Hz that overlap the second and third harmonics of the advertisement call that are often greater or equal in magnitude to that of the fundamental (Bass et al., 1999; Fig. 5.2).

Sisneros and Bass (2003) hypothesized that the increased encoding observed during the breeding season aids in the detection and localization of humming males in their nests in the shallow intertidal zone. This hypothesis rests upon the observation that higher harmonics are transmitted over a greater distance in shallow water as predicted by the influence of water depth on what is referred to as the cut-off frequency, the frequency below which sound transmission is negligible (as water depth decreases, the cut-off frequency increases; see review in Bass and Clark (2003) and Fine and Lenhardt (1983) for experimental studies in the closely related toadfish, *Opsanus tau*). Two strategies might be adopted to shift a signal's spectrum farther away from the cut-off frequency at any one depth. One solution would be to produce a signal with higher frequency components. Although the fundamental frequency of midshipman calls is near 100 Hz, which is close to the predicted cut-off frequency for calls in their habitat at night when their nests are submerged in the intertidal zone (see Bass and Clark, 2003), the advertisement hum has prominent harmonics above 100 Hz that will expand its transmission distance. Another solution to the cut-off frequency problem is to call in deeper water; midshipman males mainly call at night when water depths reach their peak at high tide which should contribute to a shift in the cut-off frequency to a lower value. Sisneros and Bass (2003) did not examine possible seasonal changes in frequency encoding in males, but there is no reason why they too would not show a similar expansion in frequency sensitivity measured in terms of phase locking. This attribute would similarly aid males in the detection of suitable nest sites where males are already humming as well as the detection of a neighbor once they have a nest of their own.

Most recently, Sisneros *et al.* (2004a) have shown that the steroid hormones testosterone and 17β-estradiol can induce a reproductive auditory phenotype in non-reproductive females (Fig. 5.2). Thus, artificially elevated steroid hormone levels can lead to an increase in the degree of temporal encoding by the eighth nerve to higher frequencies that are prominent in the advertisement calls of males. This is consistent with the seasonal shift in frequency encoding observed among wild-caught populations of females (Fig. 5.2) and with naturally elevated levels of testosterone and 17β-estradiol among females both just prior to and during the reproductive season (Sisneros *et al.*, 2004b). These studies have prompted the question as to why there is a narrowing of the window of frequency encoding during the non-reproductive months. There are at least two possible alternative, but complementary, explanations. One neural-based, mechanistic explanation might simply be that it is energetically expensive to maintain an expanded range of increased temporal encoding. A second, ecologically based explanation might be that it relates to increased detection of other sonic species. For example, during the non-reproductive periods of the fall and winter when midshipman shift to offshore sites, they may overlap the path of migrating whales (e.g. see Crane and Lashkari, 1996; McDonald et al., 2001). Some baleen whales (e.g. blue and gray) make calls with significant spectral energy below 100 Hz (e.g. see Crane and Lashkari, 1996; McDonald et al., 2001; Bass and Clark, 2003), a range that overlaps the main bandwidth sensitivity of the midshipman ear during the fall and winter. While baleen whales may not actively prey upon midshipman fish, midshipman may become part of their by-catch when they are feeding upon krill. Importantly, midshipman fish collected from trawls during the winter months have been found to have krill in their stomach contents (J. Sisneros, personal communication). Perhaps midshipman fish attend to the calls of migrating whales to avoid them while they are feeding.

5.4 Conclusion and future prospects

Together, the above studies show how studies of teleost fish are providing new insights into the hormone-dependent plasticity of both vocal motor and auditory systems. Studies of the interface between neuroendocrine, vocal and auditory systems will remain a major focus of future studies because they provide models for identifying cellular and molecular mechanisms that underlie the plasticity in auditory and vocal traits. Given the conserved pattern of organization of vocal (see Bass and Baker, 1997), auditory (see McCormick, 1999; Bass et al., 2005) and neuroendocrine (see Goodson and Bass, 2001) systems across vertebrates, it is likely that these mechanisms will be conserved among vertebrates. The most recent studies of Sisneros et al. (2004a) open up a new set of biophysical studies of the inner ear. Will steroids affect the electric resonance properties of hair cells or the filtering properties of the afferent-hair cell synapse? Other studies in teleosts have defined the spectral and temporal encoding properties of central auditory neurons (see reviews cited earlier). How will neuropeptides, steroids and other neurochemicals (e.g. serotonin; Hurley and Pollak, 1999) affect those encoding properties and thus adaptations to the detection of changing environmental stimuli? How will neurochemicals affect multi-sensory integration? For example, there is evidence to support the central integration of lateral line and auditory systems (e.g. see Weeg and Bass, 2000; Edds-Walton and Fay, 2003). How might peptides and steroids affect such interfaces depending on an animal's reproductive state? This is an especially interesting question to ask of midshipman and toadfish since Weeg and Bass (2002) showed that the lateral line system of midshipman is sensitive to the low frequency components of the advertisement call of males. These and many other exciting questions lie ahead for researchers that use teleost fish as study species.

Acknowledgments

Supported in part by NIH-NIDCD (DC00092) and NSF (9987241). Thanks to Joseph Sisneros for help with the figures.

REFERENCES

Bass AH (1996) Shaping brain sexuality. *Am Sci* 84: 352–363.
Bass AH, Baker R (1990) Sexual dimorphisms in the vocal control system of a teleost fish: morphology of physiologically identified neurons. *J Neurobiol* 21: 1155–1168.
Bass AH, Baker R (1997) Phenotypic specification of hindbrain rhombomeres and the origins of rhythmic circuits in vertebrates. *Brain, Behav Evol* 50: 3–16.
Bass AH, Clark CW (2003) The physical acoustics of underwater sound communication. In: Simmons AM, Fay RR, Popper A (eds), *Springer Handbook of Auditory Research, Acoustic Communication*, Vol. 16, pp. 15–64. Springer, New York, NY.
Bass AH, McKibben JR (2003) Neural mechanisms and behaviors for acoustic communication in teleost fish. *Prog Neurobiol* 69: 1–26.
Bass AH, Marchaterre MA, Baker R (1994) Vocal–acoustic pathways in a teleost fish. *J Neurosci* 14: 4025–4039.
Bass AH, Bodnar DA, Marchaterre MA (1999) Complementary explanations for existing phenotypes in an acoustic communication system. In: Hauser M, Konishi M (eds), *Neural Mechanisms of Communication*, pp. 493–514. MIT Press, Cambridge, Mass.
Bass AH, Rose GJ, Pritz MB (2005) Auditory midbrain of fish, amphibians and reptiles: models systems for understanding auditory function. In: Winer JA, Schreiner CE (eds), *The Inferior Colliculus*, pp. 459–492. Springer-Verlag, New York, NY.
Brantley RK, Bass AH (1994) Alternative male spawning tactics and acoustic signaling in the plainfin midshipman fish, *Porichthys notatus. Ethology* 96: 213–232.
Crane NL, Lashkari K (1996) Sound production of gray whales, *Eschrichtius robustus*, along their migration route: a new approach to signal analysis. *J Acoust Soc Am* 100: 1878–1886.
Edds-Walton PL, Fay RR (2003) Directional selectivity and frequency tuning of midbrain cells in the oyster toadfish, *Opsanus tau. J Comp Physiol* A 189: 527–543.
Fay RR (1978) Phase-locking in goldfish saccular nerve fibers accounts for frequency discrimination capacities. *Nature* 275: 320–322.
Fine ML, Lenhardt ML (1983) Shallow-water propagation of the toadfish mating call. *Comp Biochem Physiol A* 76: 225–231.
Goldberg JM, Brown PB (1969) Response of binaural neurons of dog superior olivary complex to dichotic tonal stimuli: some physiological mechanisms of sound localization. *J Neurophysiol* 32: 613–636.
Goodson JL, Bass AH (2000a) Forebrain peptide modulation of sexually polymorphic vocal motor circuitry. *Nature* 403: 769–772.
Goodson JL, Bass AH (2000b) Vasotocin innervation and modulation of vocal–acoustic circuitry in the teleost *Porichthys notatus. J Comp Neurol* 422: 363–379.
Goodson JL, Bass AH (2001) Social behavior functions and related anatomical characteristics of vasotocin/vasopressin systems in vertebrates. *Brain Res Rev* 35: 246–265.

Goodson JL, Bass AH (2002) Vocal–acoustic circuitry and descending vocal motor pathways in teleost fish: convergence with terrestrial vertebrates reveals conserved traits. *J Comp Neurol* 448: 298–321.

Goodson JL, Evans AK, Bass AH (2003) Isotocin distributions in vocal fish: relationship to vasotocin and vocal–acoustic circuitry. *J Comp Neurol* 462: 1–14.

Hurley L, Pollak GD (1999) Serotonin differentially modulates responses to tones and frequency-modulated sweeps in the inferior colliculus. *J Neurosci* 19: 8071–8082.

Knapp R, Marchaterre MA, Bass AH (2001) The relationship between courtship behavior and steroid hormone levels in parental male midshipman fish. *Horm Behav* 39: 355 (abstract).

Lu Z (2004) Neural mechanisms of hearing in fishes. In: von der Emde G, Mogdans J, Kapoor BG (eds), *The Senses of Fish: Adaptations for the Reception of Natural Stimuli*, pp. 147–172. Narosa Publishing House, New Delhi, India.

McCormick CA (1999) Anatomy of the central auditory pathways of fish and amphibians. In: Popper A, Fay RR (eds), *Springer Handbook of Auditory Research, Comparative Hearing: Fish and Amphibians*, Vol. 11, pp. 155–217. Springer, New York, NY.

McDonald MA, Calambokidis J, Teranishi AM, Hildebrand JA (2001) The acoustic calls of blue whales off California with gender data. *J Acoust Soc Am* 109: 1728–1735.

Popper AN, Fay RR (1999) The auditory periphery in fishes. In: Popper A, Fay RR (eds), *Springer Handbook of Auditory Research, Comparative Hearing: Fish and Amphibians*, Vol. 11, pp. 43–100. Springer, New York, NY.

Remage-Healey L, Bass AH (2003) Steroid hormones exert rapid, sex-specific changes in vocal communication signals in gulf toadfish, *Opsanus beta*. *Horm Behav* 44: 72 (abstract).

Remage-Healey L, Bass AH (2004) Rapid, hierarchical modulation of vocal patterning by steroid hormones. *J Neurosci* 24: 5892–5900.

Remage-Healey L, Bass AH (2005) Simultaneous, rapid, elevations in steroid hormones and vocal signaling during playback challenge: a field experiment in Gulf toadfish. *Horm Behav* 47: 297–305.

Sisneros JA, Bass AH (2003) Seasonal plasticity of peripheral auditory frequency sensitivity. *J Neurosci* 23: 1049–1058.

Sisneros J, Forlano P, Deitcher D, Bass AH (2004a) Steroid-dependent plasticity of frequency sensitivity in the adult auditory periphery. *Science* 305: 404–407.

Sisneros JA, Forlano PM, Knapp R, Bass AH (2004b) Seasonal variation of steroid hormone levels in an intertidal-nesting fish, the vocal plainfin midshipman. *Gen Comp Endocrinol* 136: 101–116.

Weeg MS, Bass AH (2000) Midbrain lateral line circuitry in a vocalizing fish. *J Comp Neurol* 418: 841–864.

Weeg MS, Bass AH (2002) Frequency response properties of lateral line superficial neuromasts in a vocal fish, with evidence for acoustic sensitivity. *J Neurophysiol* 88: 1252–1262.

6
Processing of Species-Specific Vocalizations in the Auditory Brainstem and Midbrain of Mexican Free-Tailed Bats (*Tadarida brasiliensis*)

Achim Klug, Eric E. Bauer, Joshua T. Hanson and George D. Pollak

6.1 Introduction

Hearing is a sensory modality of critical importance for the survival of many animal species. Animals listen to sounds from the world around them when hunting for prey, or conversely, avoiding predators. Animals listen to sound from the environment to identify potential mates, attract the attention of their parents, or when interacting in other social contexts. For these reasons many species, including birds and mammals, have hearing organs, which are both powerful and well tailored to the animals' specific needs.

Receiving the sound at the ears and stimulation of the auditory nerve is just the very first in a long sequence of processing steps in the brain before auditory information becomes behaviorally useful. Sounds arrive at the ears in the form of a complex amplitude-varying multi-frequency sound wave, from which the relevant information has to be extracted. For example, a sound wave does not contain information about the location of the sound source in space. Instead, spatial information is computed by the brain from time and intensity differences that a sound creates between the two ears (Rayleigh, 1907). Also, the ears or initial auditory centers simply follow the physical properties of sounds such as their frequencies and timing, but cannot identify the biological relevance of, for example, an alarm call or a mating call. This type of information emerges only after central processing and a complex series of computations (Rauschecker *et al.*, 1995; Rauschecker, 1997; Rauschecker and Tian, 2000; Wang, 2000). Exactly how the brain processes sound information to extract biologically relevant information is one of the central questions in auditory research. The purpose of this chapter is to look at the question of how species-specific communication signals are processed in the auditory brainstem and midbrain. We will discuss some recent findings from the *dorsal nucleus of the lateral lemniscus* (DNLL), an auditory brainstem nucleus (Bauer *et al.*, 2002), and the *central nucleus of the inferior colliculus* (ICC), an auditory midbrain nucleus (Klug *et al.*, 2002). Both nuclei were studied in Mexican free-tailed bats (*Tadarida brasiliensis*), a very social bat species for which a large number of vocalizations have been described previously (Gelfand and McCracken, 1986; Balcombe, 1990; Balcombe and McCracken, 1992; French and Lollar, 1998; French *et al.*, 2003). We will show, that neurons in these two areas respond in a very different way to the same set of natural stimuli, although a number of inputs to these nuclei are the same.

Many studies of the encoding of natural sounds have been conducted in the auditory forebrain and cortex (e.g. Muller and Scheich, 1987; Rauschecker *et al.*, 1995; Esser *et al.*, 1997; Hessler and Doupe, 1999; Kanwal, 1999; Kanwal *et al.*, 1999; Theunissen *et al.*, 2000; Wang, 2000; Sen *et al.*, 2001; Wang and Kadia, 2001), fewer addressed the representation of natural signals in the auditory midbrain (Scheich *et al.*, 1977; Fuzessery and Feng, 1983; Keller and Takahashi, 2000), and virtually nothing was known about how nuclei below the ICC code natural sounds. Yet the stream of auditory information passes through brainstem and midbrain nuclei before ascending to the thalamus and auditory cortex (Clarey *et al.*, 1992; Winer, 1992), and any transformations that occur in lower levels almost certainly affect the processing in higher-order areas. In particular, the ICC and DNLL are targets of ascending and descending auditory information and receive inputs from many lower level nuclei (Roth *et al.*, 1978; Brunso-Bechtold *et al.*, 1981; Glendenning *et al.*, 1981; Ross *et al.*, 1988; Oliver and Huerta, 1992; Saldana and Merchan, 1992; Huffman and Covey, 1995; Kelly *et al.*, 1998; Winer *et al.*, 1998), and thus may perform significant modifications to the auditory stream of information. Moreover, the ICC is the principle source of ascending information to the thalamus and further to the cortex so that knowing how the ICC represents and processes natural sounds is a critical step in understanding the strategies of the higher auditory system for their further processing.

6.2 The importance of hearing for bats

The reason we chose bats for our studies of auditory processing was that hearing plays an especially important role in their lives (Pollak and Casseday, 1989; Popper and Fay, 1995; Neuweiler, 2000), which is an advantage when addressing questions of the ethology, physiology, and anatomy. These nocturnal mammals use hearing in place of vision and figuratively "see" their world through their ears. Bats are able to navigate through complex environments and detect, identify and locate prey in total darkness by "echolocation" that is by emitting loud ultrasonic calls and listening to the echoes that are reflected from nearby objects (Griffin, 1958). Far less attention was given to their use of sound for social communication. Many bat species, including Mexican free-tailed bats, live at high-population densities in caves where they employ a rich repertoire of spectrally and temporally complex communication calls for a wide variety of social interactions, including mother–infant interactions, courting, agonistic encounters, and territoriality (Gelfand and McCracken, 1986; Balcombe, 1990; Balcombe and McCracken, 1992; Kanwal et al., 1994; French and Lollar, 1998; Kanwal, 1999; French and Lollar, 2000; French et al., 2003). The importance of vocal communication is well illustrated by the ability of mothers to recognize their own pup from its vocal signature (Balcombe, 1990; Balcombe and McCracken, 1992) in colonies of thousands or millions of individuals. During the night, the Mexican free-tailed bats hunt in the open skies for small insects such as moths. They spend the days in the shelter of caves where they raise their pups, which are left in large nursing colonies when mothers go on their nightly hunting expeditions. When returning, mothers have to find their own pup among possibly millions of other pups. Vocal interaction between mothers and pups has been shown to facilitate this process (Balcombe, 1990; Balcombe and McCracken, 1992).

In our view, the processing of communication signals, even in animals with the additional specialization of echolocation, is one of the primary tasks for which the auditory system was designed. Neuroethological adaptations required for echolocation were added subsequently to the bats' brains, built on top of the previous circuitry though not surplanting or interfering with the underlying processing of communication sounds. Early studies of the auditory systems of bats expected that numerous specializations would be found to explain the ability of bats to echolocate and thereby distinguish their auditory systems from those of other mammals. However, it was found that there is a lot of similarity between bats and other mammals commonly used in studies of the auditory system, such as in cell types, synaptic morphology, connections, neurochemistry, and physiology of brainstem auditory nuclei, from cochlear nucleus to inferior colliculus. Thus, the structural and mechanistic underpinnings for the processing of communication signals in the brainstem and midbrain are almost certainly features common to mammals rather than specializations unique to bats.

6.3 The structure of the auditory brainstem and midbrain: a brief overview

The most fundamental organizational feature of the auditory system is its tonotopic organization, which is created at the basilar membrane. Due to the physical properties of this membrane, every frequency is represented by a distinct place (Bekesy, 1960; Pickles, 1988). Thus, the inner ear performs a Fourier analysis of a multi-frequency sound with each frequency component exciting a different spot along the cochlea. The strength of the stimulation is proportional to the intensity of that particular frequency. This frequency separation is maintained by the auditory nerve fibers (Liberman, 1978; Liberman and Kiang, 1978; Pickles, 1986) and subsequent auditory nuclei. During these initial steps of auditory processing, the amplitude and timing of a frequency component is encoded by the number of auditory nerve fibers activated above their response threshold, and by their rate and temporal pattern of firing (Pickles, 1986, 1988). This is the input to the auditory pathways of the brain.

Auditory nerve fibers branch when entering the brain and make synaptic contacts with neurons in the divisions of the cochlear nucleus (Pickles, 1988; Brugge, 1992; Webster, 1992). These neurons project in series of parallel pathways to a myriad of further brainstem auditory nuclei (Brugge, 1992; Irvine, 1992; Webster, 1992). Each of the cell types in the cochlear nucleus extracts a different aspect from the common spike trains of auditory nerve fibers and thus conveys a different type of information to higher centers. Likewise, each of the higher center targets also processes incoming spike trains uniquely. Thus, the information contained in the discharges of auditory nerve fibers is divided up among the parallel pathways each conveying some portion of the total information.

Many of the ascending pathways give off branches to the DNLL. This nucleus is anatomically located just below the main midbrain auditory nucleus, the inferior colliculus (Adams and Mugnaini, 1984; Winer *et al.*, 1995), receives excitatory and inhibitory inputs (Fig. 6.1a) from many of the same nuclei as the inferior colliculus (Fig. 6.1b), is purely GABAergic, binaural, and provides a large-inhibitory projection to the ICC (Shneiderman and Oliver, 1989; Li and Kelly, 1992; Covey, 1993; Faingold *et al.*, 1993; Burger and Pollak, 2001). Most of the brainstem pathways converge in the ICC (Fig. 6.1b) (Brunso-Bechtold *et al.*, 1981; Ross *et al.*, 1988; Oliver and Huerta, 1992; Oliver *et al.*, 1995; Kelly *et al.*, 1998). The ICC processes and integrates almost all acoustically evoked information from lower centers and provides the main innervation to the medial geniculate body (Clarey, 1992; Winer, 1992; Winer *et al.*, 1996) and thus, indirectly, to the primary auditory cortex and to surrounding cortical fields.

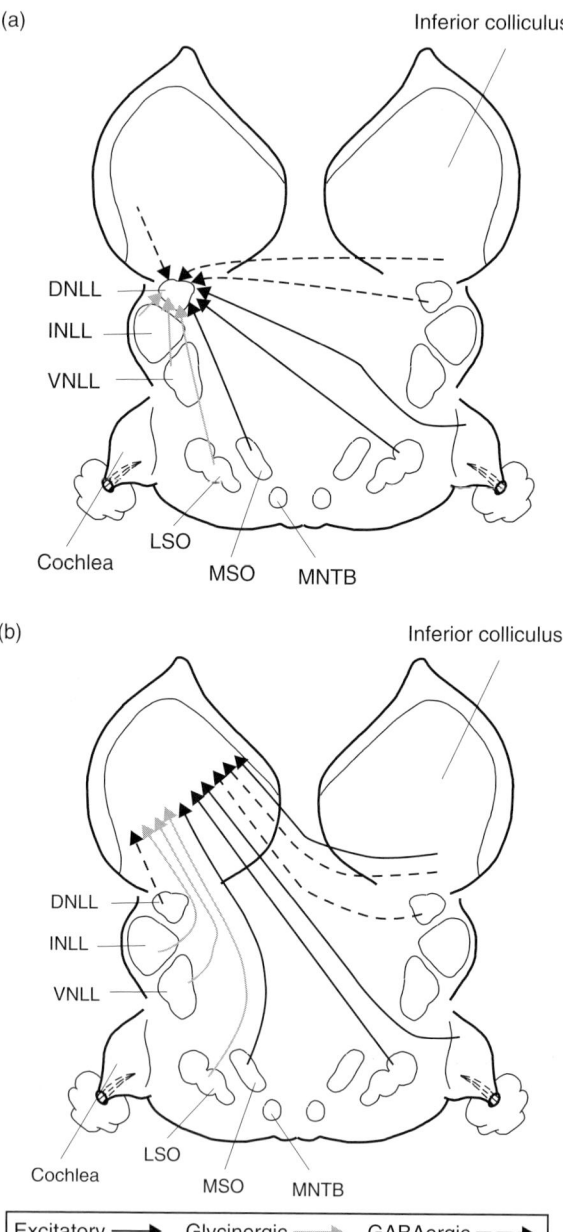

Figure 6.1. Brainstem and midbrain projections to the (a) DNLL and (b) inferior colliculus. LSO: lateral superior olive; MSO: medial superior olive; MNTB: medial nucleus of the trapezoid body; VNLL: ventral nucleus of the lateral lemniscus, INLL: intermediate nucleus of the lateral lemniscus.

The cochlear tonotopy is reestablished in the three-dimensional auditory nuclei by projections from distinct places along the cochlea into a sheet of cells of very similar best frequencies (BFs), termed isofrequency contours. For simplicity, it is assumed that inputs to the ICC isofrequency contours are processed in the same way. In other words, each isofrequency contour is regarded as a module of processing and each module is an iterated version of the unit module. The isofrequency contours or sheets are stacked one on top of each other in an orderly fashion, with low frequencies represented more dorsally and higher frequencies represented more ventrally in the ICC (Clopton and Winfield, 1973; Fitzpatrick, 1975; Semple and Aitkin, 1979; Zook et al., 1985; Schreiner and Langner, 1997).

6.4 Representation of communication sounds in lower auditory nuclei

The social calls of *Tadarida* are acoustically complex in that each one is composed of multiple harmonics with spectral components that change in both frequency and amplitude throughout the duration of each call (Gelfand and McCracken, 1986; Balcombe, 1990; Balcombe and McCracken, 1992; Bauer et al., 2002; Klug et al., 2002; French et al., 2003). A selection of only eight of the many types of *Tadarida* communication calls (SC1–8), as well as two example echolocation calls (EC9 and 10), is shown in Fig. 6.2. Note that some calls are composed of frequency modulations that sweep downward at various sweep rates (e.g., SC1, SC7, EC9 and 10), or both upward and downward frequency modulations (e.g., SC3–6, 8), or harmonic stacks of constant frequencies (SC2) or very shallow frequency modulations (SC1). During echolocation, the bats emit very rapid frequency modulated downward sweeps (EC9 and 10).

What happens when a bat's ear receives a natural communication call and how is it coded and represented in auditory brain centers? The only nucleus below the inferior colliculus, in which the processing of communication calls has been studied, is the DNLL. The processing of species-specific signals is straightforward and simple in DNLL neurons (Bauer et al., 2002) illustrated by two features: non-selectivity and homogeneity. Non-selectivity of DNLL neurons refers to a response to any signal so long as it has spectral energy in the neuron's excitatory response area, regardless of the temporal or spectral structure of the sound (Bauer et al., 2002). This non-selectivity is illustrated by the responses of five "isofrequency" DNLL neurons (Fig. 6.3) to five monaurally presented species-specific calls (compare Fig. 6.2). The neurons were unselective for these calls, in that all five neurons responded to all five signals due to

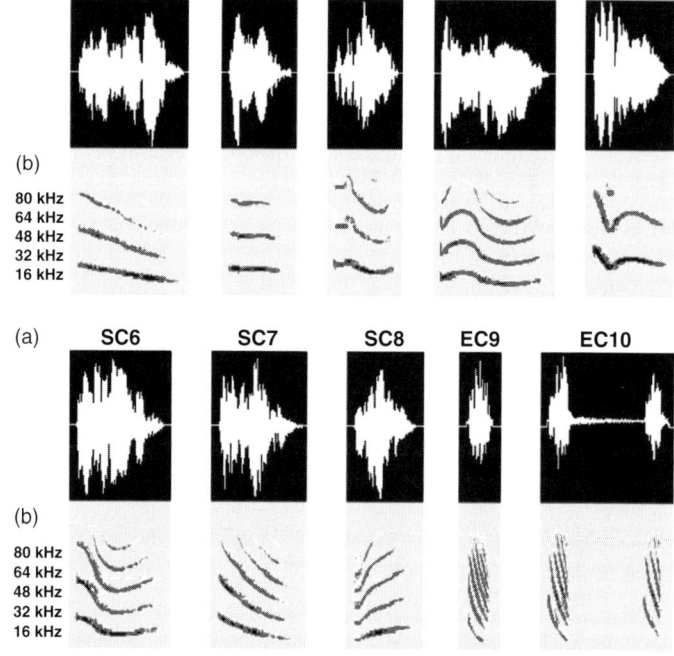

Figure 6.2. Waveforms and spectrograms of 10 vocalizations recorded from *Tadarida brasiliensis*. Calls SC1–8 are social communication calls, EC9 and 10 are echolocation calls: (a) shows waveforms and (b) shows spectrograms.

the fact that they contained energy at a frequency suitable to drive the neurons. However, even though all calls caused a response, the response to each call type was more or less unique and homogeneous, in that DNLL neurons with similar BFs responded to a given call with similar response profiles, defined by response latency, magnitude and temporal discharge pattern. Each of the five calls had a different spectro-temporal structure evoking a very similar call-specific response profile in the five neurons. SC1, for example, evoked a brief response with only one discharge peak, SC5 a more sustained response with several peaks.

What do these data suggest about the processing of complex signals by DNLL neurons? DNLL processing of monaural signals appears to be simple and determined predominantly by the integration of excitatory innervation. The specific spectro-temporal properties of a complex call do not seem to play a role in the neuron's response or non-response to a given stimulus. In other words, it seems to matter little, what other frequencies at what temporal arrangement are present in the signal so long as the signal contains some frequency component with sufficient energy to excite the neuron. The neuron's response pattern to a given

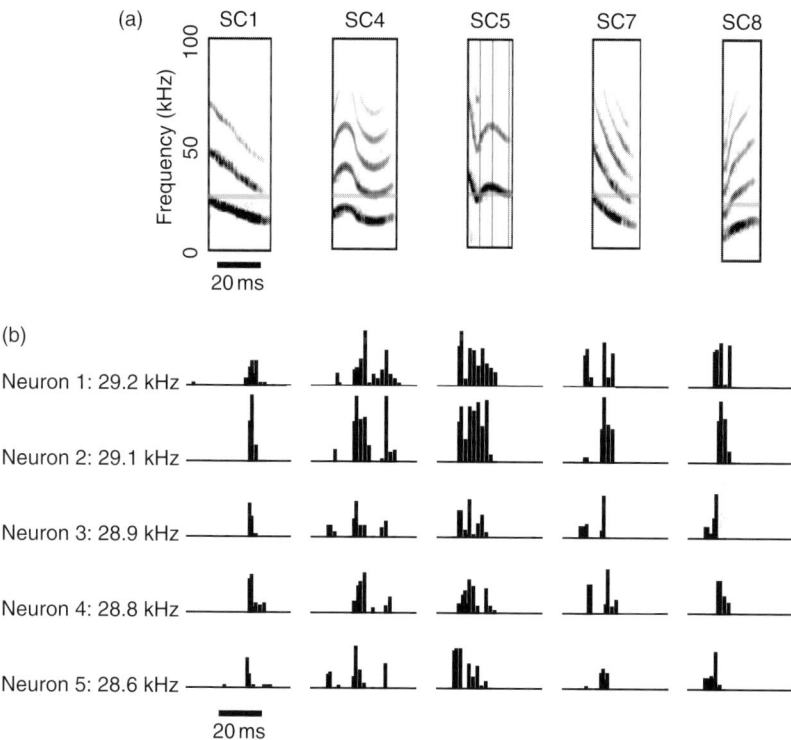

Figure 6.3. DNLL neurons with similar BF respond to vocalizations in a similar way. (a) Spectrograms of five different vocalizations and (b) response of five DNLL neurons to these vocalizations. The neurons' BFs ranged from 28.6 to 29.2 kHz. This range is indicated by the gray boxes in the spectrograms. Figure reproduced, with permission, from Bauer et al. (2002).

call depends on the temporal properties of the call, that is when and for how long the excitatory frequencies occur in the signal.

If these findings can be generalized, the responses of a DNLL neuron to one set of stimuli should be able to predict the neuron's response to an entirely different set of stimuli. We tested this idea by measuring the responses of a DNLL neuron to short tone bursts (a set of simple stimuli) and used this information to predict how the neuron should respond to another set of signals, namely the ten vocalizations described above (Fig. 6.2). The method used is to employ a convolution, which is a form of matrix multiplication, or a calculation in which two matrices are slid past each other and multiplied at each point of overlap. In our case, one of the two matrices is a spectro-temporal representation of the complex signal of interest (Fig. 6.4a) and the other matrix is a representation of

(a)

	7.68	10.24	12.80	15.36	17.92	20.48	23.04	25.60	28.16	30.72	33.28
17.18 kHz	0	0	0	0	0	1	24	27	19	4	0
17.96 kHz	0	0	0	0	0	22	30	19	8	0	0
18.74 kHz	0	0	0	0	0	26	23	6	0	0	0
19.53 kHz	0	0	0	0	19	17	0	0	0	0	0
20.31 kHz	0	0	0	17	23	15	0	0	0	0	0
21.09 kHz	0	0	17	24	15	2	0	0	0	0	0
21.87 kHz	0	5	18	21	2	0	0	0	0	0	0
22.65 kHz	4	23	12	9	0	0	0	0	0	0	0
23.44 kHz	8	25	5	0	0	0	0	0	0	0	0
24.22 kHz	17	8	0	0	0	0	0	0	0	0	0
25.00 kHz	10	0	0	0	0	0	0	0	0	0	0
25.78 kHz	6	0	0	0	0	0	0	0	0	0	0
26.56 kHz	1	0	0	0	0	0	0	0	0	0	0
27.34 kHz	0	0	0	0	0	0	0	0	0	0	0
28.12 kHz	0	0	0	0	0	0	0	0	0	0	0

(b)

	7.68	10.24	12.80	15.36	17.92	20.48	23.04	25.60	28.16	30.72	33.28
17.18 kHz	0	0	0	0	0	0	0	0	0	0	0
17.96 kHz	0	0	0	0	0	0	0	0	0	0	0
18.74 kHz	0	0	0	0	0	0	0	0	0	0	0
19.53 kHz	0	0	0	0	0	0	0	0	0	0	0
20.31 kHz	0	0	0	0	6	0	0	0	0	0	0
21.09 kHz	0	0	0	16	24	0	0	0	0	0	0
21.87 kHz	0	0	0	16	24	25	0	0	0	0	0
22.65 kHz	0	0	0	19	28	32	0	0	0	0	0
23.44 kHz	0	0	0	14	26	30	0	0	0	0	0
24.22 kHz	0	0	0	3	31	17	0	0	0	0	0
25.00 kHz	0	0	0	3	31	17	0	0	0	0	0
25.78 kHz	0	0	0	0	18	0	0	0	0	0	0
26.56 kHz	0	0	0	0	0	0	0	0	0	0	0
27.34 kHz	0	0	0	0	0	0	0	0	0	0	0
28.12 kHz	0	0	0	0	0	0	0	0	0	0	0

Figure 6.4. Demonstration of the convolution process. (a) A small segment of the matrix of the spectrogram of SC1. Amplitude at a given time and a given frequency is represented in arbitrarily scaled units. Numbers greater than zero are indicated in bold. Time is represented on x-axis; the temporal resolution is 2.56 ms per bin. Frequency is represented on the y-axis, the frequency resolution is 780 Hz. (b) Response area of a DNLL neuron. The numbers represent numbers of spikes counted within consecutive 2.56 ms time bins when short tone bursts at the indicated frequencies were use as stimuli. The units of this matrix match the units of matrix A.

the neuron's responses to short tone stimuli at various frequencies, termed the neuron's "response region" (Fig. 6.4b).

Matrix A (Fig. 6.4a) is the spectro-temporal representation of vocalization SC1 as already shown in Fig. 6.2. For illustrative purposes, the figure shows only a small segment of the entire matrix, which resembles the shallow downward sweeps of call SC1. The matrix is organized by time on the x-axis and frequency on the y-axis. The width of the time bins used in this example are 2.56 ms, of the frequency bins 760 Hz. The numbers in the matrix represent units of amplitude. High numbers indicate that a particular frequency is present at a particular time at a high amplitude. Low numbers and zeros indicate low amplitudes or absence, respectively, of a particular frequency at a given time. Note that this type of representation not only decomposes a signal into its frequency components but also carries information on the timing of those frequencies.

Matrix B (Fig. 6.4b), the neuron's response region, is a representation of a neuron's responsiveness to different frequencies. Again, Fig. 6.4b shows only a small segment of the entire matrix. The time bins and frequency channels of the response region correspond exactly to the units of the first matrix. The numbers in this second matrix represent numbers of spikes counted within a time bin, when the neuron responded to short tone bursts at different frequencies. The latency of the neuron's response is represented by the position of the field along the x-axis, the response duration by the number of consecutive time bins with values greater than zero. Figure 6.4a and b shows only a small segment of the actual matrices used in the convolution. The original matrices contained 103 frequency bands covering frequencies from 0 kHz to 80 kHz and 78 time bins covering time from 0 ms to 200 ms.

Our hypothesis was that if a DNLL neuron was stimulated with a complex call consisting of frequencies of the neuron's response region, then the neuron's response should occur at the exact time at which at least one of these frequencies occurred in the call. The convolution performs the predictive calculation in a convenient and automated way. Mathematically, one of the matrices is flipped along the time axis and both matrices are slid past each other bin by bin along the time axis. At each bin, numbers on top of each other are multiplied so that the products fill the bins corresponding to each frequency and time. The result of this process is a third matrix, which represents a raw version of the expected response of the neuron to the signal that was tested. In order to get the spike train that the neuron is predicted to fire, the matrix along the frequency axis is collapsed into one single row of numbers representing the temporal sequence and magnitude of discharges we expect the neuron to fire when stimulated with the signal. This is equivalent to a raw version of the predicted envelope of the spike train histogram. The absolute scaling in this train is arbitrary since one of the matrices (4A) already contained arbitrarily scaled numbers, and hence the results need to be normalized. Normalization leads to the envelope of the predicted spike train.

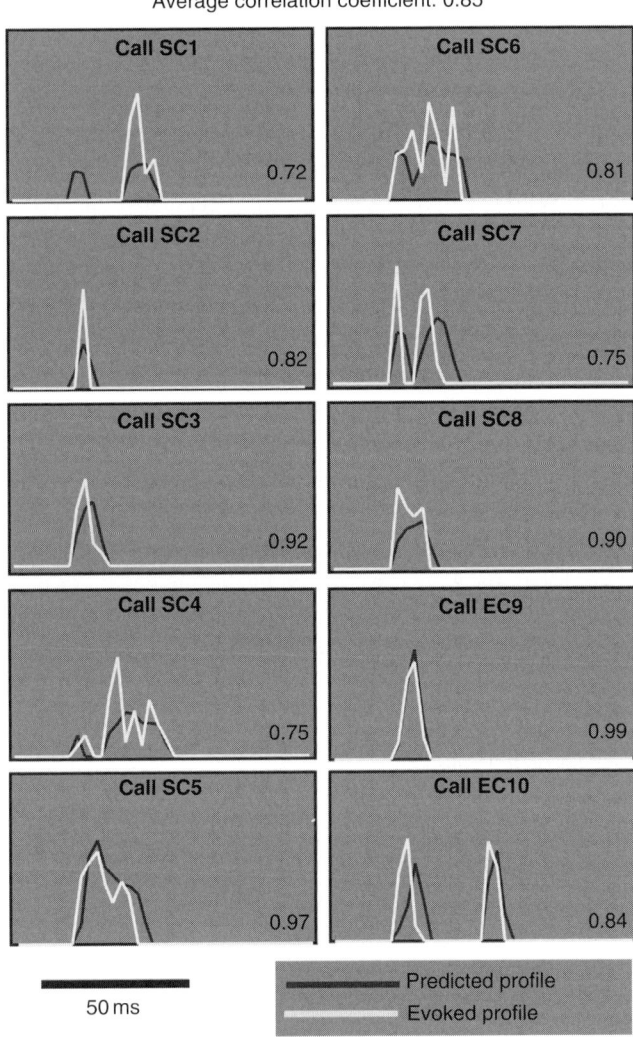

Figure 6.5. Envelopes of response profiles for predicted responses (black) and the responses actually evoked (white) by 10 species-specific calls for a DNLL neuron. Numbers in bottom right corner show the correlation coefficients for predicted versus evoked responses to each call. All signals were presented at 40 dB SPL, which was 40 dB above the neuron's threshold. Windows: 100 ms.

The result of the calculation by this convolution is a prediction that must be tested against the actual response of the neuron to the complex call. The curve of the histogram of the actual response is compared to the predicted response envelope using simple correlation calculations. The more similar the actual and predicted response profiles are, the closer the correlation gets to +1. Similarity in this context includes the overall temporal alignment of the patterns of responses (i.e. how well peaks and troughs line up with each other) as well as the relative magnitude of response at each time point. Thus, the biggest peak of the actual response must align with the biggest peak of the predicted response, and the second biggest peak must align with the second biggest peak, etc. Misalignment in timing and magnitude will decrease the correlation score.

Across DNLL neurons, convolutions predicted the neurons' responses to complex calls with very high accuracy. An example is shown in Fig. 6.5. Note the close correspondence between the predicted and obtained responses, with correlations ranging from 0.72 to 0.99 (average for this neuron = 0.85). Of 32 DNLL neurons tested, the average correlation coefficient was 0.72 (Bauer et al., 2002). Thus, the response profiles of DNLL neurons to a set of complex communication sounds are in general accurately predicted from their responses to a set of simple tones. This implies that integrative mechanisms determining the response of a DNLL neuron to single frequencies and to complex stimuli are similar.

As an extension of the hypothesis of similar integrative mechanisms and as further evidence of the homogeneity of DNLL cells, consider that the response of a given DNLL neuron can, by convolving it with the species-specific calls, accurately predict the response to those calls of any other DNLL neuron with the same BF (Bauer et al., 2002). This is illustrated in Fig. 6.6, which shows the predicted and actual responses to three species-specific calls for six "isofrequency" DNLL neurons. The important feature in this figure is that the predicted responses for all neurons are based on the response area of neuron 4 marked by the gray box. These predictions are compared to and correlated with the actual responses of each one of the six neurons. The correlations of actual and predicted response profiles were, of course, high for neuron 4, since it was correlated with its own response. More interestingly, the correlations were just as high for the five other neurons, emphasizing the point that the responses of DNLL neurons are very homogeneous.

6.5 Processing of communication sounds is much more complex in the inferior colliculus

The ICC is only one synapse above the DNLL, and receives many of the same inputs (Fig. 6.1b), but the processing in the ICC is far more complex than in the DNLL. While DNLL neurons are unselective and homogeneous as explained above, ICC neurons are highly selective and heterogeneous (Klug et al., 2002). Response selectivity and heterogeneity of five ICC neurons are illustrated in Fig. 6.7. The neurons are tuned to about the same frequency and may belong to

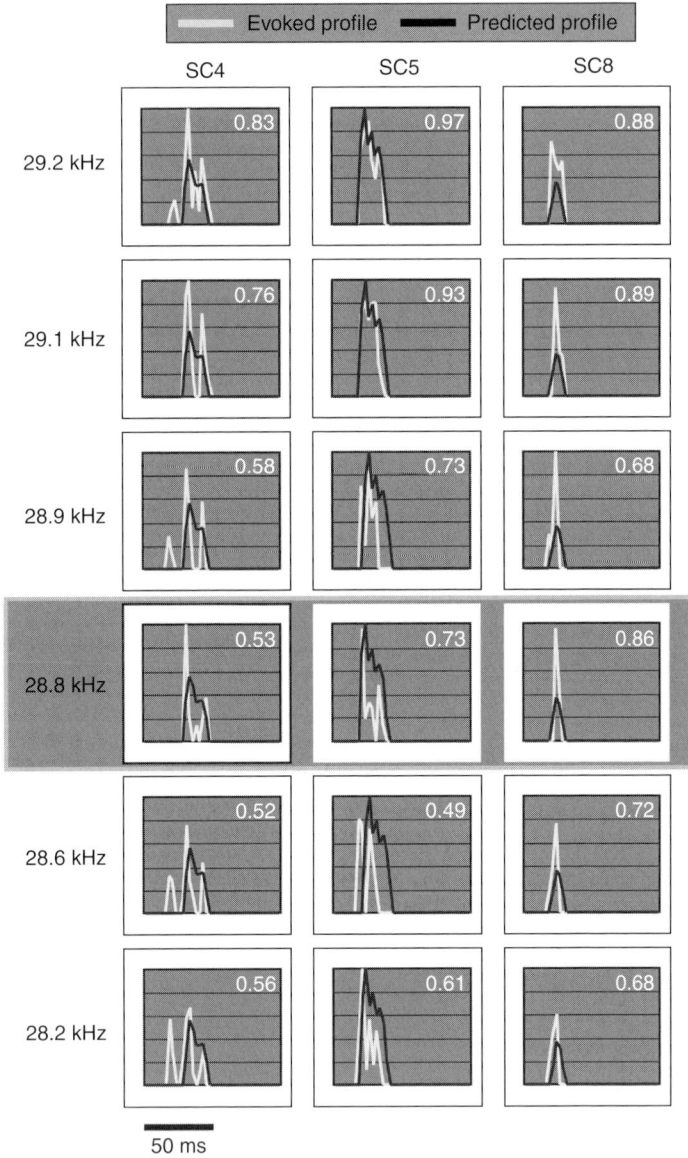

Figure 6.6. Predicted responses for DNLL neurons predict responses to other DNLL neurons with similar BFs. All predicted responses shown were based on the convolution results of neuron 4 (marked by the gray box). The correlation between the predicted responses of neuron 4 and the evoked responses of each neuron are shown in the upper right of each panel. All neurons had similar BFs ranging from 28.2 to 29.2 kHz and the correlations of predicted and evoked responses were high. All signals were between 30 and 40 dB above threshold.

6.5 Communication sounds in the inferior colliculus

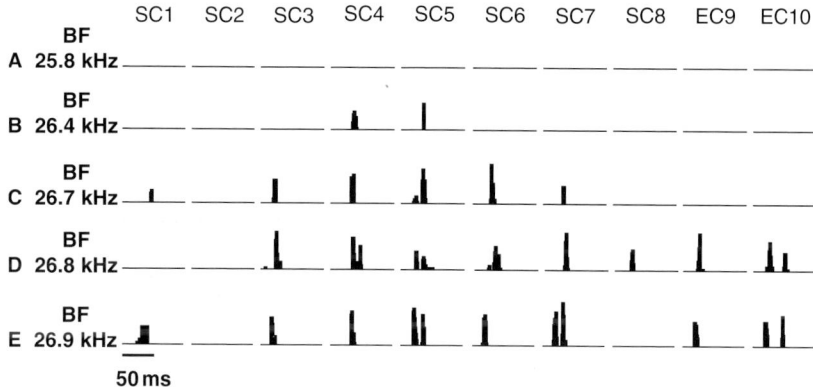

Figure 6.7. Different response selectivities displayed by five ICC neurons to the 10 species-specific calls shown in Fig. 2. The BFs of the neurons were closely aligned and ranged from 25.8 kHz to 26.9 kHz. None of the cells responded to call SC2, which was composed of harmonic stacks of constant frequency components. The lowest constant frequency component was about 20 kHz and did not encroach upon the response areas of any of the neurons. Signal levels above threshold for neurons A, B, C, D, E: 40, 50, 40, 30, 40 dB, respectively.

the same isofrequency contour, yet each neuron responded to a different subset of the 10 calls (Fig. 6.2). Rather trivial is that none of the neurons responded to call SC2 because its frequencies did not enter the neurons' response regions. The other nine calls, however, had spectral components within the neurons' excitatory response regions. The five neurons displayed selectivity, since each failed to respond to some of the calls. Their responses were heterogeneous since the particular subset of calls to which each neuron responded was different. Neuron A did not respond to any of the calls, neuron B responded only to two calls (SC4, SC5), neuron D eight calls, while neuron E responded to a different subset of eight calls compared to neuron D. On average, 145 ICC neurons responded to only 3.19 of the 10 calls presented (Klug et al., 2002).

This response selectivity is largely due to inhibitory innervation at the ICC. This can be shown by reversibly inactivating the inhibitory inputs to an ICC neuron by iontophoretically applying antagonists of inhibitory transmission, in this case the $GABA_A$ antagonist bicuculline and the glycine antagonist strychnine. When cells were retested for selectivity while inhibitory inputs were inactivated, they responded on average to twice as many calls than they did with inhibition intact (+2.7 calls, n = 50; Klug et al., 2002).

As a consequence of the effects of inhibition, convolutions calculated in the ICC with only with excitatory responses evoked by tonal stimuli must provide inaccurate predictions of responses to species-specific calls. An example is shown

146 6 *Processing of species-specific vocalizations*

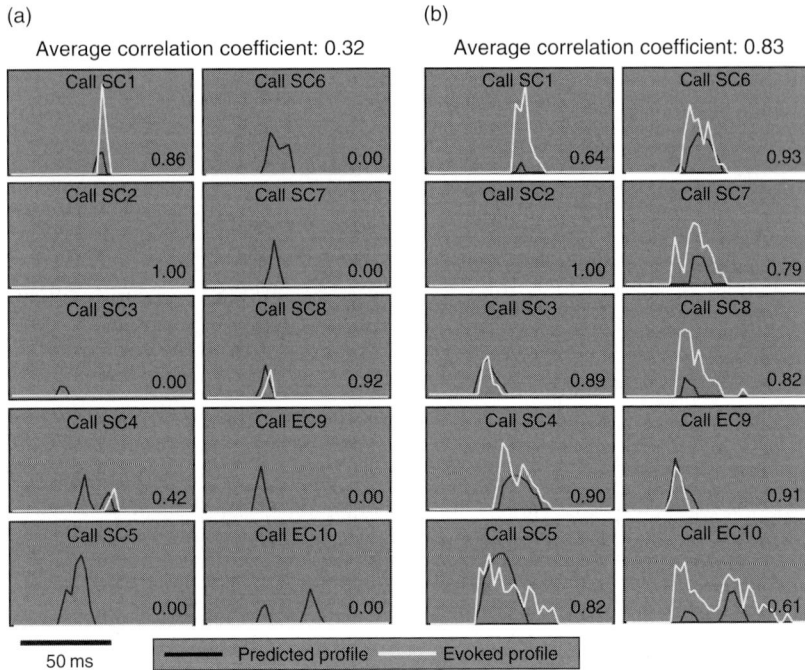

Figure 6.8. Envelopes of response profiles for predicted responses (black) and the responses actually evoked (white) by 10 species-specific calls for an ICC neuron. Numbers in bottom left corner show the correlation coefficients for predicted versus evoked responses to each call. All signals were presented at 40 dB SPL, which was 30 dB above threshold. Windows: 100 ms. (a) Response profiles when inhibition was intact, (b) response profiles of the same cell when inhibition was functionally disabled by iontophoretically applying antagonists to inhibitory synaptic transmission.

in Fig. 6.8a. The convolutions predicted that the neuron responded to nine of the calls having energy within its excitatory response region. However, it only responded to three of the nine calls (SC1, SC4 and SC8). Thus, the neuron displayed selectivity and more importantly, that selectivity was poorly predicted by the responses to tonal stimulation. An indicator of the poor predictability is the low-correlation coefficient of 0.32 obtained for this neuron. This was a finding typical for 104 ICC neurons. The average correlation coefficient was 0.43 (Klug *et al.*, 2002) compared to 0.72 for the DNLL (see above).

When inhibition was blocked in this neuron (Fig. 6.8b), it responded to nine of the 10 calls, exactly as predicted by the convolution. The average correlation coefficient improved markedly from 0.32 to 0.83, confirming the view that inhibition played a significant role in shaping the response to natural calls.

Among the population of ICC neurons, the effects of removing inhibition were diverse in that some neurons showed very little or no improvement in predictability, while others showed a moderate or dramatic (Fig. 6.8) improvement in predictability. On average, removing inhibition significantly improved the predictability of responses to calls among ICC neurons by 0.15 (Klug et al., 2002). However, even when inhibition was inactivated, the overall predictability of responses was still lower than for DNLL neurons. In summary, ICC neurons appear to be diverse and highly selective when inhibitory innervation is active. When inhibition is turned off, they are less diverse and become more like DNLL neurons. The transformation, however, is not complete. ICC isofrequency neurons never display the homogeneity of DNLL neurons and fire with a variety of response profiles to any particular signal, even after inhibition is blocked.

6.6 Communication sounds elicit a unique spatio-temporal pattern of activity in the ICC

It appears that one of the principal transformations in the ICC is a change from processing that emphasizes similarity of responses among the neuronal population to one that emphasizes diversity and selectivity. The increased call selectivity of individual neurons in the ICC relative to the DNLL is at least partially due to the additional contralaterally evoked-inhibitory inputs to the ICC. Heterogeneity of responses across an isofrequency population of neurons is another important aspect in which ICC processing is different from that in lower nuclei. Lower nuclei typically seem to display more similar response properties within one isofrequency contour. Response properties differ between nuclei but neurons within the same nucleus are usually rather similar. The ICC, in contrast, is very heterogeneous. Different ICC neurons, even within the same isofrequency contour, have very different response properties when tested with the same sound stimulus (e.g. Pollak et al., 1986; Yang et al., 1992; Park and Pollak, 1993a; Saitoh and Suga, 1995; Wenstrup and Leroy, 2001).

We suggest that the transformation from relative homogeneity to heterogeneity is largely a consequence of the anatomical distribution of excitatory and inhibitory inputs along isofrequency contours in the ICC. The DNLL, for example, is innervated by a number of lower nuclei, but most high-frequency DNLL neurons presumably receive the same complement of projections (Glendenning et al., 1981; Shneiderman et al., 1988; Yang et al., 1996) and have a similar complement of voltage-gated channels (Fu et al., 1996, 1997a, b; Wu and Kelly, 1996). This can explain why a given signal evokes a similar response profile in neurons along one isofrequency contour. A similar argument is advanced for the neuronal population in each lower auditory nucleus. The innervation pattern of the ICC, however, is markedly different. Inputs from most lower nuclei form bands innervating restricted regions of an isofrequency contour (Shneiderman and Henkel, 1987; Oliver and Beckius, 1992; Oliver *et al.*, 1995). The band from a particular lower nucleus only partially overlaps with the bands from other nuclei. Thus, any segment of an isofrequency contour receives a unique complement of inputs from a subset of lower nuclei, which differs from the subset innervating adjacent regions of the same contour (Ross and Pollak, 1989; Peruzzi *et al.*, 1997). Such differences in excitatory inputs can account for the response diversity among ICC neurons even when inhibition is blocked. This diversity is further amplified as responsiveness of ICC neurons undergoes substantial modification by inhibition (Faingold *et al.*, 1991; Vater *et al.*, 1992; Yang *et al.*, 1992; Park and Pollak, 1993a, b; Pollak and Park, 1993; Casseday *et al.*, 1994, 2000; Fuzessery and Hall, 1996; Le Beau *et al.*, 1996; Kuwada *et al.*, 1997). The inhibitory innervation, which modifies response profiles and creates selectivity, ascends from lower nuclei, of which different subsets project differently along a frequency contour. The result is a large variety of innervation patterns resulting in a large diversity of response types that is not present in the ascending auditory system up to this point. The diversity is even further enhanced when the mixed and matched inputs synapse on ICC neurons with different complements of voltage-gated channels (Peruzzi *et al.*, 2000; Sivaramakrishnan and Oliver, 2001). The consequence is that neurons of isofrequency contours can respond to complex signals with far more diversity and selectivity as lower brainstem nuclei innervating the ICC.

What is the functional advantage of creating diversity and selectivity in the ICC? We suggest that the population of ICC neurons can respond to any signal with a unique spatio-temporal pattern of activity. Any one of the "simpler" lower nuclei would also respond with a unique spatio-temporal pattern to any given signal, based on the principles of tonotopy and rate-coding. However, the large variation of input patterns and channel complements among ICC neurons is well suited to amplify any differences in the population response to different complex sounds. Communication calls are emitted in specific social situations and may convey specific informations about the senders and their behavioral context. In order to respond adequately, the perceiving animals have to precisely discriminate the calls. With the results presented in this chapter we suggest that any call will, for the first time in the ascending auditory system, generate a unique and distinct population response in the ICC, which then is transferred to higher regions in the forebrain for further processing and recognition.

We illustrate the qualitative nature of pattern exaggeration in Fig. 6.9, showing the responses of 13 neurons to calls SC4 and SC6 of very similar spectro-temporal

6.6 Communication sounds in the ICC 149

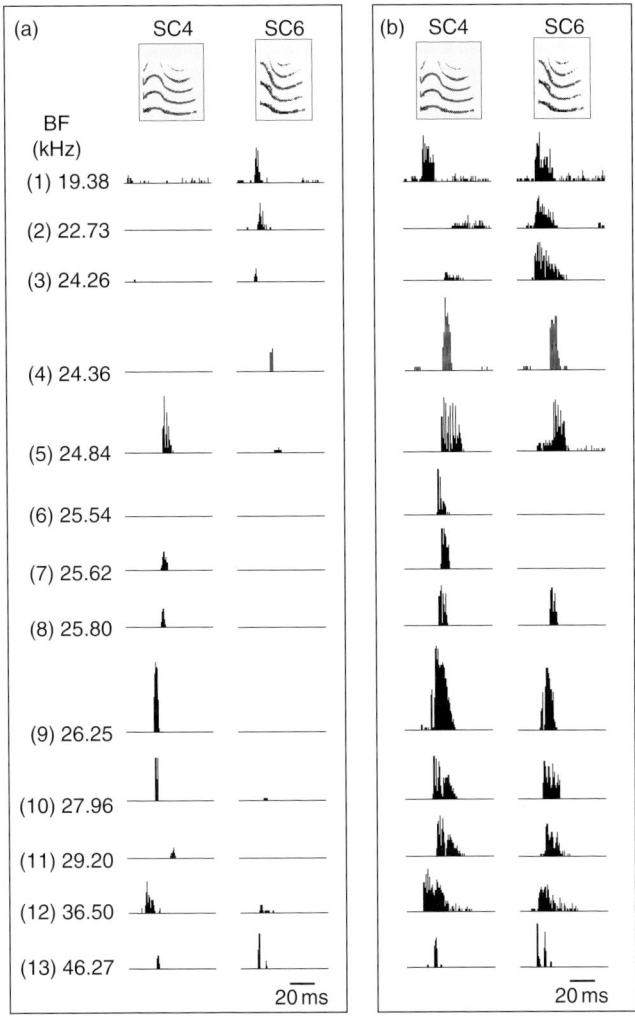

Figure 6.9. Responses of 13 neurons to two calls, SC4 and SC6, (a) before and (b) while inhibition was blocked. The BFs of the neurons are arranged from low to high to simulate the tonotopic organization of the ICC. Both signals were presented to all neurons at 30–50 dB above threshold. Note that the calls had similar spectro-temporal features but evoked different responses from the neurons. Before inhibition was blocked, neurons responded with different selectivities to the two calls (eight neurons responded to one but not the other call). Also differed the response profiles of the four neurons responding to both calls. Blocking inhibition greatly reduced selectivity and allowed 11 of the 13 neurons to respond to both calls. Additionally, the response profiles evoked by the two calls were now similar in many neurons. Figure reproduced, with permission, from Klug et al. (2002).

features. The neurons are stacked from dorsal to ventral in the ICC according to their increasing BFs along the tonotopic organization. Eight neurons had BFs of about 22–26 kHz, which is a frequency range over-represented in the Mexican free-tailed bat's auditory system (Vater and Siefer, 1995; Bauer et al., 2000). Eight neurons responded to one call but not the other one, while one neuron did not respond to either call. Four neurons responded to both calls, but with different response profiles. In short, there was a pronounced difference in the "population" response to the two calls. The difference was less pronounced when inhibition was blocked, leading 11 neurons to respond to both calls whereas only four did so when inhibition was intact. Moreover, both calls evoked similar response profiles in many neurons when inhibition was blocked. In short, pronounced disparities in the ICC population responses evoked by the two call signals occurred both within and across isofrequency contours and those disparities were enhanced considerably by inhibition.

6.7 Conclusion

In this chapter we discussed the question of how complex sounds are initially coded and represented by the peripheral auditory system and how this representation looks like and changes at two higher order nuclei, namely the DNLL of the auditory brainstem and the ICC of the auditory midbrain. Although both nuclei are several synapses upstream of the auditory nerve, and receive excitatory and inhibitory projections from a number of same lower nuclei, the DNLL and the ICC respond to complex and natural signals in a very different way. Neurons in the DNLL respond in a simplistic and stereotypic way, with a high degree of similarity between DNLL neurons. The responses of these neurons to complex signals can be predicted with a high degree of accuracy simply by convolving the neurons' response regions as obtained from tone bursts with the complex signal of interest. This indicates that the processing of complex stimuli in DNLL neurons follows the same rules as when responding to simple stimuli. By contrast, neurons in the ICC respond to complex signals much more selectively and there is a much larger degree of variation in the responses even between neurons of similar BFs. Moreover, the convolution process that yielded high levels of predictability of responses to complex sounds in the DNLL yielded much lower predictability in the ICC. We showed that neural inhibition involved in the ICC processing of complex sounds is important for decreasing the predictability of ICC responses. We suggest that differences of innervation patterns between DNLL and ICC are responsible for an overall exaggeration of differences in the neural population response to different sounds in the ICC. Such a magnification of differences between acoustic properties of species-specific vocalizations by the ICC processing seems to be useful and important for sound recognition at higher levels of the auditory system.

Acknowledgments

We thank Frederic Theunissen for help with implementing the mathematical procedures and Laura Hurley and John Meitzen for help with the data collection. Supported by a grant from the NIH to G.D.P.

REFERENCES

Adams JC, Mugnaini E (1984) Dorsal nucleus of the lateral lemniscus: a nucleus of GABAergic projection neurons. *Brain Res Bull* 14: 585–590.

Balcombe JP (1990) Vocal recognition of pups by mother Mexican free-tailed bats, *Tadarida brasiliensis mexicana*. *Anim Behav* 39: 960–966.

Balcombe JP, McCracken GF (1992) Vocal recognition in Mexican free-tailed bats: do pups recognize mothers? *Anim Behav* 43: 79–87.

Bauer EE, Klug A, Pollak GD (2000) Features of contralaterally evoked inhibition in the inferior colliculus. *Hear Res* 141: 80–96.

Bauer EE, Klug A, Pollak GD (2002) Spectral determination of responses to species-specific calls in the dorsal nucleus of the lateral lemniscus. *J Neurophysiol* 88: 1955–1967.

Bekesy V (1960) *Experiments in Hearing*, McGraw-Hill, New York.

Brugge JF (1992) An overview of central auditory processing. In: Popper AN, Fay RR (eds), *The Mammalian Auditory Pathway: Neurophysiology*, 1st Edition, pp. 1–33. Springer-Verlag, New York.

Brunso-Bechtold JK, Thompson GC, Masterton RB (1981) HRP Study of the organization of auditory afferents ascending to central nucleus of inferior colliculus in cat. *J Comp Neurol* 197: 705–722.

Burger RM, Pollak GD (2001) Reversible inactivation of the dorsal nucleus of the lateral lemniscus reveals its role in the processing of multiple sound sources in the inferior colliculus of bats. *J Neurosci* 21: 4830–4843.

Casseday JH, Ehrlich D, Covey E (1994) Neural tuning for sound duration: role of inhibitory mechanisms in the inferior colliculus. *Science* 264: 847–850.

Casseday JH, Ehrlich D, Covey E (2000) Neural measurement of sound duration: control by excitatory–inhibitory interactions in the inferior colliculus. *J Neurophysiol* 84: 1475–1487.

Clarey JC, Barone P, Imig TJ (1992) Physiology of thalamus and cortex. In: Webster DB, Popper AN, Fay RR (eds), *The Mammalian Auditory Pathway: Neurophysiology*, pp. 232–334. Springer-Verlag, New York.

Clopton BM, Winfield JA (1973) Tonotopic organization in the inferior colliculus of the rat. *Brain Res* 56: 355–358.

Covey E (1993) Response properties of single units in the dorsal nucleus of the lateral lemniscus and paralemniscal zone of an echolocating bat. *J Neurophysiol* 69: 842–859.

Esser KH, Condon CJ, Suga N, Kanwal JS (1997) Syntax processing by auditory cortical neurons in the FM–FM area of the mustached bat *Pteronotus parnellii*. *Proc Natl Acad Sci USA* 94: 14019–14024.

Faingold CL, Boersma Anderson CA, Caspary DM (1991) Involvement of GABA in acoustically-evoked inhibition in inferior colliculus neurons. *Hear Res* 52: 201–216.

Faingold CL, Anderson CA, Randall ME (1993) Stimulation or blockade of the dorsal nucleus of the lateral lemniscus alters binaural and tonic inhibition in contralateral inferior colliculus neurons. *Hear Res* 69: 98–106.

Fitzpatrick KA (1975) Cellular architecture and topographic organization of the inferior colliculus of the squirrel monkey. *J Comp Neurol* 164: 185–207.

French B, Lollar A (1998) Observations on the reproductive behavior of captive *Tadarida brasiliensis mexicana* (Chiroptera: Molossidae). *Southwest Nat* 43: 484–490.

French B, Lollar A (2000) Communication among Mexican free-tailed bats. *Bats: Bat Conserv Int* 18: 1–4.

French B, Lollar A, Ma T, Page R, Steinberg R, Xie R (2003) Library of Social Communication Calls of the Mexican Free-tailed Bat (*Tadarida brasiliensis*). *Intern Bat Rehab J* 1: 1–13.

Fu XW, Wu SH, Brezden BL, Kelly JB (1996) Potassium currents and membrane excitability of neurons in the rat's dorsal nucleus of the lateral lemniscus. *J Neurophysiol* 76: 1121–1132.

Fu XW, Brezden BL, Wu SH (1997a) Hyperpolarization-activated inward current in neurons of the rat's dorsal nucleus of the lateral lemniscus *in vitro*. *J Neurophysiol* 78: 2235–2245.

Fu XW, Brezden BL, Kelly JB, Wu SH (1997b) Synaptic excitation in the dorsal nucleus of the lateral lemniscus: whole-cell patch-clamp recordings from rat brain slice. *Neuroscience* 78: 815–827.

Fuzessery ZM, Feng AS (1983) Mating call selectivity in the thalamus and midbrain of the leopard frog (*Rana pippens*): single and multiunit analyses. *J Comp Physiol A* 150: 333–344.

Fuzessery ZM, Hall JC (1996) Role of GABA in shaping frequency tuning and creating FM sweep selectivity in the inferior colliculus. *J Neurophysiol* 76: 1059–1073.

Gelfand DL, McCracken GF (1986) Individual variation in the isolation calls of Mexican free-tailed bat pups (*Tadarida brasiliensis mexicana*). *Anim Behav* 34: 1078–1086.

Glendenning KK, Brunso-Bechtold JK, Thompson GC, Masterton RB (1981) Ascending auditory afferents to the nuclei of the lateral lemniscus. *J Comp Neurol* 197: 673–703.

Griffin DR (1958) *Listening in the Dark*. Yale University Press, New Haven, CT.

Hessler NA, Doupe AJ (1999) Social context modulates singing-related neural activity in the songbird forebrain. *Nat Neurosci* 2: 209–211.

Huffman RF, Covey E (1995) Origin of ascending projections to the nuclei of the lateral lemniscus in the big brown bat, *Eptesicus fuscus*. *J Comp Neurol* 357: 532–545.

Irvine DRF (1992) Physiology of the auditory brainstem. In: Popper AN, Fay RR (eds), *The Mammalian Auditory Pathway: Neurophysiology*, pp. 153–231. Springer-Verlag, New York.

Kanwal JS (1999) Processing Species-specific Calls by Combination-sensitive Neurons in an Echolocating Bat. In: Hauser MD, Konishi M (eds), *The Design of Animal Communication*, pp. 133–157. Bradford/MIT Press, Cambridge, Massachusetts.

Kanwal JS, Matsumura S, Ohlemiller K, Suga N (1994) Analysis of acoustic elements and syntax in communication sounds emitted by mustached bats. *J Acoust Soc Am* 96: 1229–1254.

Kanwal JS, Fitzpatrick DC, Suga N (1999) Facilitatory and inhibitory frequency tuning of combination-sensitive neurons in the primary auditory cortex of mustached bats. *J Neurophysiol* 82: 2327–2345.

Keller CH, Takahashi TT (2000) Representation of temporal features of complex sounds by the discharge patterns of neurons in the owl's inferior colliculus. *J Neurophysiol* 84: 2638–2650.

Kelly JB, Liscum A, van Adel B, Ito M (1998) Projections from the superior olive and lateral lemniscus to tonotopic regions of the rat's inferior colliculus. *Hear Res* 116: 43–54.

Klug A, Bauer EE, Hanson JT, Hurley L, Meitzen J, Pollak GD (2002) Response selectivity for species-specific calls in the inferior colliculus of Mexican free-tailed bats is generated by inhibition. *J Neurophysiol* 88: 1941–1954.

Kuwada S, Batra R, Yin TC, Oliver DL, Haberly LB, Stanford TR (1997) Intracellular recordings in response to monaural and binaural stimulation of neurons in the inferior colliculus of the cat. *J Neurosci* 17: 7565–7581.

Le Beau FE, Rees A, Malmierca MS (1996) Contribution of GABA- and glycine-mediated inhibition to the monaural temporal response properties of neurons in the inferior colliculus. *J Neurophysiol* 75: 902–919.

Li L, Kelly JB (1992) Inhibitory influence of the dorsal nucleus of the lateral lemniscus on binaural responses in the rat's inferior colliculus. *J Neurosci* 12: 4530–4539.

Liberman MC (1978) Auditory-nerve response from cats raised in a low-noise chamber. *J Acoust Soc Am* 63: 442–455.

Liberman MC, Kiang NY (1978) Acoustic trauma in cats. Cochlear pathology and auditory-nerve activity. *Acta Oto-laryngol Suppl* 358: 1–63.

Muller CM, Scheich H (1987) GABAergic inhibition increases the neuronal selectivity to natural sounds in the avian auditory forebrain. *Brain Res* 414: 376–380.

Neuweiler G (2000) *The Biology of Bats*. Oxford University Press, New York.

Oliver DL, Beckius GE (1992) Fine structure of GABA-labeled axonal endings in the inferior colliculus of the cat: immunocytochemistry on deplasticized ultrathin sections. *Neuroscience* 46: 455–463.

Oliver DL, Huerta MF (1992) Inferior and superior colliculi. In: Webster DB, Popper AN (eds), *The Mammalian Auditory Pathway: Neuroanatomy*, pp. 168–222. Springer-Verlag, New York.

Oliver DL, Beckius GE, Shneiderman A (1995) Axonal projections from the lateral and medial superior olive to the inferior colliculus of the cat: a study using electron microscopic autoradiography. *J Comp Neurol* 360: 17–32.

Park TJ, Pollak GD (1993a) GABA shapes a topographic organization of response latency in the mustache bat's inferior colliculus. *J Neurosci* 13: 5172–5187.

Park TJ, Pollak GD (1993b) GABA shapes sensitivity to interaural intensity disparities in the mustache bat's inferior colliculus: implications for encoding sound location. *J Neurosci* 13: 2050–2067.

Peruzzi D, Bartlett E, Smith PH, Oliver DL (1997) A monosynaptic GABAergic input from the inferior colliculus to the medial geniculate body in rat. *J Neurosci* 17: 3766–3777.

Peruzzi D, Sivaramakrishnan S, Oliver DL (2000) Identification of cell types in brain slices of the inferior colliculus. *Neuroscience* 101: 403–416.

Pickles JO (1986) The neurophysiological basis of frequency selectivity. In: Moore BCJ (ed.), *Frequency Selectivity in Hearing*, Academic Press, London.

Pickles JO (1988) *An Introduction to the Physiology of Hearing*. Academic Press, London.

Pollak GD, Casseday JH (1989) *The Neural Basis of Echolocation in Bats*, 1st Edition. Springer-Verlag, Berlin-Heidelberg, NY.

Pollak GD, Park TJ (1993) The effects of GABAergic inhibition on monaural response properties of neurons in the mustache bat's inferior colliculus. *Hear Res* 65: 99–117.

Pollak GD, Wenstrup JJ, Fuzessery ZM (1986) Auditory processing in the mustache bat's inferior colliculus. *Trends Neurosci* 9: 556–561.

Popper AN, Fay RR (eds) (1995) *Hearing by Bats*, 1st Edition. Springer-Verlag, New York.

Rauschecker JP (1997) Processing of complex sounds in the auditory cortex of cat, monkey, and man. *Acta Oto-laryngol Suppl* 532: 34–38.

Rauschecker JP, Tian B (2000) Mechanisms and streams for processing of "what" and "where" in auditory cortex. *Proc Natl Acad Sci USA* 97: 11800–11806.

Rauschecker JP, Tian B, Hauser M (1995) Processing of complex sounds in the macaque nonprimary auditory cortex. *Science* 268: 111–114.

Rayleigh L (1907) On our perception of sound direction. *Philos Mag* 13: 214–232.

Ross LS, Pollak GD (1989) Differential ascending projections to aural regions in the 60 kHz contour of the mustache bat's inferior colliculus. *J Neurosci* 9: 2819–2834.

Ross LS, Pollak GD, Zook JM (1988) Origin of ascending projections to an isofrequency region of the mustache bat's inferior colliculus. *J Comp Neurol* 270: 488–505.

Roth GL, Aitkin LM, Andersen RA, Merzenich MM (1978) Some features of the spatial organization of the central nucleus of the inferior colliculus of the cat. *J Comp Neurol* 182: 661–680.

Saitoh I, Suga N (1995) Long delay lines for ranging are created by inhibition in the inferior colliculus of the mustached bat. *J Neurophysiol* 74: 1–11.

Saldana E, Merchan MA (1992) Intrinsic and commissural connections of the rat inferior colliculus. *J Comp Neurol* 319: 417–437.

Scheich H, Langner, G., Koch, R. (1977) Coding of narrow-band and wide-band vocalizations in the auditory midbrain nucleus (MLD) of he Guinea fowl (*Numida melearis*). *J Comp Physiol A* 117: 245–265.

Schreiner CE, Langner G (1997) Laminar fine structure of frequency organization in auditory midbrain. *Nature* 388: 383–386.

Semple MN, Aitkin LM (1979) Representation of sound frequency and laterality by units in central nucleus of cat inferior colliculus. *J Neurophysiol* 42: 1626–1639.

Sen K, Theunissen FE, Doupe AJ (2001) Feature analysis of natural sounds in the songbird auditory forebrain. *J Neurophysiol* 86: 1445–1458.

Shneiderman A, Henkel CK (1987) Banding of lateral superior olivary nucleus afferents in the inferior colliculus: a possible substrate for sensory integration. *J Comp Neurol* 266: 519–534.

Shneiderman A, Oliver DL (1989) EM autoradiographic study of the projections from the dorsal nucleus of the lateral lemniscus: a possible source of inhibitory inputs to the inferior colliculus. *J Comp Neurol* 286: 28–47.

Shneiderman A, Oliver DL, Henkel CK (1988) Connections of the dorsal nucleus of the lateral lemniscus: an inhibitory parallel pathway in the ascending auditory system? *J Comp Neurol* 276: 188–208.

Sivaramakrishnan S, Oliver DL (2001) Distinct k currents result in physiologically distinct cell types in the inferior colliculus of the rat. *J Neurosci* 21: 2861–2877.

Theunissen FE, Sen K, Doupe AJ (2000) Spectral-temporal receptive fields of nonlinear auditory neurons obtained using natural sounds. *J Neurosci* 20: 2315–2331.

Vater M, Siefer W (1995) The cochlea of *Tadarida brasiliensis*: specialized functional organization in a generalized bat. *Hear Res* 91: 178–195.

Vater M, Habbicht H, Kossl M, Grothe B (1992) The functional role of GABA and glycine in monaural and binaural processing in the inferior colliculus of horseshoe bats. *J Comp Physiol A* 171: 541–553.

Wang X (2000) On cortical coding of vocal communication sounds in primates. *Proc Natl Acad Sci USA* 97: 11843–11849.

Wang X, Kadia SC (2001) Differential representation of species-specific primate vocalizations in the auditory cortices of marmoset and cat. *J Neurophysiol* 86: 2616–2620.

Webster DB (1992) An Overview of the Mammalian Auditory Pathways with an Emphasis on Humans. In: Webster DB, Popper AN, Fay RR (eds), *The Mammalian Auditory Pathway: Neuroanatomy*, 1st Edition p. 485. Springer-Verlag, New York.

Wenstrup JJ, Leroy SA (2001) Spectral Integration in the Inferior Colliculus: Role of Glycinergic Inhibition in Response Facilitation. *J Neurosci* 21: RC124.

Winer JA (1992) Functional architecture of the medial geniculate body and primary auditory cortex. In: Webster DB, Popper AN, Fay RR (eds),*The Mammalian Auditory Pathway: Neuroanatomy*, pp. 222–409. Springer-Verlag, New York.

Winer JA, Larue DT, Pollak GD (1995) GABA and glycine in the central auditory system of the mustache bat: structural substrates for inhibitory neuronal organization. *J Comp Neurol* 355: 317–353.

Winer JA, Saint Marie RL, Larue DT, Oliver DL (1996) GABAergic feedforward projections from the inferior colliculus to the medial geniculate body. *Proc Natl Acad Sci USA* 93: 8005–8010.

Winer JA, Larue DT, Diehl JJ, Hefti BJ (1998) Auditory cortical projections to the cat inferior colliculus. *J Comp Neurol* 400: 147–174.

Wu SH, Kelly JB (1996) *In vitro* brain slice studies of the rat's dorsal nucleus of the lateral lemniscus. III. synaptic pharmacology. *J Neurophysiol* 75: 1271–1282.

Yang L, Pollak GD, Resler C (1992) GABAergic circuits sharpen tuning curves and modify response properties in the mustache bat inferior colliculus. *J Neurophysiol* 68: 1760–1774.

Yang L, Liu Q, Pollak GD (1996) Afferent connections to the dorsal nucleus of the lateral lemniscus of the mustache bat: evidence for two functional subdivisions. *J Comp Neurol* 373: 575–592.

Zook JM, Winer JA, Pollak GD, Bodenhamer RD (1985) Topology of the central nucleus of the mustache bat's inferior colliculus: correlation of single unit properties and neuronal architecture. *J Comp Neurol* 231: 530–546.

7
A Distributed Cortical Representation of Social Communication Calls

Jagmeet S. Kanwal

7.1 Introduction

The mammalian auditory system has been studied extensively over the last two to three decades. A large majority of these studies have focused on understanding either the basic properties of auditory neurons within different centers of the auditory pathways of the brain and/or their role in either sound localization or echolocation (e.g. Griffin *et al.*, 1960; Suga and Jen, 1976; Suga *et al.*, 1978, 1987; Konishi and Knudsen, 1982; Jenkins and Merzenich, 1984; Fuzessery *et al.*, 1985; Rajan *et al.*, 1990; Schreiner and Mendelson, 1990; Sutter and Schreiner, 1991, 1995; Schreiner and Sutter, 1992; Schreiner *et al.*, 1992; Middlebrooks *et al.*, 1994; Mittmann and Wenstrup, 1995; Fitzpatrick *et al.*, 1997; Sutter *et al.*, 1999; Recanzone *et al.*, 2000). With few exceptions (e.g. Ehret and Moffat, 1985; Ehret, 1987; Wang *et al.*, 1995; Ohlemiller *et al.*, 1996; Esser *et al.*, 1997; Kanwal, 1999; Nagarajan *et al.*, 2002; Šuta *et al.*, 2003; Geissler and Ehret, 2004; Medvedev and Kanwal, 2004a), the study of call processing in mammals has generally lagged behind because it entails analyzing and synthesizing complex sounds that are more difficult to study systematically. Furthermore, call synthesis and editing requires extensive computing power that was not readily available before the emergence of high-speed desktop computers. Also, unlike either constant frequency (CF) or simple frequency-modulated (FM) tones, social communication calls are often complex in the frequency, amplitude and time domains, so that a multidimensional framework is required to both characterize the acoustics of perception and understand the representation of calls in the activity of the auditory system. The processing of social communication calls is an important common function of the auditory system. Social communication calls are used to coordinate various social behaviors, such as sexual behavior, care of the young, territoriality and establishment of social hierarchies. Some calls, such as alarm calls, play a communicative role also across individuals of different species as a biologic indicator of the presence of a predator in the environment (Searcy and Caine, 2003; Rainey *et al.*, 2004). In a more general context, call types may also be considered as independent "auditory objects" (Husain *et al.*, 2004; Nelken, 2004), although this designation does not emphasize the special neural processing mechanisms and nonlinearities for processing calls. This chapter will provide a brief overview of the organizational features of the auditory cortex (AC) in mammals and focus on an interpretation of what we know about the representation of social communication calls in the AC of mustached bats.

7.2 The mammalian AC

In the mammalian brain, the AC harbors an elaborate spatial representation for the perception of complex sounds. In this cortical sheet, different acoustic features can be represented in an organized fashion within topologically defined areas. Cortical areas upstream of the AC, such as the frontal cortex and the anterior cingulate, are largely involved in multisensory integration that defines the context within which to extract the meaning of complex sounds. These areas are also more closely involved in the production of both vocal and behavioral responses. To begin to understand the role of AC in auditory processing as a whole, it is important to know (1) how a representation of complex sounds is created in the AC and (2) what does this representation look like? The answer to the first question is complex and requires studies at all levels of the auditory system since auditory representations in the AC are the outcome of multiple steps of processing at levels both below and above the AC. The second question can be addressed relatively easily by presenting a set of complex sounds, such as species-specific social communication calls, in a systematic manner and examining the responses of hundreds of single neurons to the same set of sounds (Ohlemiller *et al.*, 1994). Alternative strategies may involve examining the responses of populations of neurons by recording local field potentials at various locations or by simply studying the basic response properties of neurons and then formulating a computational framework within which the coding and representation of the responses to complex sounds can be estimated (Nelken *et al.*, 1994; Medvedev *et al.*, 2002; Medvedev and Kanwal, 2004a). A non-neurophysiologic strategy that can also tell us something about the possible representation of complex sounds is to study the statistics of the acoustical parameters of natural sounds in order to derive the natural distribution of orthogonal versus correlated parameters that must be extracted, encoded and "mapped" in order to obtain the necessary information for sound discrimination (e.g. Herrnberger *et al.*, 2002; Lewicki, 2002; Karklin and Lewicki, 2003). This approach, of course, cannot provide information about what is biologically important and about the neural mechanisms that must be implemented for the representation of complex sounds. It can, however, provide the upper and lower bounds of the complexity in the acoustical parameter space that needs to be encoded to have a useful representation of complex sounds. All of these approaches have been used in various fashions to identify the parameters that are important for generating a neural representation of complex communication sounds. The different methodologic approaches and the use of several species are complimentary for a general understanding of the representation of complex communication calls in the mammalian AC.

7.2.1 Organization of the AC

By its cytoarchitecture and its connectivity with partitions of the medial geniculate body and other thalamic nuclei, the mammalian AC is divided into primary and higher-order auditory cortical fields (for details, see e.g. Winer *et al.*, 1992; de Ribaupierre, 1997; Rouiller, 1997). In several mammal species, especially primates, primary fields are referred to as the core areas and higher-order fields as the belt and parabelt areas of the AC (Kaas and Hackett, 1998). Functionally, each primary area is characterized by a tonotopic organization reflecting the frequency gradient along the basilar membrane of the cochlea. Neurons in primary areas usually show clear and brisk responses to tone bursts with only a small tendency to habituate. An across-species comparison on the basis of thalamocortical projections and cytoarchitecture of different cortical areas in *Rhinolophus* suggests that bat brains can be equated to those of primates in terms of a core-belt–parabelt organization (Radtke-Schuller, 2001, 2004; Radtke-Schuller *et al.*, 2004). In mustached bats, the division into core and belt areas has not been established. However, the central tonotopic axis that includes the anterior primary auditory (AIa), the Doppler-shifted constant frequency (DSCF) and the posterior primary auditory (AIp) areas likely constitute the primary AC or the core. The CF/CF, DM and TE and a portion of the surrounding gray zone ("l"and "m" in Fig. 7.1a) are likely analogous to the belt and parabelt areas. The status of the dorsally located FM–FM area is unclear as it represents frequencies that are largely absent in the primary tonotopic axis and has reciprocal connections with the medial geniculate body (Fitzpatrick *et al.*, 1998). Yet, the FM–FM area does not lie within the main tonotopic axis. From a physiologic perspective, a separation into core and belt areas is less relevant and the AC of the mustached bat is best divided into primary (AI) and nonprimary areas. The cortical areas (DSCF AIp, CF/CF and FM–FM) discussed in this

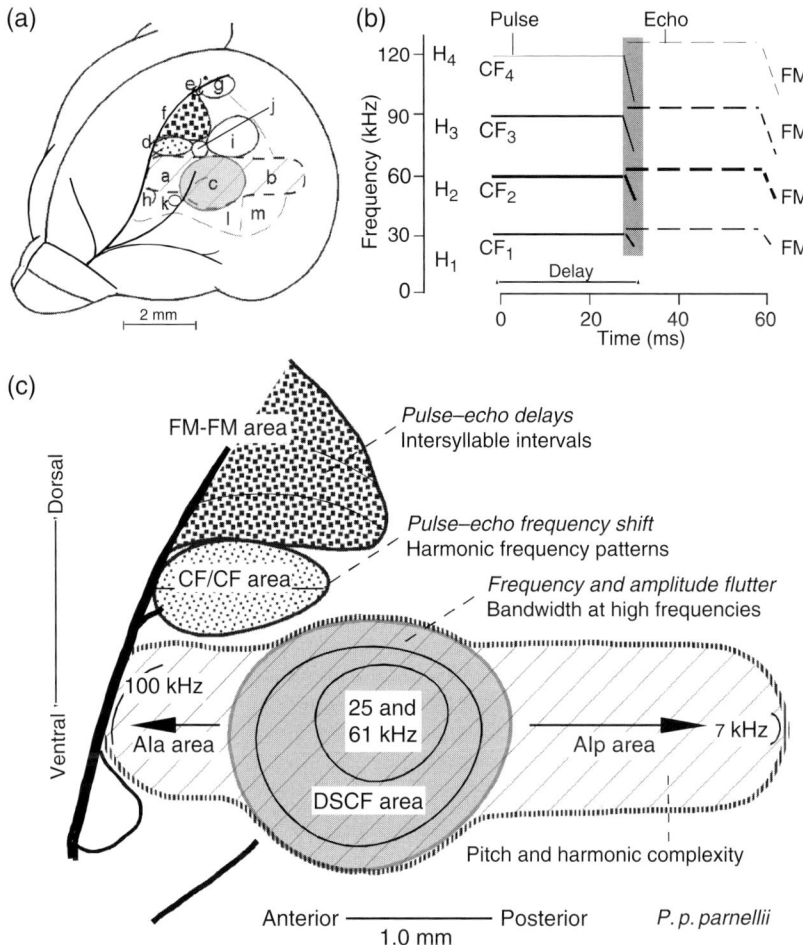

Figure 7.1. Organization of the AC in mustached bats. (a) Lateral view of the mustached bat's brain showing the functionally defined subdivisions of the AC. The primary AC (AI) is shown as hatched. "a": AI-anterior, "b": AI-posterior, "c": DSCF area (shown in gray), "d": CF/CF area (fine stippled), "e": DIF area, "f": FM–FM area (coarse stippled), "g": DF area, "h": VF area, "i": DM area, "j": TE area, "k": H_1–H_2 area, "l": VA area, "m": VP area (adapted from Suga, 1984). (b) Schematized spectrogram of the mustached bat's echolocation signal. H_{1-4} refer to harmonics 1 through 4 of the echolocation pulse or echo. The shaded region shows the overlap of time and frequency components in the pulse and echo signal. (c) Detailed representation of the FM–FM, CF/CF, DSCF and AIp areas in the AC. The DSCF area occupies nearly 30% of the area of the tonotopically organized primary AC and represents a small range (60.6–62.3 kHz for a resting CF_2 of 61 kHz) frequencies. Labels with dashed lines indicate echolocation parameters (noted in italics) and social call parameters that are represented in each area.

chapter are considered to co-represent the perception of social calls since the range of initial response latencies to tones and calls in each area is roughly equivalent and relatively short (5–25 ms).

7.2.2 Response properties of primary auditory cortical neurons

The tonotopy, often reflecting the cochlear frequency representation, provides the basic auditory cortical map in the primary auditory cortex. This map is one that is "hard wired" from the inner ear through centers of the ascending auditory pathways up to primary auditory cortical fields. The neurons within the primary AC (AI), however, do not simply represent single sound frequencies in a tonotopic fashion. There are various other acoustic parameters and neural response parameters that have been shown to be represented in an organized fashion (maps) along with tonotopy (see e.g. Ehret, 1997). These maps may represent the basis functions or substrate for generating the perception of the acoustical gestalt of a sound. Which parameters are mapped and how their gradients are spatially organized at the level of the AI may vary with the species. For example, in several species, bands of binaural excitatory (E–E) versus contralateral excitatory and ipsilateral inhibitory (E–I) neural responses are found superimposed on the tonotopy (e.g. Imig *et al.*, 1986; Liu and Suga, 1997). In cats, tone-response threshold minima and best amplitudes as well as latencies are all represented in a systematic manner perpendicular to the tonotopy of AI (Schreiner and Sutter, 1992; Schreiner *et al.*, 1992; Schreiner, 1995). Periodicity pitch is represented along a horseshoe-shaped pattern in the AI of Mongolian gerbils (Schulze *et al.*, 2002) and orthogonal to the frequency axis in the AC of humans (Langner *et al.*, 1997). Mustached bats show an amplitopic representation and binaural bands within the central region of their AI, which is known as the DSCF processing area (Suga and Jen, 1976; Liu and Suga, 1997).

Within AI, neurons may be tuned best to a single frequency, but the response areas of the neurons are found to be complex in terms of conventional tuning curves or spectrotemporal receptive fields. In several species, the excitatory response areas are surrounded by inhibitory response areas (e.g. ferrets: Shamma *et al.*, 1993; bats: Kanwal *et al.*, 1999; cats: Sutter *et al.*, 1999). In ferrets, the organization of the inhibitory areas is asymmetric and shifts with location in the cortex. In bats, excitatory response areas are partially masked by inhibitory response areas leading to an increase in the sharpness of tuning to one or more parameters, such as frequency, pulse–echo delay, etc. (Suga, 1995). Asymmetric inhibitory response areas also skirt and frequently overlap the boundaries of the facilitatory response areas in the cortex of bats (Kanwal *et al.*, 1999), but their general organization and shape varies from that in cats and ferrets (Sutter and Schreiner, 1991; Shamma *et al.*, 1993). Both in cats and bats, multi-peaked harmonic frequency tuning is present within AI, but at different locations; in the dorsal part in cats (Sutter and Schreiner, 1991), within the DSCF and AIp areas in bats (Kanwal *et al.*, 1999; Peng *et al.*, submitted). Thus, a description of AI in terms of frequency selectivity alone is inadequate considering the multiple parameters that are represented and sometimes mapped in an orderly fashion here (Read *et al.*, 2002). Some have even argued that the primary function of AI is not to map parameters or features, but to integrate sounds on different time scales (Nelken *et al.*, 2003; Ulanovsky *et al.*, 2004b).

The remaining part of this chapter will focus on the organization and response properties of neurons in the AC of mustached bats. In particular, the acoustic structure of complex communication calls or simply "calls" will be described in relation to the response properties of neurons in different cortical areas and in relation to the representation of biosonar information. These data will be used to propose a distributed representational scheme that, in principle, may apply to the representation of complex sounds in most species.

7.3 The acoustic structure of complex sounds in mustached bats

7.3.1 The structure of echolocation signals

During echolocation, mustached bats, *Pteronotus parnellii*, emit brief (~30 ms long) sound pulses that consist of a fundamental frequency and three harmonics. These are labeled as H_1–H_4. The fundamental and each harmonic has a CF component (CF_1-CF_4) and a downward-sweeping FM component (FM_1–FM_4) (Fig. 7.1b). The CF components are harmonics of an approximately 30 kHz fundamental. The flying bat hears both, its own pulses and the returning echoes. The latter are Doppler-shifted upward in frequency, and may partially overlap in time with the original pulse if the target is nearby. The auditory periphery is exquisitely sensitive to the CF_2 echo (near 61 kHz in *P. parnellii parnellii*) because of a well-defined cochlear resonance, but it is relatively insensitive to the emitted CF_2 pulse (near 59 kHz). Hence, masking of the echo by the pulse is minimal.

7.3.2 The structure of social communication calls

Mustached bats emit a variety of highly structured sounds or calls for social communication (Kanwal *et al.*, 1994). Calls consist of either simple syllables, or of composites of two or more simple syllables produced together without a silent interval between them, and of phrases consisting of both simple syllables and composites. A few syllable types may also be vocalized in sequences of repeated simple syllables separated by more-or-less fixed silent intervals. Calls have many features in common with echolocation signals, such as varying duration of call and inter-call intervals, presence of harmonics, unidirectional and bidirectional FMs, pulse–echo versus syllable pairs, predominant harmonics, etc. An important difference from the rather stereotypic sequences of echolocation sounds is the large variety and high structural variability of social communication calls. Also, calls are richer in their acoustic features than echolocation pulses and the meaning of calls can change with the behavioral context since most calls are associated with emotions (see also the chapter by Clement *et al.*, Chapter 3, this volume). In some situations, even echolocation pulses can take on a communicative function (Fenton *et al.*, 2004). Thus, for social communication to be effective, the animal must perceive each sound by its intended meaning together with the context in which it is produced in order to release appropriate behavioral responses.

Based on their predominant acoustic patterns, calls can be roughly divided into CF-type, FM-type and NB-type (noise burst) calls. A multidimensional scaling analysis of simple syllabic calls of all three types revealed a discrete progression from one call type to another across two to three dimensions representing the multiparametric boundaries for each call type (Kanwal *et al.*, 1994). The dimensional reduction explains over 90% of the variability in the calls, thus simplifying the number of features that need to be mapped or represented within the AC without an inherent loss of perceptual information. Although some calls have distinct amplitude patterns, these patterns are not of major importance as they can easily vary because of the directionality of high-frequency sounds and the irregular dissipation and absorption of sound energy at relatively short distances (Simmons, 1969). To minimize the dissipation and absorption effects, bats frequently emit calls at high intensities (near about 100 dB SPL) and aim their head in the direction of the receiver and sometimes may even "speak" into the receiver's ear.

7.4 The AC in mustached bats

7.4.1 Functional organization for echolocation

The AC contains neurons tuned to combinations of the pulse H_1 and various components in the echo H_2–H_4 (Fig. 7.1b). The DSCF area is the largest portion of the primary AC. It contains frequency-versus-amplitude coordinates for mapping the amplitude of the CF_2 component of the echo (Manabe *et al.*,

1978) (Fig. 7.1a and c). This region occupies nearly 30% of the AI and contains a magnified representation of a narrow range of frequencies between 60.6 and 62.3 kHz (normalized to a resting pulse CF_2 frequency of 61 kHz) in combination with frequencies of 23–29 kHz in the FM_1 component of the pulse (Fig. 7.1b).

In the CF/CF area (Fig. 7.1c), neurons tuned to CF combinations (e.g. pulse CF_1 and echo CF_2 or CF_3) are sensitive to the amount of Doppler shift between pulse and echo, and thus process the relative velocity between the bat and the structure generating the echo by reflection (Suga, 1984). The CF/CF area is functionally organized into columns each consisting of neurons that respond best to a particular combination of CF_1/CF_2 or CF_1/CF_3 tones. The frequency differences in the paired tones to which the neurons respond best vary in a systematic manner along the surface of the cortex. Other cortical neurons tuned to FM combinations in the FM–FM area (Fig. 7.1c) (e.g. pulse FM_1 and echo FM_2) are selective for certain echo delays, and thus process target range information (O'Neill and Suga, 1982; Suga, 1984). The combination-sensitivity of CF/CF and FM–FM neurons is manifested in the facilitation of the response when both stimulus components are presented either simultaneously or with a specific delay. This facilitation, encompasses both spectral (CF/CF neurons) and temporal domains (FM–FM neurons). Similar facilitation mechanisms have been reported for neurons in the midbrain of the leapord frog (Fuzessery and Feng, 1983) and the forebrain of the zebra finch (Margoliash and Fortune, 1992).

7.4.2 Neural mechanisms for pulse–echo representation

Cortical neurons map multiple target characteristics for prey recognition and capture via echolocation behavior (Suga, 1984). The characteristics of neural responses include (i) combination-sensitivity to different acoustical parameters of the pulse and echo, (ii) tuning to the fundamental of the pulse and sharp tuning to harmonics CF_2, CF_3 or CF_4 in the echo, (iii) specialization to respond only to downward FMs and (iv) sharp tuning to pulse–echo delays. Exactly how bats, in general, use these response characteristics and maps for echolocation is not clear as different species of bats emit different types of echolocation pulses that generate different types of representations for the same ultimate goal of navigation, tracking and prey capture (Wong and Shannon, 1988; Shannon-Hartman et al., 1992; Dear et al., 1993). For example, big brown bats that use FM pulses for echolocation may reconstruct an auditory scene not by comparisons of single pulse–echo pairs, but by integrating the information in multiple echoes (Dear et al., 1993). They could also use "auditory streaming" to parse the pulse stream from the echo stream for tracking changes in their environment or use other integrative properties of neurons that have not been systematically examined (Kanwal et al., 2003). These same response properties of cortical neurons that are used to construct an image from an echo can also be used to generate unique percepts for different types of calls during audiovocal communication.

7.4.3 Neural mechanisms for call representation

Calls enable social interactions among conspecifics via audiovocal communication. Cortical neurons likely encode calls within a multidimensional framework. As for echolocation, combination-sensitivity is exploited as a mechanism to represent acoustic features within calls (Kanwal, 1999). In the last few years, neurophysiologic experiments have generated a database of responses to the same set of 14 (of 19 observed) simple syllabic calls and their variants emitted by mustached bats. This includes responses from over 300 neurons in different regions of the mustached bat's cortex (Kanwal et al., 1992; Ohlemiller et al., 1994, 1996; Esser et al., 1997; Kanwal, 1999; Medvedev and Kanwal, 2004a). Single unit call response data that have been examined include peak firing rates and response magnitudes (total number of spikes generated), spike timing as well as some temporal response parameters, such as response duration, time of peak response and initial response latencies. For details of the stimulation paradigm and data analysis in these experiments, see Kanwal et al. (1999). More recently, the temporal response pattern of neural ensembles is also implicated in call encoding and discrimination (Medvedev and Kanwal, 2004a).

Most studies on call processing have to grapple with the important question of deciding on the "basis functions" that play a primary role in call representation. Some studies pose this question explicitly, whereas others make implicit assumptions about the role of single versus multiple parameters. In either case, the receptive fields of cortical neurons need to be carefully and comprehensively tested. Single parameter representations include mappings of the receptive fields of single frequencies (tonotopy), amplitude (amplitopy), latency, bandwidth and pitch periodicity (periodotopy) along some axis. The concept here is that the representation of a complex sound is based on the sum of several linear transformations. The spatiotemporal pattern of activity of a set of neurons, whose frequency–amplitude response areas match the spectrographic structure of a complex sound, provides a unique representation of that sound (Wang et al., 1995; Machens et al., 2004). This scheme is highly generalizable, but does not attach any importance to neural specializations, such as combination-sensitivity in the spectral and time domains. Alternate hypotheses that take neural specializations into consideration, may invoke the representation of basic acoustic patterns (e.g. FMs, or combinations of patterns) that constitute the minimal excitatory elements in calls. Due to the large number of possibilities of such patterns and the variations within them, these representations have not been rigorously tested via neurophysiologic experiments. The responses of single cortical neurons to whole calls, however, have been systematically tested in the AC of mustached bats (Ohlemiller *et al.*, 1994, 1996; Esser *et al.*, 1997; Kanwal, 1999). Whereas the use of whole calls is neuroethologically valid, the drawback of this approach is that by itself it does not necessarily identify the "basis functions" that may be mapped for call representation. Regardless, it provides the quickest means to identify neurons that respond best to one or more calls. The variability of this response to variants of call types at different call amplitudes can also be obtained. Furthermore, stimulation with single as well as combinations of dissected call components can reveal the information-bearing elements (IBEs) that must be locally extracted without being necessarily mapped across the cortical surface to uniquely represent a call type.

Finally, for a complete understanding, one must be able to relate call responses to tonotopic representations, and at the same time resolve response ambiguity emerging from an overlap of common frequencies in different sounds as well as effects of amplitude and small natural variations in the frequency and time domains. The role of inhibitory receptive fields that span large parts of the audiogram within single neurons also need to be adequately addressed. To test a representational scheme for calls, it is also useful to know the statistically invariant acoustic features in calls and their relationship to presumptive IBEs.

7.5 Call responses in mustached bats

From studies on call processing in the FM–FM, CF/CF and DSCF areas thus far (Kanwal *et al.*, 1992; Ohlemiller *et al.*, 1994, 1996; Esser *et al.*, 1997; Kanwal, 1999), a few pieces of the puzzle are in place. First, we know that the same neurons that are highly specialized for processing echolocation signals also respond to whole calls and exhibit specializations for call processing. Second, most simple syllabic calls are represented in multiple areas of the AC (Fig. 7.2). Third, from an analysis of the systematic digital dissection of calls, it appears that neurons in each cortical region are "tuned" to a particular acoustic feature within calls, rather than the calls themselves. That is to say, cortical areas are not organized by call type, but presumably by key acoustic features within calls. This is explained below in further detail by an examination of call responses in different areas identified on the bais of responses to echolocation signals (see Fig. 7.1). Each of the four sections below identify a key acoustic feature that is presumably extracted by neurons in a particular area.

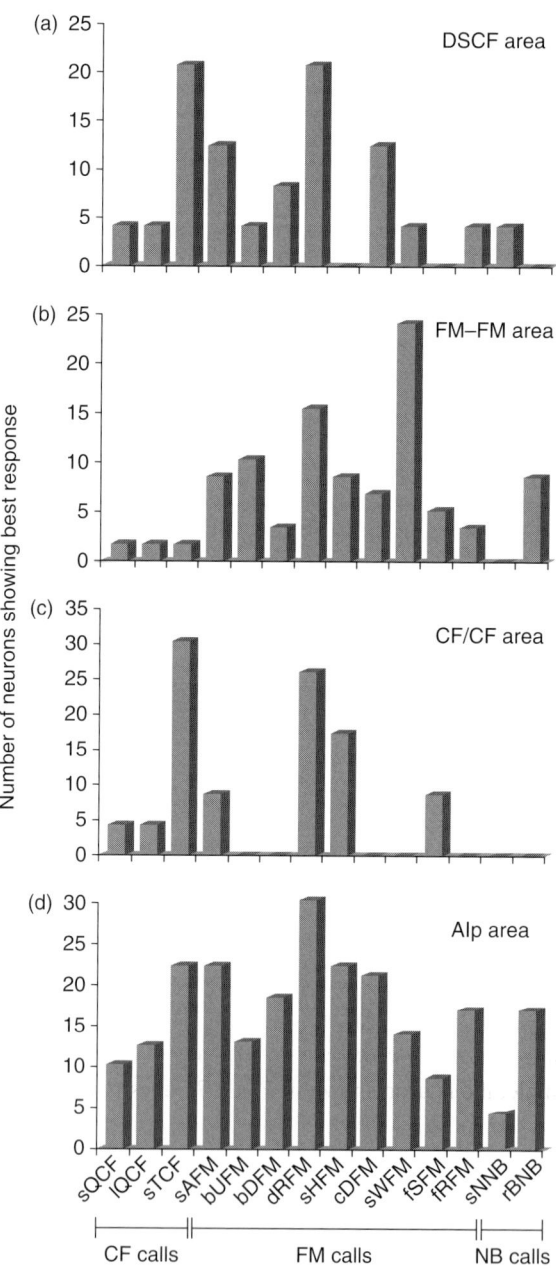

Figure 7.2. (a–d) Bar graphs showing the relative number of neurons in the DSCF, FM–FM, CF/CF and AIp areas that preferred particular call types indicated on the X-axis. A call is labeled as a preferred call if its response is within 50% of the response to the best call. CF/CF neurons are the most selective, whereas AIp neurons show

7.5.1 FM–CF combinations

Neurons in the DSCF area are sensitive not just to combinations of CFs, but also to FM components of calls. A sensitivity to specific FM components results from the neurons' excitatory and facilitatory response areas that overlap with two or more narrow bands of FM within calls (Fig. 7.3a). For many neurons with a relatively wide bandwidth (>2 kHz) of frequency tuning at best amplitudes, the response to a simple syllabic call or a call segment is greater than to the best tone pair corresponding to components in the echolocation signal (Fig. 7.3b and c). In other, more sharply tuned neurons, the response to a call segment often leads to suppression of the response compared to stimulation with a tone in the CF_2 (~60 kHz) region. This latter category of neurons can be considered to be highly specialized for processing only the echolocation signal. There are more of such neurons in the right hemisphere compared to the left hemisphere resulting in a left hemispheric dominance for call processing in the AC (Kanwal and Suga, 1995).

The data available so far suggest that the variation in call preference across a local population of neurons is related to the match between the dominant frequencies in the calls and the shapes of the excitatory, facilitatory and inhibitory response areas of the neurons (Kanwal *et al.*, 1992). Data obtained from neurons in the DSCF area illustrate three important points:

1. Selective, high-magnitude responses to calls are based primarily on spectral facilitation. Thus, when call components fall in the best-frequency (BF) areas of the excitatory response area of a neuron, it shows clear facilitation with a response that is better than the sum of the responses to each component alone (Kanwal, 1999).
2. A combination of bandpass filtered call components may elicit a larger response magnitude than the whole natural call because of the absence of acoustic energy in the inhibitory areas. Hence, the important conclusion that responses to calls can be greatly influenced by the organization of the inhibitory frequency-response areas.
3. DSCF neurons are sensitive to the direction of FM components within calls. As an example, the neuron shown in Fig. 7.3c does not respond well to the "bent upward FM" (bUFM) syllable if the facilitatory call components are

Caption for fig. 7.2. (cont.)
the least amount of selectivity. Data was obtained from 60 to 100 neurons in each area. Abbreviations: short, quasi CF (sQCF); Long, quasi CF (lQCF); short, true CF (sTCF); single arched FM (sAFM); bent upward FM (bUFM); bent downward FM (bDFM); descending rippled FM (dRFM); single humped FM (sHFM); checked downward FM (cDFM); short wrinkled FM (sWFM); fixed sinusoidal FM (fSFM); fixed rippled FM (fRFM); short, narrowband NB (sNNB); rectangular broadband NB (rBNB).

168   7 A distributed cortical representation

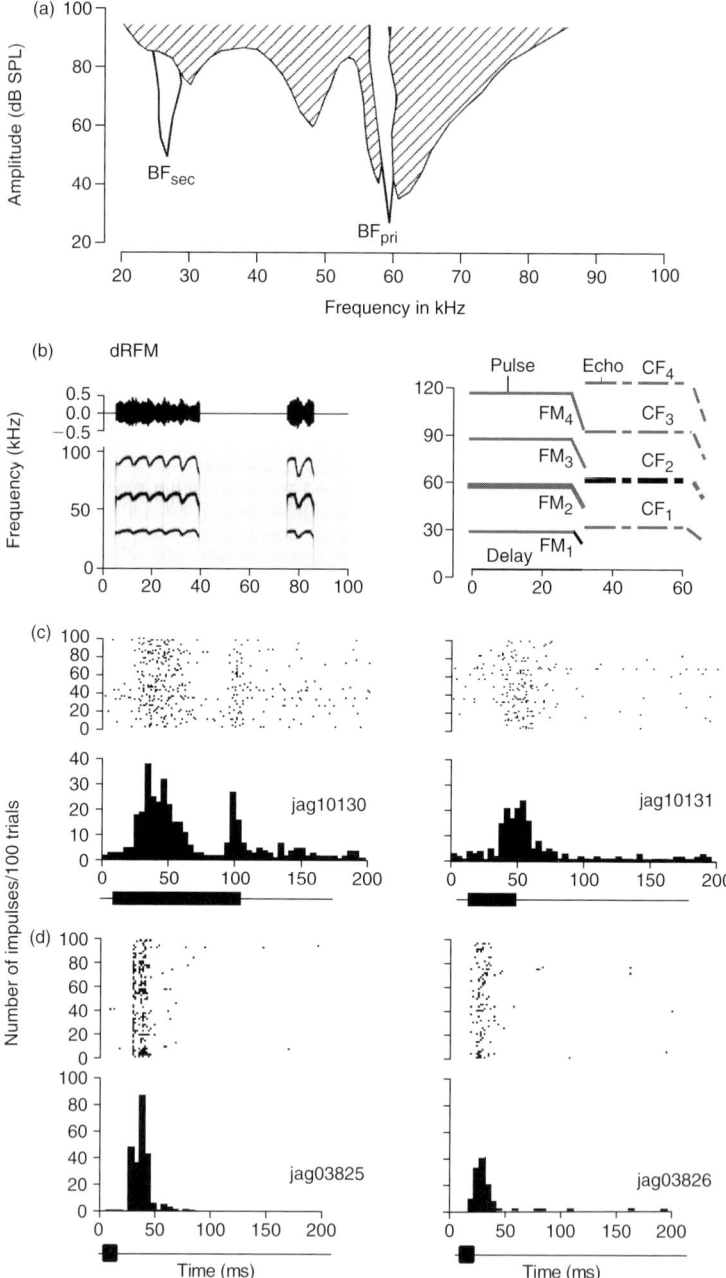

Figure 7.3. Frequency tuning and neural responses in the DSCF area. (a) Facilitatory and inhibitory (shaded) frequency–amplitude response areas with a low (~25 kHz: BF_{sec}) and a high (~60 kHz: BF_{pri}) BF for a neuron in the DSCF area. (b) Amplitude

presented as temporally reversed (Fig. 7.4a). This is explained by the asymmetrical inhibitory areas surrounding the excitatory response areas (see Fig. 7.3a) for the primary BF (BF_{pri}) that create an FM direction selectivity in many DSCF neurons. Figures 7.4b and c show examples of two other DSCF neurons, which show a decline in response to the presentation of the temporally reversed calls compared to the presentation of the same call in the forward (normal) direction. The whole call was presented to the neuron shown in Fig. 7.4a and a combination of two excitatory bands carved out of the call spectrum was presented in the forward and reverse direction for neurons shown in Fig. 7.4b and c. In short, recent studies show that for DSCF neurons, call responses can be generally explained on the basis of the shape and asymmetry in a neuron's inhibitory frequency-response areas that make it sensitive to the direction of FMs. For speech sounds of humans, this would translate to a sensitivity to the direction of the "formant transitions". As inhibitory areas are known to vary somewhat with depth within a single column, these variations in inhibitory tuning may be used to extract small variations within acoustic features that encode the identity and emotional status of the emitter.

7.5.2 FM–FM combinations: the time domain

Neurons sensitive to time domain processing of calls are uncommon in the DSCF area. In the FM–FM area, however, neurons are tuned to FMs and are highly sensitive to variations in the time domain. More than half of the neurons studied show facilitative interactions to specific temporal relationships between spectral and temporal components of pairs and trains of simple syllables (Ohlemiller *et al.*, 1996). Temporal facilitation is indicated if the response to the whole call is >120% of the sum of responses to the iso- or hetero-syllabic parts. For analysis in the time domain, the silent interval between syllables is gradually increased

Caption for fig. 7.3 (cont.)
envelop (above) and spectrogram of the dRFM call (left) and the echolocation signal (right) consisting of the pulse (solid lines) and its echo (dashed lines). Other harmonics that are normally present in the signal are shown as gray. (c) A single neuron's response to a dRFM call (shown above) in (b) versus an FM_1 and a CF pair (shown above as bold in (b)). The raster plot is shown on the top and the peri-stimulus time histogram (PSTH) below for 100 repetitions of the stimulus. Initial response latency (from stimulus onset) to the call was 9 ms and to the pulse–echo components was 16 ms. Response duration for the first syllable and the pulse–echo components were 52 and 48 ms, respectively. (d) Raster plot (above) and PSTHs (below) show the effect of call reversal (bUFM) on a single unit's response. The timing of stimuli is shown in the solid bars at the bottom. Bin size: 5 ms.

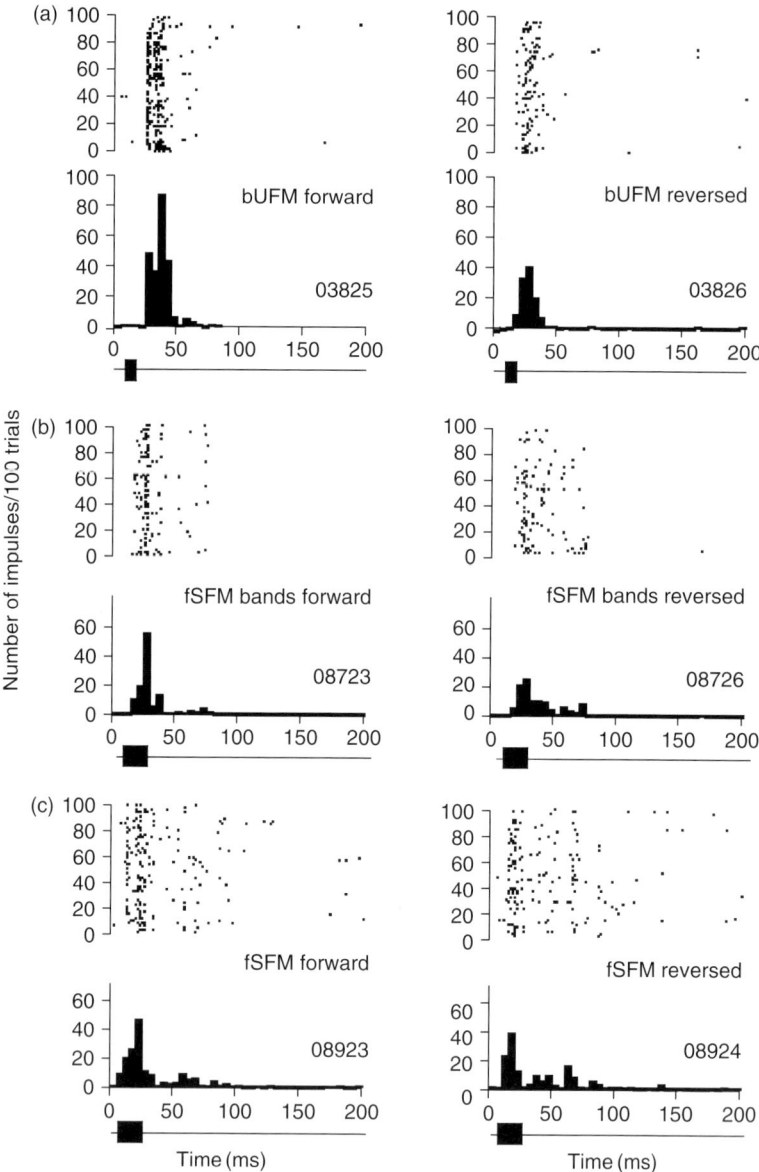

Figure 7.4. (a) Raster plot (above) and PSTH of a single neuron in the DSCF area to show the effect of presenting the bUFM call in the forward (normal) and reverse directions. (b and c) Raster and PSTH plots of the responses of two other neurons to the presentation of a combination of spectral notches or bands (b) from the fSFM bands (b) and whole calls (c) presented in the forward and reverse directions. The spectral notches or bands were carved out of the power spectrum to match the width of

from 0 ms through a range such that the naturally occurring interval falls approximately in the middle. Also in the case of syllable pairs, the first syllable is presented by itself in the beginning and the second one is presented by itself at the end of the scan. This type of scan tests for both the tuning of the neuron to various silent intervals as well as the magnitude of facilitation of a combination-sensitive neuron. The responses to these manipulations can then be compared with the neuron's response to the original syllable pair. As a result, the response to a syllable pair usually declines in a graded fashion when the duration of the silent interval extends on either side of the normal variation in the calls. Data on tuning to silent intervals show that FM–FM neurons are tuned to intersyllable intervals matching those in the naturally emitted syllable pairs or trains. These silence intervals can be an order of magnitude different from pulse–echo delays in echolocation used for ranging (Ohlemiller *et al.*, 1994).

The FM pattern in a syllable can be "destructured" such that predominant peaks in the power spectrum remain unchanged. Experiments in which the modulation pattern either of the first or second syllable in a syllable pair was destructured, showed that the FM pattern in the second syllable was most important in elucidating a neuron's response in the FM–FM area. These data together with experiments varying the silent interval between syllables suggest that the first syllable acts to prime a neuron's response to the second syllable (Ohlemiller *et al.*, 1996), which must have the correct FM pattern and the appropriate delay in order to then evoke the neuron's maximum response rate. Both, modification of the optimal time delay and FM pattern of the second syllable reduce the responsiveness of the neuron. Thus, temporal combination-sensitivity of FM–FM neurons for syllable pairs is found to co-exist with spectral combination-sensitivity. In other words, neuronal responses to natural syllable pairs are highly vulnerable to (i) a destruction of the FMs in the second syllable, (ii) to the introduction of artificially short or long silent periods between the syllables and (iii) to playing the syllables in reverse.

These observations provide independent lines of evidence at the single-unit level that "phonologic" syntax in calls is represented in the neural responses of the bat's AC. These findings, together with statistical evidence of structural syntax in the calls and the occurrence of both facilitative and suppressive intersyllable interactions, clearly point to the importance of silent intervals as an acoustic feature or information-bearing parameter as defined by Suga (1988) that is important for the identification of communication sounds in bats.

Caption for fig. 7.4. (cont.)
excitatory frequency tuning at best amplitude for each neuron. Each stimulus was repeated 100 times. Bin width: 5 ms.

7.5.3 CF/CF combinations: high-frequency domain

Frequency tuning curves of auditory-nerve fibers typically have a tail at their low-frequency end (Sachs and Abbas, 1974). At increasingly higher levels of the nervous system including the AC, the shapes of frequency tuning curves usually become more diverse, including very sharp or rather broad tuning near a neuron's BF, or even tuning to multiple best frequencies (e.g. Sutter and Schreiner, 1991; de Ribaupierre, 1997; Ehret, 1997). Very sharp tuning is a characteristic of neurons in auditory cortical areas of the mustached bat that are highly specialized for echolocation (e.g. DSCF area, CF/CF area). In addition, neurons in the DSCF and CF/CF areas show two (sometimes three) peaks (BFs) of tuning which correspond to the fundamental of the pulse and a higher harmonic in the echo. These well-separated (by 30 kHz or more) BFs have been considered as extreme specializations for the sole purpose of echolocation. Other mammals that neither echolocate nor use ultrasonic communication calls seem not to have such a large separation between peaks of frequency tuning curves in AC neurons. Call response data from the mustached bat show that many of these neurons also respond well to calls, although response magnitudes are generally smaller than those to echolocation stimuli. The sharp tuning to high frequencies, also makes CF/CF neurons more selective to calls than neurons in other cortical areas. Neurons in the CF/CF area respond better to calls containing CF components compared to broadband FM components.

7.5.4 Harmonic complexity: low-frequency domain

Virtually all neurons in the AIp area (Fig. 7.1c) of mustached bats show multiple peaks of tuning to low frequencies (<50 kHz) (Fig. 7.5a and b), but do not exhibit combination-sensitivity to call components as observed in the DSCF and FM–FM areas (Peng et al., submitted). Since AIp neurons are unlikely to be involved in echolocation, I propose that multi-peaked frequency tuning, especially to harmonics, may be considered a common feature of auditory cortical neurons that may also be present at subcortical levels (Wenstrup, 1999; Portfors and Wenstrup, 2001, 2002). A harmonic relationship in the peaks of frequency tuning of AIp neurons is likely based on the tonal structure of natural sounds (i.e. most naturally produced sounds contain multiple harmonics based on principles of physics). Social communication calls have relatively low fundamental frequencies and are also tonal so that a functional adaptation or plasticity via learning favored by evolution and natural selection may use this feature for their perception. Furthermore, the simultaneous activation of two or more neurons would lead to their synchronous firing that could be selected for by higher-level neurons onto which the simultaneously activated neurons converge.

Thus, neurons with harmonic, multi-peaked tuning would result from simultaneous and consistent stimulation with harmonically structured sounds such as echolocation pulses and calls used in social communication. Fine tuning of multi-peaked response areas to Doppler-shifted harmonics in the echo (for echolocation) and to near-perfect harmonics in calls (for communication) is likely guided by the sounds an animal is exposed to and their functional importance during critical periods of its development (Knudsen, 1985). This neuronal response feature can be stabilized by associative (Hebbian) learning and is postulated to be an important mechanism for the perception of pitch, a presumably universal feature of auditory perception in mammals, including humans (Medvedev *et al.*, 2002). The "motivation" to extract pitch, which contains information about the identity, size and/or emotional state of the emitter might have been so strong that natural selection has endowed humans and many animal species the capability of perceiving a virtual pitch (i.e. a fundamental frequency that by itself is absent in the sound and is perceived only by the presence of its harmonics) (e.g. Pantev *et al.*, 1989; Preisler and Schmidt, 1995). In the AIp area of mustached bats, multidimensional scaling of the temporal pattern of call responses shows that each call type can be uniquely identified by its temporal pattern of response (Medvedev and Kanwal, 2004a). The segregation of call types is highly correlated with the harmonic complexity of calls, which includes the shape of a call's normalized power spectrum as well as its pitch.

Based on peak response magnitudes, the AIp area in mustached bats is fairly nonselective for call types compared to other cortical areas (see Fig. 7.2d). Part of the reason for a lack of call selectivity is a relative insensitivity to the direction of FMs, which is an acoustic feature present in many calls (Fig. 7.5c and d). Therefore, these neurons may contribute to the distinction of a call from other sounds. Recordings from mustached bats and mice indicate that multi-peaked frequency tuning emerges in the inferior colliculus (Mittman and Wenstrup 1995; Egorova *et al.*, 2001). In cats, neurons tuned to multiple frequencies are present in the dorsal field of AI (Sutter and Schreiner, 1991) and probably in other secondary auditory cortical areas. The comparably wide separation of the multiple BFs (BF_{low} and BF_{high}) in the tuning of auditory neurons in the mustached bat may result simply from the nearly one order of magnitude larger shift towards high frequencies in the fundamental frequency of its vocalizations and its audiogram compared to those of humans and many other mammalian species. For similar reasons, neural tuning to Doppler-shifted frequencies rather than to the exact harmonics in many cortical areas may be considered an additional specialization of neurons in the central auditory system of mustached bats. Since bats must attend to the fundamental of the pulse and the Doppler-shifted frequencies in the echo for extracting the relevant information to carry out

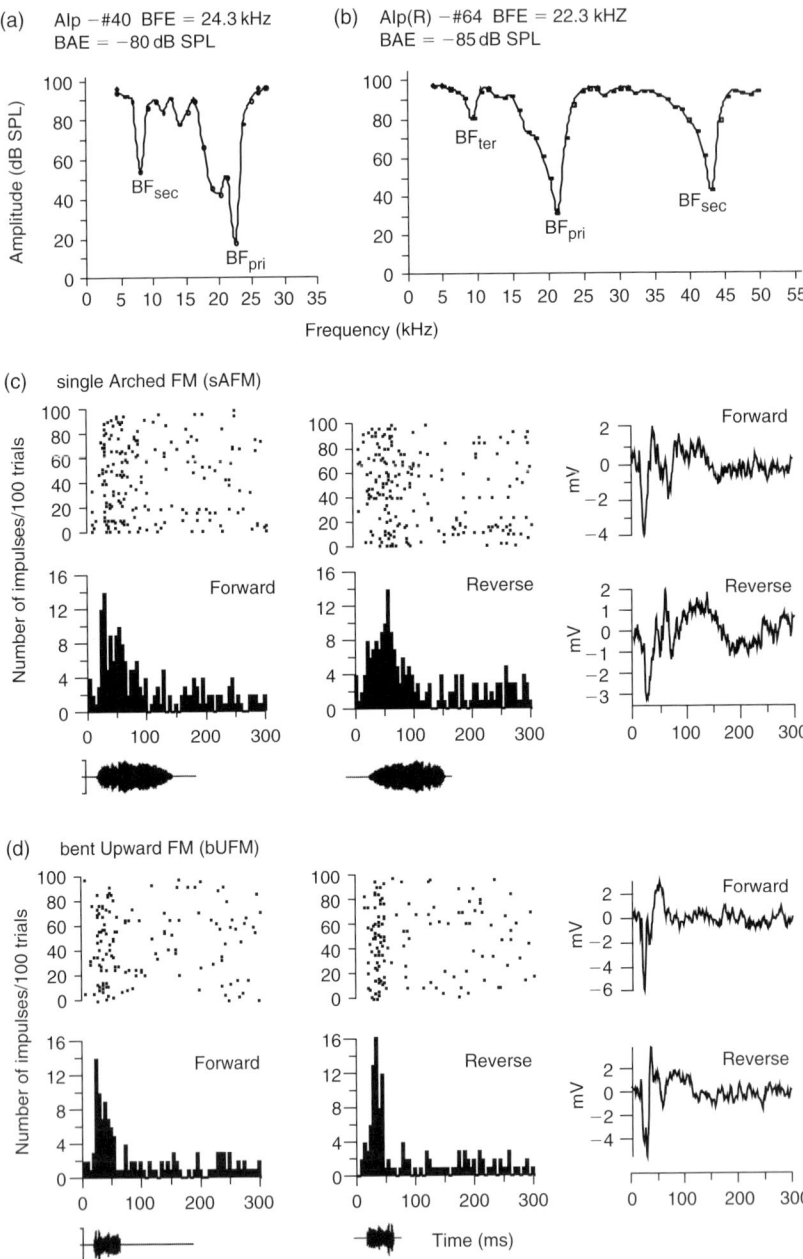

Figure 7.5. Frequency tuning and neural responses in the AIp area. (a and b) Two examples of neurons with a multi-peaked excitatory response area. Nearly all neurons in the AIp area exhibit this type of multi-peaked frequency tuning with a primary (BF_{pri}), a secondary (BF_{sec}) and, sometimes, a tertiary best frequency (BF_{ter}).

survival–critical auditory tasks, such as orientation, navigation and especially target tracking for prey capture, their cortical neurons become tuned to and/or can dynamically shift their tuning to off-harmonic combinations of frequencies. Developmental studies are needed to test whether this should be considered an evolutionary adaptation or an example of plasticity by learning. The same principle would explain the representation of formants in speech sounds in the human AC (i.e. the formant structure rather than the individual harmonics carry information about the identity of the sound). Whether the center frequency at peaks of energy in the sound spectrum is a pure tone or a relatively noisy frequency band is irrelevant. What is important is that the frequency peak of one formant is clearly separate from the frequency peaks of other formants so that the formants can be resolved by the critical band mechanism in the frequency domain. In bats that only use FM signals for orientation, etc., mapping of individual Doppler-shifted frequencies is not an efficient method to extract target velocity information because of the ambiguity in Doppler-shift created by multiple frequencies arriving within a very short time.

7.6 A hypothesis for the cortical representation of calls

7.6.1 Multiparametric distributed representations

The central theme of this chapter is the claim that complex communication sounds are represented in a distributed fashion in the AC of mammals. To test this hypothesis, neural responses to a set of complex communication sounds were obtained from different areas of the AC in separate sets of experiments. The same set of frequency-shifted variants of call types were presented at several intensities as stimuli for three specialized and one relatively unspecialized (AIp) area of the AC of mustached bats. The selectivity for FM (FMs and/or frequency sweeps), bandwidth and CF parameters determined the call preference of specialized neurons in the DSCF area of the primary AC, whereas FMs and intersyllabic silence intervals appeared to determine the response magnitudes and call selectivity of specialized neurons in the FM–FM area of the AC. Similarly, other spectral features of calls such as combinations of CFs in the high-frequency domain (>50 kHz) together with ~30 kHz frequencies triggered

Caption for fig. 7.5. (cont.)
Inhibitory areas were not mapped. (c) Single unit responses (raster and PSTH plots in left panel) and local field potentials (right panel) generated by two calls. There was no significant effect of call reversal on the temporal pattern and peak response in single unit (left versus right for PSTH and raster plots) and LFP (upper versus lower traces) responses.

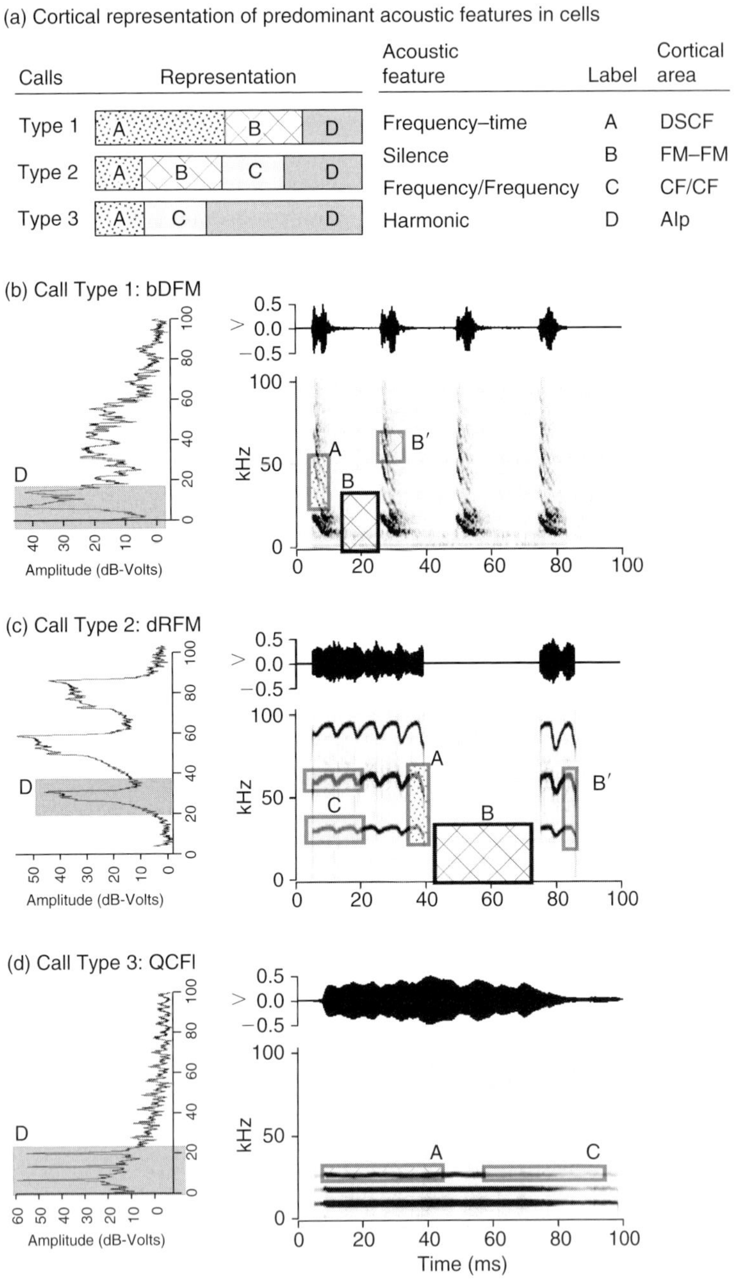

176   *7 A distributed cortical representation*

(a) Cortical representation of predominant acoustic features in cells

Calls	Representation	Acoustic feature	Label	Cortical area
Type 1	A B D	Frequency–time	A	DSCF
Type 2	A B C D	Silence	B	FM–FM
Type 3	A C D	Frequency/Frequency	C	CF/CF
		Harmonic	D	Alp

(b) Call Type 1: bDFM

(c) Call Type 2: dRFM

(d) Call Type 3: QCFI

Figure 7.6. (a) Scheme indicating that different call types consist of different combinations of acoustic features that are represented within different cortical areas. Left panel: acoustic features of three call types labeled as "A" (stippled), "B"

7.6 A hypothesis for the cortical representation of calls

activity in sharply tuned neurons in the CF/CF area. Harmonically complex frequency bands in the low-frequency range (<50 kHz) excited neurons with broad, multi-peaked frequency tuning in the AIp area. Accordingly, each call type seems to be dissected into common acoustic features as encoded within subcortical networks and represented within a combinatorial scheme in different cortical areas. The dominance of certain acoustic features is reflected in the relatively higher-peak response magnitudes or rates to a feature combination or call element in neurons in one area compared to the rest of the cortex. This concept is summarized for three hypothetical call types by the scheme presented in Fig. 7.6a. This global representational scheme is based largely on peak firing rates of AC neurons, which are presumed to transmit the bulk of information about sensory signals. According to this scheme, the coordinated firing of many neurons determines the perceived identity of a call type. Since every feature combination is distributed in a spatially segregated manner (i.e. within different cortical areas), such a map may be categorized as a "feature map". This map is made up of complex acoustic features that are distributed across the whole AC and together characterize a particular call type.

The coding scheme proposed in Fig. 7.6a is illustrated in Fig. 7.6b with examples of three call types. The bent downward FM (bDFM) call type consists of a multiharmonic, downward hyperbolic frequency sweep with a lower boundary of the fundamental at about 8 kHz. A stack of higher-harmonics results in an overall broadband signal that extends upwards to nearly 100 kHz with maximal energy at approximately 8 and 16 kHz. This call is always emitted as a train of four or more syllables with a fairly constant silent interval of about 15 ms. The number of syllables can be as many as 10–12. The spectrographic pattern in successive syllables is gradually modified. Figure 7.6b shows the power spectrum in the left panel aligned to the frequency axis of the spectrogram on the right with the amplitude contour above the latter. The upper-case letters and rectangles adjacent to them are positioned to highlight acoustic features of the calls that are encoded by cortical neurons residing in different cortical areas. The

Caption for fig. 7.6. (cont.)
(cross-hatched), "C" (light gray) and "D" (dark gray). The length of the marked area indicates the relative dominance of a feature in each call type. (B–D) Power spectrum (left) and spectrographic (right) plots illustrating the different acoustic features that are present within each of three different call types. The acoustic features are outlined by rectangles and labeled as "A", "B", "B'", "C" and "D". The frequency power spectrum of each call type is aligned to the frequency axis of its spectrogram on the right. Gray rectangles highlight the bandwidth in the power spectrum that is represented within a cortical area. The amplitude envelop (oscillogram) is shown on the top. The FM–FM area represents both frequency contours and temporal delays between syllables in a call.

stippled area "A" includes 25 through 61 kHz frequencies in the FM components whose harmonics span a range of ~8–100 kHz. The temporally overlapping 25 and 61 kHz frequencies are well suited to trigger activity in some DSCF neurons even though this does not represent the predominant frequency band in the call. Cross-hatched areas labeled as "B" and "B'" represent, respectively, the silent interval between the FM patterns in successive syllables that is encoded by FM–FM neurons. Finally, AIp neurons respond to the low-frequency peaks of high energy present at <25 kHz in the syllable – an acoustic domain that is not represented in any of the other cortical areas. This is illustrated by the gray area labeled as "D". Neurons in the CF/CF area do not respond well to this multisyllabic call (see also Fig. 7.1) as there are no distinct CF components in any of the syllables.

The descending Rippled FM (dRFM) call is acoustically similar to the multiharmonic echolocation signal with a fundamental of near 30 kHz (Kanwal *et al.*, 1994). The CF–FM echolocation pulse in mustached bats is followed by the echo with varying silent intervals between them. The dRFM call consists of a similar spectral structure, except that the CF is replaced by a rippled FM and the silence interval of >30 ms between syllables is nearly twice as long as the longest pulse–echo delays (~17 ms) perceived and mapped in the FM–FM area of mustached bats for echolocation and an order of magnitude longer than the short pulse–echo delays of <5 ms to which most of the FM–FM neurons are tuned. This call does not trigger a best response in most AIp neurons, but frequently is a best stimulus in the DSCF, FM–FM and CF/CF areas (see also Fig. 7.2). This is explained by the structural features highlighted in Fig. 7.6c, and labeled as "A", "B", "B'", "C" and "D". The two rectangles above and below "C" jointly represent an echolocation-like narrowband (approximating CF/CF) combination that activate neurons in the DSCF and CF/CF areas. The intersyllable silence interval labeled as "B" within the FM sweeps (B') triggers activity in FM–FM neurons. A peak of energy at low frequencies, labeled as "D" in the power spectrum, is present in the last descending limb of the FM fundamental (Figs. 7.3b and 7.6c). This peak overlaps with the excitatory frequency tuning of neurons in the AIp area. Therefore, when presented at high intensities (>85 dB SPL), this call is also relatively well represented in the AIp area.

Finally, hypothetical calls of type 3 in Fig. 7.6a are illustrated by the long quasi-CF (QCFl) call shown in Fig. 7.6d. This call of *P. parnellii* has frequencies largely below 30 kHz. Calls of this type are represented mainly in the AIp area. DSCF neurons that respond well to frequencies in the neighborhood of 25 kHz may also contribute to its cortical representation and hence its perception. FM–FM neurons do not contribute to the processing and perception of this call because it lacks features of silence gaps and FM patterns as well as

7.6 A hypothesis for the cortical representation of calls

high-frequency harmonics in most variants. Extreme upwardly frequency-shifted variants of QCFl calls and those that contain 4–5 harmonics can be well represented in the response of CF/CF neurons.

In summary, simultaneous activation of many neurons at multiple locations in the cortex can represent the identity of a call or syllable and possibly also the emitter's identity and emotional state (see Clement et al., Chapter 3, this volume). According to Fig. 7.6a, neurons in a given area of the AC can contribute to different percepts (the perception of different calls) in the functional context of social communication as well as echolocation (Kanwal et al., 2004). In other words, a set of neurons in a particular area encode acoustic features more uniquely than specific-call types. The simultaneous representation of different acoustic features is likely based on a parallel, hierarchical processing of communication sounds as has been shown for the case of echolocation signals (Suga, 1989). In the scheme presented here, an acoustic feature may be shared by several calls some of which may have very different perceived qualities. Every call may stimulate hundreds or thousands of neurons in several areas of the AC. For every call, these neurons representing "hot spots" of activity are considered to constitute "call constructs" in neural space in very much the same way as responses of neuronal clusters in different areas of the mustached bat AC may reconstruct from a series of pulse–echo pairs the auditory scene with information about the distance, size and shape and other unique characteristics of a target. Interestingly, the spatiotemporal representational scheme (Fig. 7.6a) for neural representation of calls parallels that demonstrated for encoding molecular features of odorants in the olfactory bulb (Shepherd, 1992a; Guthrei et al., 1993; Buck, 1996). This suggests that such a scheme may provide a universal solution to the problem of categorizing a large diversity of multidimensional stimuli in the natural environment – a problem that nearly all sensory systems have to solve if the organism is to behave appropriately and according to the biologic and environmental demands. In the olfactory bulb of mammals, such a representational scheme has been referred to as a "consensus" map (Shepherd, 1992b).

Our local field potential data from AIp neurons further suggest that complex sounds generate complex patterns of temporal activity at the recording sites (Medvedev and Kanwal, 2004a). These temporal patterns also contain oscillations that extend into the gamma frequency range (40–100 Hz). The exact role of these oscillations is unclear. They can be used, however, to segregate the mustached bat's communication calls into three groups with 4–5 call types per group, since each group is characterized by either two or one or no peak of gamma activity (Medvedev and Kanwal, 2004a; 2004b). In guinea pigs, optical recordings show that responses to vocalizations are represented by complex spatiotemporal patterns of activity (see Horikawa et al., Chapter 8, this volume). These spatiotemporal response patterns spread via three streams from the core (primary) to the belt areas of the AC. The concept of a spatial pattern of hot spots of activation has also been alluded to by Ehret (1997) and is based mainly on data of topographies of neuronal response characteristics in cats and time-coordinated firing of groups of neurons (Eggermont, 1994; deCharms and Merzenich, 1996). All of these ideas purport a dynamical framework for encoding complex sounds within AI.

One way to consider these dynamics is to postulate that the mapping of calls and possibly other "auditory objects" in the AC is fuzzy both in the spatial and time domains. Higher-order cortical areas can act as feedback controllers that accept noisy, imprecise input. This mechanism would enhance and improve the adaptability of the recognition system. This idea is formalized in fuzzy logic. Systems based on an implementation of this concept are being increasingly used in artificial intelligent-based task control systems (Golay *et al.*, 1998).

7.6.2 Representation versus perception

The ultimate goal of all sensory processing is perception and the coordination of goal-directed behaviors that involve either instantaneous action or adaptive behaviors via learning and memory. Knowledge of the neural representations of sensory stimuli can tell us something about how perception may be created by the neural machinery. For vertebrates, we can assume that representations in higher brain centers are dynamic and adaptive rather than static and fixed. To examine such representations, it is useful to combine studies of sensory stimulation with those of perception and behavior. This, however, is not easily accomplished because we have to study the activation of neurons to a set of sounds when an animal is performing a particular task. Monitoring immediate-early gene activation (e.g. of c-fos) in neurons in response to ultrasound elicited retrieving of pups in mice (Geissler and Ehret, 2004) and studying learning and memory in a task-directed paradigm either in primates who are free to move their arms or in freely moving songbirds with tethered electrodes are examples of two approaches where this can be accomplished (Margoliash *et al.*, 1994; Rauske *et al.*, 2003). Such approaches help us to understand how cortical representations of sensory stimuli, such as calls are shaped by learning and how higher-order cortical structures can modify the representations based on attention (Shea and Margoliash, 2003).

The approach taken in our studies is to present natural stimuli to awake animals in which the behavioral context is missing as the animal is physically constrained. We do, however, use a relatively good representative set of call stimuli that are presented at many sound pressure levels. Since we are dealing with sensory and perceptual representations in the AC, not within centers of motor control, we are confident that the distributed representations of call-specific neural activity that we observe reflect the basic format by which social calls are encoded in the brain. The type of multidimensional representation discussed above has the potential to easily "map" many and also very new stimuli on the same neural substrate (i.e. it does not constrain the neural machinery by an exhaustive mapping of all dimensions of a single call type). Instead it "maps" common acoustic features that may be shared by many different call types. Thus, we have an open-ended, flexible and adaptable representational scheme for all kinds of sound signals, which is totally different from the idea that each sound or call is represented as a percept best or only at a specific locus in the AC. It is also different from the idea that a sound is represented by the sequential activation or coordinated activity of all neurons whose excitatory frequency tuning includes one or more frequencies contained in the spectrogram of a complex sound (Kilgaard and Merzenich, 1999; Nagarajan et al., 2001).

The multifocal and distributed spatiotemporal representational scheme proposed here may be sharpened both with experience and when an animal is involved in various dedicated tasks that require attention and awareness. In other tasks, that are more automatic and driven by the motivational status of the animal, a cortical representation may not be directly engaged as these behaviors are probably driven by the highly conserved limbic areas of the brain. Many of these limbic structures, such as the amygdala and hippocampus, are well connected to the cortex, thalamus, hypothalamus and reticular and motor nuclei in the brainstem (Price, 2003). These limbic structures are likely governed by an independent set of rules for perception (see Ehret, Chapter 4, this volume) and may act as extracortical feedback controllers of the boundaries and nature of the cortical representation of natural stimuli. However, because there is a strong reciprocal coupling of cortical areas with respective thalamic areas which convey motivational information, motivations certainly reach cortical representations and may modify them.

Echolocation is a sensory-motor task with many non-cognitive, reflexive components, such as orientation, pursuit and capture of prey. A major cognitive (i.e. learned), component is the discrimination between prey and non-prey. Other uses of echolocation include exploration of the physical environment, awareness and discrimination of conspecifics and other bat species. For example, bats may change their echolocation pulse parameters for 1–2 pulses to "inspect" objects; they can start an approach sequence and then abort it in the middle, suggesting cognitive processing (Ulanovsky et al., 2004a). They also learn the location of objects in their everyday flight path for foraging. Bats can transfer echolocation information about their space into memory per the experiments of bats flying through wires (Griffin, 1958). Sometimes, bats may fly into unexpected objects as when taking a familiar path to their home cave after foraging, which suggests that they rely on memory transfered from echolocation rather than automated echolocation behavior, which would detect those objects (Kanwal, personal observation). For most of these tasks, echolocation information must be very well defined and secure compared to the less well-defined representation of complex acoustic features within social communication calls that vary from individual to individual. For both sets of stimuli, however, the auditory system extracts the basic stimulus parameters in a parallel–hierarchical manner (Suga, 1989; Kanwal *et al.*, 2004), and our data show that many of the common parameters of complex acoustic features are represented within the same cortical areas and neurons, though with different levels of precision.

7.7 Conclusions

Our data from the AC of the mustached bat show that various types of social communication calls are represented within a consensus map in the AC such that different combinations of acoustic parameters in the calls trigger activity of different sets of neurons mapped according to their response selectivities in specialized areas of the AC. The orderly organization of acoustic parameters within its auditory cortical fields is obviously necessary to support a very fast and efficient echolocation behavior in mustached bats. This organization of specialized auditory cortical fields has enabled me to propose a well-defined framework for call representation via a "mapping" of acoustic features in calls. According to this framework, neural activity distributed across different cortical areas and determined by complex acoustic features may constitute the functional unit for mapping a call type within the cortex of mustached bats. A similar strategy may be employed by many other species with the caveat that neurons responsive to different acoustic features may be less well segregated than in mustached bats and, therefore, may be more difficult to identify. It is also possible that basic acoustic parameters, such as pitch and bandwidth, are systematically mapped within different spatial and temporal axes of the cortex so that calls become represented within the spatiotemporal activity in overlapping populations of neurons. In the absence of specialized neurons, this latter scheme would result in a lot of ambiguity. In either case, a neural consensus signals either the identity of the call type or the mood (emotional/motivational status) and/or the size of the emitter. In humans, this could translate into the recognition of either a speech sound or of pitch/prosody in the speech stream. More work is clearly needed to further define and test this hypothesis in bats and other mammalian species.

Acknowledgments

I thank Farouk Muradali for collecting the bats and the Departments of Agriculture and the Natural Resource Conservation Authority of Jamaica and the Department for Wildlife in Trinidad for permission to export bats. Drs. N. Suga, K.K. Ohlemiller, K.-H. Esser, J.P. Peng and A.V. Medvedev collaborated for the acquisition of data on call responses in mustached bat's AC. I thank Dr. G. Ehret for a critical review that helped to improve the manuscript. Zhicheng Lai assisted with software development for the analysis of some of the data. This work was supported in part by a grant DC02054 to J.S.K. from the National Institutes of Health.

REFERENCES

Buck LB (1996) Information coding in the vertebrate olfactory system. *Annu Rev Neurosci* 19: 517–544.

Dear SP, Simmons JA, Fritz J (1993) A possible neuronal basis for representation of acoustic scenes in auditory cortex of the big brown bat. *Nature* 364: 620–623.

deCharms RC, Merzenich MM (1996) Primary cortical respresentation of sounds by the coordination of action potential timing. *Nature* 381: 610–613.

de Ribaupierre F (1997) Acoustical information processing in the auditory thalamus and cerebral cortex. In: Ehret G, Romand R (eds), *The Central Auditory System*, pp. 317–397. Oxford University Press, New York.

Doupe AJ, Kuhl PK (1999) Birdsong and human speech: common themes and mechanisms. *Ann Rev Neurosci* 22: 567–631.

Eggermont JJ (1994) Neural interaction in cat primary auditory cortex. II. Effects of sound stimulation. *J Neurophysiol* 71: 246–270.

Egorova M, Ehret G, Vartanian I, Esser KH (2001) Frequency response areas of neurons in the mouse inferior colliculus. I. Threshold and tuning characteristics. *Exp Brain Res* 140: 145–161.

Ehret G (1987) Left hemisphere advantage in the mouse brain for recognizing ultrasonic communication calls. *Nature* 325: 249–251.

Ehret G (1997) The auditory cortex. *J Comp Physiol A* 181: 547–557.

Ehret G, Moffat AJM (1985) Inferior colliculus of the house mouse. III. Response probabilities and thresholds of single units to synthesized mouse calls compared to tone and noise bursts. *J Comp Physiol A* 156: 637–644.

Esser KH, Condon CJ, Suga N, Kanwal JS (1997) Syntax processing by auditory cortical neurons in the FM–FM area of the mustached bat *Pteronotus parnellii*. *Proc Natl Acad Sci USA* 94: 14019–14024.

Fenton MB, Jacobs DS, Richardson EJ, Taylor PJ, White E (2004) Individual signatures in the frequency-modulated sweep calls of African large-eared, free-tailed bats *Otomops martiensseni* (Chiroptera: Molossidae). *J Zool* 262: 11–19.

Fitzpatrick DC, Batra R, Stanford TR, Kuwada S (1997) A neuronal population code for sound localization. *Nature* 388: 871–874.

Fitzpatrick DC, Olsen JF, Suga N (1998) Connections among functional areas in the mustached bat auditory cortex. *J Comp Neurol* 391: 366–396.

Fuzessery ZM, Feng AS (1983) Mating call selectivity in the thalamus and midbrain of the leopard frog (*Rana p. pipiens*): single and multiunit analyses. *J Comp Physiol A* 150: 333–344.

Fuzessery ZM, Wenstrup JJ, Pollak GD (1985) A representation of horizontal sound location in the inferior colliculus of the mustache bat (*Pteronotus p. parnellii*). *Hear Res* 20: 85–89.

Geissler DB, Ehret G (2004) Auditory perception vs. recognition: representation of complex communication sounds in the mouse auditory cortical fields. *Eur J Neurosci* 19: 1027–1040.

Golay X, Kollias S, Stoll G, Meier D, Valavanis A, Boesiger P (1998) A new correlation-based fuzzy logic clustering algorithm for fMRI. *Magn Reson Med* 40: 249–260.

Griffin DR (1958) Success and Failure of Bats' Acoustic Orientation. In: Listening in the dark, pp. 148–167. Yale University Press, Inc.

Griffin DR, Webster FA, Michael CR (1960) The echolocation of flying insects by bats. *Anim Behav* 141–154.

Guthrie KM, Anderson AJ, Leon M, Gall C (1993) Odor-induced increases in c-fos mRNA expression reveal an anatomical "unit" for odor processing in olfactory bulb. *Proc Natl Acad Sci USA* 90: 3329–3333.

Hernnberger B, Kempf S, Ehret G (2002) Basic maps in the auditory midbrain. *Biol Cybern* 87: 231–240.

Hofstetter KM, Ehret G (1992) The auditory cortex of the mouse: connections of the ultrasonic field. *J Comp Neurol* 323: 370–386.

Husain FT, Tagamets M-A, Fromm SJ, Braun AR, Horwitz B (2004) Relating neuronal dynamics for auditory object processing to neuroimaging activity: a computational modeling and an fMRI study. *Neuroimage* 21: 1701–1720.

Imig TJ, Reale RA, Brugge JF, Morel A, Adrian HO (1986) Topography of cortico-cortical connections related to tonotopic and binaural maps of cat auditory cortex. In: *Two Hemispheres-One Brain: Functions of the Corpus Callosum*, pp. 103–115. Alan R. Liss, Inc.

Jenkins WM, Merzenich MM (1984) Role of cat primary auditory cortex for sound-localization behavior. *J Neurophysiol* 52: 819–847.

Kaas JH, Hackett TA (1998) Subdivisions of auditory cortex and levels of processing in primates. *Audiol Neurootol* 3: 73–85.

Kanwal JS (1999) Processing species-specific calls by combination-sensitive neurons in an echolocating bat. In: Hauser MD, Konishi M (eds), *The Design of Animal Communication*, pp. 135–157. The MIT Press, Cambridge, MA.

Kanwal JS, Suga N (1995) Left hemispheric specializations for processing species-specific communication sounds in the primary auditory cortex of mustached bats. In: *Proceedings of the 4th International Congress of Neuroethology*. Burrows M, Matheson T, Newland PL, Schuppe H (eds), *Nervous Systems and Behavior*, p. 304. Georg Thieme Verlag, Stuttgart, Germany.

Kanwal JS, Ohlemiller KK, Suga N (1992) Selective representation of call-syllables in the DSCF area in the primary auditory cortex of the mustached bat. *Soc Neurosci* 18: 884.

Kanwal JS, Matsumura S, Ohlemiller K, Suga N (1994) Analysis of acoustic elements and syntax in communication sounds emitted by mustached bats. *J Acoust Soc Am* 96: 1229–1254.

Kanwal JS, Fitzpatrick DC, Suga N (1999) Facilitatory and inhibitory frequency tuning of combination-sensitive neurons in the primary auditory cortex of mustached bats. *J Neurophysiol* 82: 2327–2345.

Kanwal JS, Medvedev AV, Micheyl C (2003) Neurodynamics for auditory stream segregation: tracking sounds in the mustached bat's natural environment. *Network* 14: 413–435.

Kanwal JS, Peng JP, Esser K-H (2004) Vocal communication and echolocation in the mustached bat: computing dual functions within single neurons. In: Thomas JA, Vater M, Moss CJ (eds), *Echolocation in Bats and Dolphins*, pp. 201–214. University of Chicago Press, Chicago.

Karklin Y, Lewicki MS (2003) Learning higher-order structures in natural images. *Network* 14: 483–499.

Kilgard MP, Merzenich MM (1999) Distributed representation of spectral and temporal information in rat primary auditory cortex. *Hear Res* 134: 16–28.

Knudsen EI (1985) Experience alters the spatial tuning of auditory units in the optic tectum during a sensitive period in the barn owl. *J Neurosci* 5: 3094–3109.

Konishi M, Knudsen EI (1982) A theory of neural auditory space. In: *Cortical Sensory Organization, Vol. 3., Multiple Auditory Areas*, pp. 219–229.

Langner G, Sams M, Heil P, Schulze H (1997) Frequency and periodicity are represented in orthogonal maps in the human auditory cortex: evidence from magnetoencephalography. *J Comp Physiol A* 181: 665–676.

Lewicki MS (2002) Efficient coding of natural sounds. *Nat Neurosci* 5: 356–363.

Liu W, Suga N (1997) Binaural and commissural organization of the primary auditory cortex of the mustached bat. *J Comp Physiol A* 181: 599–605.

Machens CK, Wehr MS, Zador AM (2004) Linearity of cortical receptive fields measured with natural sounds. *J Neurosci* 24: 1089–1100.

Manabe T, Suga N, Ostwald J (1978) Aural representation in the Doppler-shifted-CF processing area of the auditory cortex of the mustache bat. *Science* 200: 339–342.

Margoliash D, Fortune ES (1992) Temporal and harmonic combination-sensitive neurons in the zebra finch's HVc. *J Neurosci* 4309–4326.

Margoliash D, Fortune ES, Sutter ML, Yu AC, Wren-Hardin BD, Dave A (1994) Distributed representation in the song system of oscines: evolutionary implications and functional consequences. *Brain Behav Evol* 247–264.

Marler P, Peters S (1981) Birdsong and speech: evidence for special processing. In: Eimas PD (ed.), *Perspectives on the Study of Speech*, pp. 75–113. Lawrence Erlbaum Associates, Publishers, Hillsdale.

Medvedev AV, Kanwal JS (2004a) Local field potentials and spiking activity in the primary auditory cortex in response to social calls. *J Neurophysiol* 92: 52–65.

Medvedev AV, Kanwal JS (2004b) Evoked gamma activity is sensitive to the spectrotemporal structure of communication sounds. Congress of the Intl. Soc. for Neuroethology 7: PO41.

Medvedev AV, Chiao F, Kanwal JS (2002) Modeling complex tone perception: grouping harmonics with combination-sensitive neurons. *Biol Cybern* 86: 497–505.

Middlebrooks JC, Clock AE, Xu L, Green DM (1994) A panoramic code for sound location by cortical neurons. *Science* 264: 842–844.

Mittmann DH, Wenstrup JJ (1995) Combination-sensitive neurons in the inferior colliculus. *Hear Res* 90: 185–191.

Nagarajan SS, Cheung SW, Bedenbaugh P, Beitel RE, Schreiner CE, Merzenich MM (2002) Representation of spectral and temporal envelope of twitter vocalizations in common marmoset primary auditory cortex. *J Neurophysiol* 87: 1723–1737.

Nelken I (2004) Processing of complex stimuli and natural scenes in the auditory cortex. *Curr Opin Neurobiol* 14: 474–480.

Nelken I, Prut Y, Vaadia E, Abeles M (1994) Population responses to multifrequency sounds in the cat auditory cortex: one- and two-parameter families of sounds. *Hear Res* 72: 206–222.

Nelken I, Fishbach A, Las L, Ulanovsky N, Farkas D (2003) Primary auditory cortex of cats: feature detection or something else? *Biol Cybern* 89: 397–406.

O'Neill WE, Suga N (1982) Encoding of target range and its representation in the auditory cortex of the mustached bat. *J Neurosci* 2: 17–31.

Ohlemiller K, Kanwal JS, Butman JA, Suga N (1994) Stimulus design for auditory neuroethology: synthesis and manipulation of complex communication sounds. *Audit Neurosci* 1: 19–37.

Ohlemiller KK, Kanwal JS, Suga N (1996) Facilitative responses to species-specific calls in cortical FM–FM neurons of the mustached bat. *Neuroreport* 7: 1749–1755.

Pantev C, Hoke M, Lutkenhoner B, Lehnertz K (1989) Tonotopic organization of the auditory cortex: pitch versus frequency representation. *Science* 246: 486–488.

Portfors CV, Wenstrup JJ (2001) Responses to combinations of tones in the nuclei of the lateral lemniscus. *J Assoc Res Otolaryngol* 2: 104–117.

Portfors CV, Wenstrup JJ (2002) Excitatory and facilitatory frequency response areas in the inferior colliculus of the mustached bat. *Hear Res* 168: 131–138.

Preisler A, Schmidt S (1995) Virtual pitch formation in the ultrasonic range. *Naturwissenschaften* 82: 45–47.

Price JL (2003) Comparative aspects of amygdala connectivity. *Ann N Y Acad Sci* 985: 50–58.

Radtke-Schuller S (2001) Neuroarchitecture of the auditory cortex in the rufous horseshoe bat *(Rhinolophus rouxi)*. *Anat Embryol (Berl)* 204: 81–100.

Radtke-Schuller S (2004) Cytoarchitecture of the medial geniculate body and thalamic projections to the auditory cortex in the rufous horseshoe bat (*Rhinolophus rouxi*). I. Temporal fields. *Anat Embryol*, 209: 59–76.

Radtke-Schuller S, Schuller G, O'Neill WE (2004) Thalamic projections to the auditory cortex in the rufous horseshoe bat (*Rhinolophus rouxi*) II. Dorsal fields. *Anat Embryol*, 209: 77–91.

Rainey HJ, Zuberbuhler K, Slater PJ (2004) Hornbills can distinguish between primate alarm calls. *Proc R Soc Lond B Biol Sci* 271: 755–759.

Rajan R, Aitkin LM, Irvine DRF (1990) Azimuthal sensitivity of neurons in primary auditory cortex of cats. I. Types of sensitivity and the effects of variations in stimulus parameters. *J Neurophysiol* 64: 872–887.

Rauschecker JP (1997) Processing of complex sounds in the auditory cortex of cat, monkey, and man. *Acta Otolaryngol (Stockh)* 34–38.

Rauschecker JP, Tian B, Hauser M (1995) Processing of complex sounds in the macaque nonprimary auditory cortex. *Science* 268: 111–114.

Rauske PL, Shea SD, Margoliash D (2003) State and neuronal class-dependent reconfiguration in the avian song system. *J Neurophysiol* 89: 1688–1701.

Read HL, Winer JA, Schreiner CE (2002) Functional architecture of auditory cortex. *Curr Opin Neurobiol* 12: 433–440.

Recanzone GH, Guard DC, Phan ML, Su TK (2000) Correlation between the activity of single auditory cortical neurons and sound-localization behavior in the macaque monkey. *J Neurophysiol* 83: 2723–2739.

Rouiller EM (1997) Functional organization of the auditory pathways. In: Ehret G, Romand R (eds), *The Central Auditory System*, pp. 3–96. Oxford University Press, New York.

Sachs MB, Abbas PJ (1974) Rate versus level functions for auditory-nerve fibers in cats: tone-burst stimuli. *J Acoust Soc Am* 56: 1835–1847.

Schreiner CE (1995) Order and disorder in auditory cortical maps. *Curr Opin Neurobiol* 4: 489–496.

Schreiner CE, Mendelson JR (1990) Functional topography of cat primary auditory cortex: distribution of integrated excitation. *J Neurophysiol* 64: 1442–1459.

Schreiner CE, Sutter ML (1992) Topography of excitatory bandwidth in cat primary auditory cortex: single-neuron versus multiple-neuron recordings. *J Neurophysiol* 68: 1487–1502.

Schreiner CE, Mendelson JR, Sutter ML (1992) Functional topography of cat primary auditory cortex: representation of tone intensity. *Exp Brain Res* 92: 105–122.

Schulze H, Hess A, Ohl FW, Scheich H (2002) Superposition of horseshoe-like periodicity and linear tonotopic maps in auditory cortex of the Mongolian gerbil. *Eur J Neurosci* 15: 1077–1084.

Scott JW, Wellis DP, Riggott MJ, Buonviso N (1993) Functional organization of the main olfactory bulb. *Microsc Res Tech* 24: 142–156.

Searcy YM, Caine NG (2003) Hawk calls elicit alarm and defensive reactions in captive Geoffroy's marmosets (*Callithrix geoffroyi*). *Folia Primatol (Basel)* 74: 115–125.

Shamma SA, Fleshman JW, Wiser PR, Versnel H (1993) Organization of response areas in ferret primary auditory cortex. *J Neurophysiol* 69: 367–383.

Shannon-Hartman S, Wong D, Maekawa M (1992) Processing of pure-tone and FM stimuli in the auditory cortex of the FM bat, *Myotis lucifugus*. *Hear Res* 61: 179–188.

Shea SD, Margoliash D (2003) Basal forebrain cholinergic modulation of auditory activity in the zebra finch song system. *Neuron* 40: 1213–1226.

Shepherd GM (1992a) Neurobiology: modules for molecules. *Nature* 358: 457–458.

Shepherd GM (1992b) Toward a consensus working model for olfactory transduction. *Soc Gen Physiol Ser* 47: 19–37.

Simmons JA (1969) Acoustic radiation patterns for the echolocating bats, *Chilonycteris rubigenosa* and *Eptiscus fuscus*. *J Acoust Soc Am* 44: 1054–1056.

Stiebler I, Neulist R, Fichtel I, Ehret G (1997) The auditory cortex of the house mouse: left–right differences, tonotopic organization and quantitative analysis of frequency representation. *J Comp Physiol A* 181: 559–571.

Suga N (1984) The extent to which biosonar information is represented in the bat auditory cortex. In: Edelman GM, Gall WE, Cowan WM (eds), *Dynamic Aspects of Neocortical Function*, pp. 315–373. Wiley, New York.

Suga N (1988) Auditory neuroethology and speech processing: complex-sound processing by combination-sensitive neurons. In: Edelman GM, Gall WE, Cowan WM (eds), *Auditory Function: Neurobiological Bases of Hearing*, pp. 679–720, John Wiley & Sons, Inc., New York.

Suga N (1989) Principles of auditory information-processing derived from neuroethology. *J Exp Biol* 146: 277–286.

Suga N (1995) Sharpening of frequency tuning by inhibition in the central auditory system: tribute to Yasuji Katsuki. *Neurosci Res* 21: 287–299.

Suga N, Jen PH (1976) Disproportionate tonotopic representation for processing CF–FM sonar signals in the mustache bat auditory cortex. *Science* 194: 542–544.

Suga N, O'Neill WE, Manabe T (1978) Cortical neurons sensitive to combinations of information-bearing elements of biosonar signals in the mustache bat. *Science* 200: 778–781.

Suga N, Niwa H, Taniguchi I, Margoliash D (1987) The personalized auditory cortex of the mustached bat: adaptation for echolocation. *J Neurophysiol* 58: 643–654.

Šuta D, Kvašňák E, Popelář J, Syka J (2003) Representation of species-specific vocalizations in the inferior colliculus of the guinea pig. *J Neurophysiol* 90: 3794–3808.

Sutter ML, Schreiner CE (1991) Physiology and topography of neurons with multipeaked tuning curves in cat primary auditory cortex. *J Neurophysiol* 65: 1207–1226.

Sutter ML, Schreiner CE (1995) Topography of intensity tuning in cat primary auditory cortex: single-neuron versus multiple-neuron recordings. *J Neurophysiol* 73: 190–204.

Sutter ML, Schreiner CE, McLean M, O'connor KN, Loftus WC (1999) Organization of inhibitory frequency receptive fields in cat primary auditory cortex. *J Neurophysiol* 82: 2358–2371.

Thomas H, Tillein J, Heil P, Scheich H (1993) Functional organization of auditory cortex in the Mongolian gerbil (*Meriones unguigulatus*). I. Electrophysiological mapping of frequency representation and distinction of fields. *Eur J Neurosci* 5: 882–897.

Tian B, Rauschecker JP (1998) Processing of frequency-modulated sounds in the cat's posterior auditory field. *J Neurophysiol* 79: 2629–2642.

Ulanovsky N, Fenton MB, Tsoar A, Korine C (2004a) Dynamics of jamming avoidance in echolocating bats. *Proc Biol Sci.* 271: 1467–1475.

Ulanovsky N, Las L, Farkas D, Nelken I (2004b) Multiple time scales of adaptation in auditory cortex neurons. *J Neurosci* 24: 10440–10453.

Wang X, Merzenich MM, Beitel R, Schreiner CE (1995) Representation of a species-specific vocalization in the primary auditory cortex of the common marmoset: temporal and spectral characteristics. *J Neurophysiol* 74: 2685–2706.

Wenstrup JJ (1999) Frequency organization and responses to complex sounds in the medial geniculate body of the mustached bat. *J Neurophysiol* 82: 2528–2544.

Winer JA, Wenstrup JJ, Larue DT (1992) Patterns of GABAergic immunoreactivity define subdivisions of the mustached bat's medial geniculate body. *J Comp Neurol* 319: 172–190.

Wong D, Shannon SL (1988) Functional zones in the auditory cortex of the echolocating bat, *Myotis lucifugus*. *Brain Res* 453: 349–352.

8
Spatiotemporal Processing in the Guinea Pig Auditory Cortex

Junsei Horikawa, Andreas Hess, Yutaka Hosokawa and Ikuo Taniguchi

8.1 Introduction

With the development of optical recording, it became possible to monitor spatiotemporal patterns of neural activity *in vivo*. Using fast voltage-sensitive dyes, the developments of neural responses can be directly observed as real-time or as slowly played-back images. This method is most advantageous as we can study how sensory information is represented and processed by means of activity patterns of the neural population of the cortex. This article describes optical recording of the responses to various sounds in guinea pig auditory cortex using voltage-sensitive dyes. The results show that the primary auditory cortex (AI) represents sound information as spatiotemporal activity patterns. Optical recording also shows distinct activity propagations from AI to the higher auditory fields (belt areas). We discuss how auditory information is represented and processed spatiotemporally in AI, and how representation in the higher auditory fields is different from that in AI.

8.2 Optical recording methods

Methods of the optical recording technique using voltage-sensitive dyes have been described in many studies (Orbach *et al.*, 1985; Blasdel and Salama, 1986; Cohen and Lesher, 1986; Grinvald *et al.*, 1986; Kamino *et al.*, 1989; Bonhoeffer and Grinvald, 1991), so they are only briefly outlined here. Specific details are given in our previous papers (Taniguchi *et al.*, 1992; Horikawa *et al.*, 1996, 1998). The experiments described here were conducted in accordance with the provisions of the animal care committees of Tokyo Medical and Dental University and Toyohashi University of Technology. Guinea pigs were anesthetized with Nembutal combined with a neuroleptic. The auditory cortex was exposed and

190   8 *Spatiotemporal processing in the guinea pig auditory cortex*

stained with a voltage-sensitive dye RH795 for 1 to 1.5 h. It was epi-illuminated by a light (480–580 nm) from a 150 W halogen lamp and fluorescent signals (>620 nm) emitted from the cortex were detected with a 12 × 12 photodiode array (PDA). The signals were sampled with 16 bit A/D converters at 0.576 ms/frame (Fig. 8.1). The recording area was a 3 mm square and thus each pixel of the PDA received signals from a 250 μm square cortical area as shown in the right panel of Fig. 8.1. The recordings with and without stimulation synchronized with the electrocardiogram and with the artificial respiration were made to cancel the noise caused by pulsation (Fig. 8.1 bottom). The recording was averaged over five trials. In order to record over the whole auditory cortex

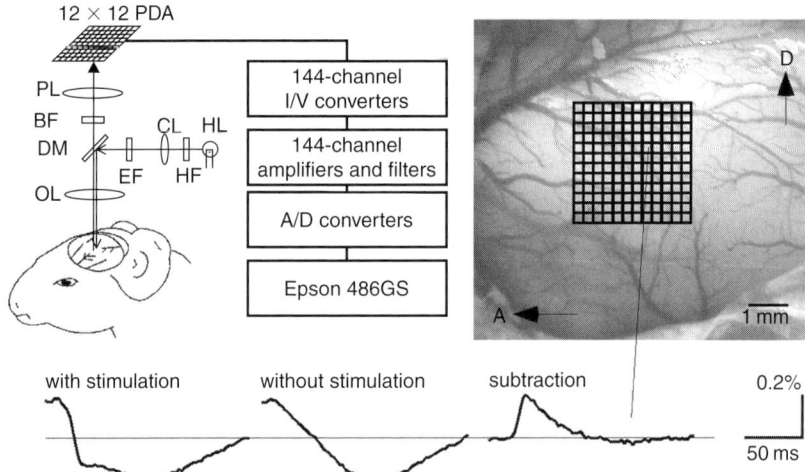

Figure 8.1. Left panel: The auditory cortex is epi-illuminated by a 150 W halogen lamp (HL) through a heat filter (HF), collective lens (CL), excitation filter (EF), dichroic mirror (DM) and objective lens (OL). The optical signal is projected to a 12 × 12 channel PDA via OL, DM, a barrier filter (BF) and a projection lens (PL). The optical signals are I/V converted, amplified, filtered and sampled by 16 bit A/D converters at 0.576 ms/frame using a personal computer. Right panel: The recording location is superimposed on a surface picture of the auditory cortex of a guinea pig. The PDA records from a 3 mm cortical square and each pixel records from a 250-μm cortical square. A: anterior; D: dorsal. The traces at the bottom show optical signals with and without tone-burst stimulation (14 kHz, 100 ms duration, 10 ms linear ramps, 75 dB SPL) recorded from a photodiode cell (indicated by a thin line), and the response signal obtained after subtraction between them. The response signal was inverted because the optical intensity decreased with a depolarizing potential. Percentage at the vertical bar means the ratio of $\Delta F/F$, where ΔF is the change in intensity and F is the background intensity of the fluorescence. The horizontal thin line indicates the level at rest.

with the high spatial resolution noted above, recordings were made from multiple locations and a composite response map was created by combining the individual recording areas in an overlapping fashion.

8.3 Tonotopic organization in auditory cortical fields

Tonotopic organization is a well-known property of mammalian auditory cortex. In the guinea pig auditory cortex, microelectrode studies have shown tonotopic organization in the core fields consisting of AI and the dorsocaudal field (DC), (after Redies *et al.*, 1989 and Wallace *et al.*, 1999, 2000; DC is also called the area II (AII) by Hellweg *et al.*, 1977). In AI, the lower frequencies are represented rostrally and the higher frequencies caudally, in DC they show a mirror-image tonotopic organization with a common high frequency border of the two fields.

Tonotopic organization has also been revealed by optical recording techniques (Taniguchi *et al.*, 1992; Uno *et al.*, 1993; Horikawa *et al.*, 1996, 1998, 2001). Optically recorded tonotopic organization in AI and DC is shown in Fig. 8.2. The individual areas responding to 2, 4, 8, 12 and 16 kHz are clipped at 60% of the maximal response for each stimulus and superimposed. There is a clear frequency gradient from anterior-low to posterior-high in AI, and a reverse frequency gradient in DC. These frequency arrangements are essentially the same as those observed in electrophysiological studies (Hellweg *et al.*, 1977; Redies *et al.*, 1989; Tanaka *et al.*, 1994; Wallace *et al.*, 1999, 2000).

However, there were some differences between the results obtained by microelectrodes and optical recording. In the optical maps, overlaps of the areas

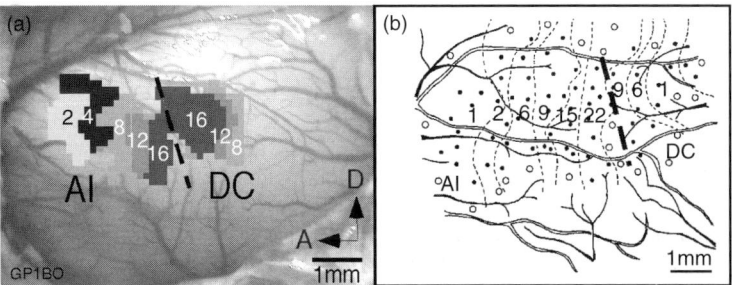

Figure 8.2. Tonotopic maps of the primary (AI) and the DC auditory cortex of guinea pigs obtained from optical recording (left panel, a). The areas responding to 2, 4, 8, 12 and 16 kHz are superimposed. The response area to each stimulation frequency was clipped at 60% of their peak activity. The numbers in the figures indicate frequencies in kHz. Right panel (b) shows a tonotopic map obtained by microelectrode recordings (modified from figures of Yamaguchi *et al.*, 2001 and Tanaka *et al.*, 1994).

responding to different frequencies were observed, while electrophysiological maps with spatially sparse sampling revealed only contours of characteristic frequencies (CFs). These electrophysiological maps represent the CFs at which the stimulus level is the lowest to elicit a response. The overlaps in the optical map partly result from the suprathreshold stimulus levels used, at which many neurons with different CF's respond to the stimulus. Another reason for the overlaps seen in the optical maps is that optical recordings may show signals from wide spread dendritic arborization (Grinvald, 1994), which has been shown to extend horizontally up to 300–500 μm in the cat auditory cortex (Ojima et al., 1991).

Using our new multiple-field optical recording scheme, the activity over wide auditory areas can be visualized (Fig. 8.3). One result of this new approach is that

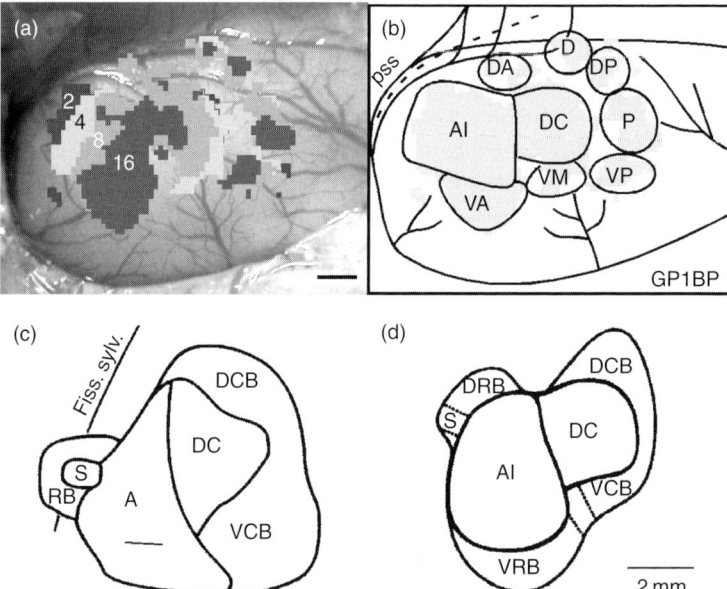

Figure 8.3. Tonotopic map of multiple auditory fields obtained by optical recording using a multiple fields recording scheme. (a) a composite response map for 2 (black), 4 (light gray), 8 (gray) and 16 kHz (dark gray) pure ton stimulations (50 ms duration, 10 ms linear ramps, 75 dB SPL) (b) multiple auditory fields obtained from the response patterns in (a). (c) and (d) multiple auditory fields obtained by microelectrode studies. AI: primary auditory field; DC: dorsocaudal auditory core field; D: dorsal field; DA: dorsoanterior field; DP: dorsoposterior field; P: posterior field; VA: ventroanterior field; VM: ventromedial field; VP: ventroposterior field; A: anterior field; DC: dorsocaudal field; DCB: dorsocaudal belt; DRB: dorsorostral belt; VCB: ventrocaudal belt; VRB ventrorostral belt; RB: rostral belt; S: small field. pss: pseudosylvian sulcus; Fiss. sylv.: sylvian fissure. ((a) and (b) from Horikawa et al., 2001; (c) from Redies et al., 1989; (d) from Rutkowski et al., 2000).

several auditory fields are discernible in the belt areas surrounding the core fields of AI and DC. These fields can be defined based on their tonotopic organization, and response latency and duration. They are the dorsoanterior (DA), dorsal (D), dorsoposterior (DP), posterior (P), ventroposterior (VP), ventromedial (VM), ventroanterior (VA) and ventral (V) fields (Horikawa et al., 2001). The shapes and sizes, and even the existence of these fields differed among individuals; AI, DC, D, DP, P, VA, VP were observed in all animals, however VM and V in 75% and DA in just 35% of animals. Tonotopic organization was observed in all areas apart from DA, D, DP and V, although in the higher fields it was not so clear as in AI. In fields D and DP, the tonotopic organization might have concentric components, as seen in the gerbil auditory cortex (Tillein et al., 1993). The overlap of the response bands was slightly larger in DC than in AI, and was much larger in the belt areas. This increase of response overlap suggests an increase in frequency integration in the higher fields, especially in the belt areas. The latency of the belt areas was 6–13 ms longer than that of the core fields, indicating that the belt areas are the higher-order fields in auditory information processing.

Figure 8.3b–d show a comparison between the multiple auditory fields in guinea pigs obtained by optical and electrophysiological recordings. The electrophysiologically observed dorsocaudal belt area (DCB) was further divided into D and DP, and the ventrocaudal belt area (VCB) into VM and VP, based on the optical data. The ventrorostral belt area (VRB) corresponded to VA, and the dorsorostral belt area (DRB) to DA. Field S was not observed in optical recording, the most likely reason being its position in the pseudosylvian sulcus and below large vessels making it difficult to obtain optical signals.

8.4 Spatiotemporal representation of constant and frequency-modulated tones

A new finding obtained by optical recording is that the response to a tone of constant frequency systematically propagates along a frequency band. Figure 8.4 shows an example of such a propagation of a response to a 16 kHz tone of 50 ms duration (10 ms linear ramps). The response mainly propagated ventrally along the isofrequency band, in the shown case (Fig. 8.4) by 1.6 mm in about 20 ms. If a lower frequency and/or a high intensity tone was used for stimulation, the response additionally spread across the higher frequency bands.

How are time-varying tones represented in AI? To study this, we used frequency-modulated (FM) tones sweeping from 4 to 16 kHz (ascending FM: FMa) or from 16 to 4 kHz (descending FM: FMd) (Horikawa et al., 1998). The response to FMa first appeared at the 4 kHz frequency band, and then a spot-like

194   8 *Spatiotemporal processing in the guinea pig auditory cortex*

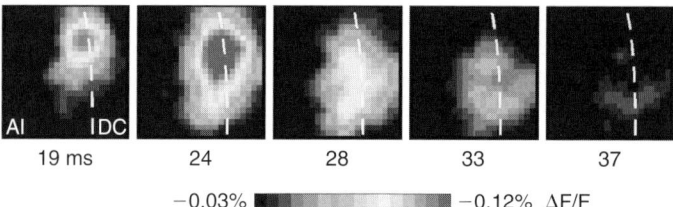

Figure 8.4. Propagation of the response to the onset of a pure tone (16 kHz, 75 dB SPL, 50 ms duration, 10 ms linear ramps) along isofrequency bands. The same phasic response was observed for longer tones (not shown). Numbers indicate time after the onset of the stimulus. Broken lines indicate the border between AI and DC (from Yamaguchi *et al.*, 2001).

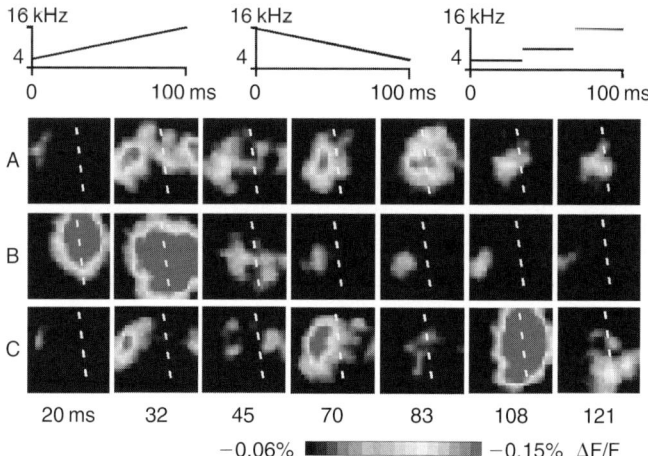

Figure 8.5. Responses to ascending FM tones (4–16 kHz, A), to descending FM tones (16–4 kHz, B) and to sound with a stepwise frequency change (SF: 4–8–16 kHz, C). Schematic sonograms of the stimuli are shown at the top. Duration of the stimuli is 100 ms. Numbers indicate time after the onset of the sounds in milliseconds. Dotted lines: the border between AI (left) and DC (right) (modified from Horikawa *et al.*, 1998).

response moved toward the border between AI and DC where the higher frequencies are represented (Fig. 8.5). The response to FMd was vise versa; it appeared first at the 16 kHz band and then moved toward the 4 kHz band. The response moved in the same period of time as the frequency sweep although it was delayed 20 ms due to response latency, for example if the sweep duration was 50 ms, it moved during 50 ms. This temporal behavior was the same for FMd except for the direction of the response movement. These results indicate that AI represents FM sounds in a spatiotemporal, that is spectrotemporal, manner.

Are responses to FM tones different from responses to a sound with stepwise frequency changes, for example from 4 to 8 and 16 kHz? The responses to the latter appeared band-like at the 4, 8 and 16 kHz bands. The responses were the same as that to each frequency alone. The difference between the response to FM sounds and that to pure tones was that the FM response in a given frequency band was spot-like rather than band-like. These shaping differences could be achieved by inhibitory interactions. Without considering inhibition, a band-like response changing its location in correspondence to instantaneous frequencies in the FM should be deduced from the constant-tone responses. But this was not the case and the existence of inhibitory interactions was confirmed by experiments applying an inhibitory receptor blocker (bicuculline) to the cortex (Horikawa *et al.*, 1996). The inhibitory areas were found to be circumscribing or sandwiching an isofrequency band.

8.5 Responses in AI to vocalized sounds

Considering spatial dynamics and excitatory–inhibitory interactions, responses in AI to vocalized sounds are expected to be complicated because such sounds often consist of complex spectral components and temporal modulations. As an example, responses to a mating call of a male guinea pig consisting of several ascending FM harmonics is shown in Fig. 8.6. The response changed spatially and temporally as the frequencies of the components changed over time. Strong responses seemed to occur in the frequency bands corresponding to prominent instantaneous frequency components. However, the responses were too complex to be analyzed in detail. Therefore, we measured responses to a synthesized sound with two FM harmonics that imitated temporal modulation of the vocalizations. The results showed that inhibitory interactions occurred between the areas corresponding to instantaneous and previous frequencies of the two FM components. In the response to vocalizations, such interactions are expected to occur between the many frequency components simultaneously.

8.6 Functional significance of the spatiotemporal representation of sounds

Based on the results obtained with pure tones, FM and vocalized sounds, we hypothesized that AI represents sounds by spatiotemporal activity patterns,

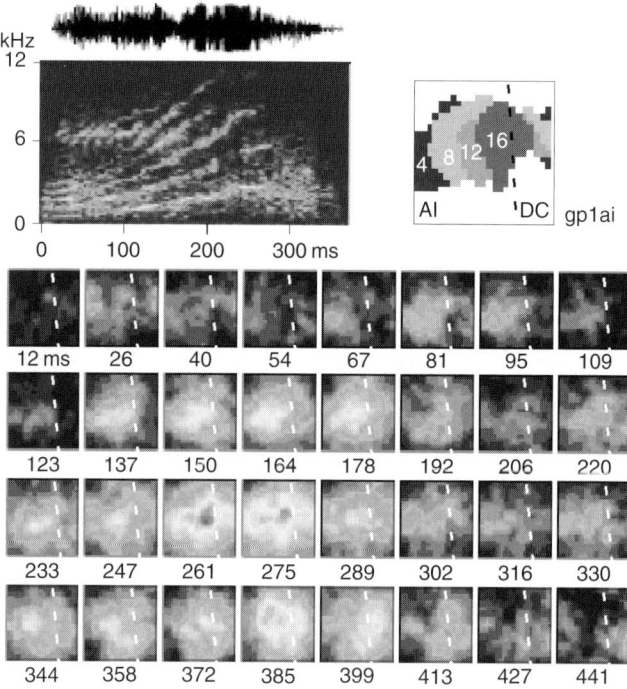

Figure 8.6. Responses to a guinea pig vocalization. Upper left: waveform and sonogram. Upper right: tonotopic organization of AI and DC. Lower panels: response to the vocalization. Numbers indicate time after the onset of the sound in milliseconds.

though it is not known what roles these patterns play in AI with regard to sound analysis, representation and perception. To evaluate neural mechanisms behind spatiotemporal activity patterns, a neural network model using neural oscillators representing localized neural populations was established (Yamaguchi et al., 2001). It is only briefly reviewed here. In this model, a neural oscillator represented mutually connected excitatory and inhibitory neural populations at a local position of AI. The oscillators with excitatory and inhibitory interconnections were arranged in a two-dimensional manner. This network model could reconstruct the fundamental characteristics of response wave propagation in AI. Hence, the model underlined the important contribution of the above described lateral inhibition phenomena to locate and shape active sites in AI. The model further suggested a novel way of feature binding and articulation in auditory recognition (Yamaguchi and Taniguchi, 1996; Yamaguchi, 1997; Kanzaka et al., 1999).

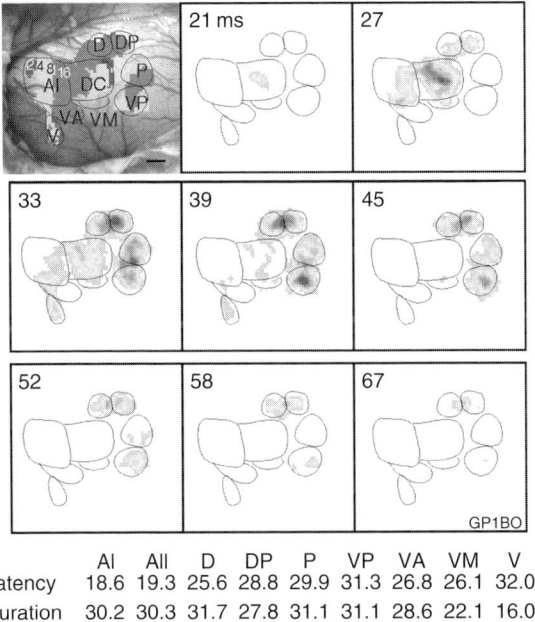

	AI	AII	D	DP	P	VP	VA	VM	V
Latency	18.6	19.3	25.6	28.8	29.9	31.3	26.8	26.1	32.0
Duration	30.2	30.3	31.7	27.8	31.1	31.1	28.6	22.1	16.0

ms n = 11 for AI–VA, 9 for VM and 8 for V

Figure 8.7. Activity developments in auditory fields in response to a pure tone (8 kHz, 75 dB SPL, 50 ms duration, 10 ms linear ramps). Upper left: auditory fields shown by a composite map of responses to 2, 4, 8 and 16 kHz. The outlines of the fields are superimposed on each response map. Numbers indicate time after the onset of the stimulus. Left side of each figure corresponds to the anterior of the animal (modified from Horikawa et al., 2001). Mean latencies and durations of the responses are also presented.

8.7 Activity spreads to the higher auditory fields

Optical recording has shown spread of neural activity from AI and DC to the higher auditory fields (Horikawa et al., 2001) which is shown in a time series of responses to 8 kHz stimulation (Fig. 8.7). The response first appeared in the core fields (AI and DC), then in fields D and DP, next in VA and V, and last in P and VP. The response remained for long in fields D, DP, P and VP. The mean latencies and durations of the responses in the fields obtained from 11 animals are shown in the table of Fig. 8.7. Animation of the response helped to see the dynamic patterns of the response as it spread from the core fields to the surrounding belt fields via mainly three distinct pathways: the dorsocaudal, ventrorostral and caudal pathways (Fig. 8.8). Anatomical studies in guinea pigs have confirmed that AI and DC are interconnected and receive direct

198   8 *Spatiotemporal processing in the guinea pig auditory cortex*

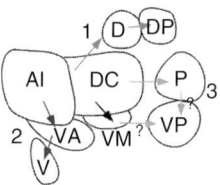

Figure 8.8. Schematic drawing of directions of response spreads from AI and DC to surrounding belt areas. 1: dorsocaudal pathway; 2: ventrorostral pathway; 3: caudal pathway.

afferent inputs from the medial geniculate nuclei (MGN, Redies *et al.*, 1989). They also showed reciprocal corticocortical projections between auditory core and belt fields (Wallace *et al.*, 2002), as observed in cats (Imig and Reale, 1980), monkeys (Morel and Kaas, 1992; Morel *et al.*, 1993) and Mongolian gerbils (Budinger *et al.*, 2000a). The long latency in the belt fields and the principal component analysis (PCA) (see section 9 below) suggest that the activity propagation is caused via corticocortical connections between the core and belt fields. However, DP and VP areas located caudal to AI in Mongolian gerbils, and the cortical region lateral to AI in macaque monkeys, were reported to receive sets of projections from dorsal (MGd) and ventral nuclei of MGN (MGv) different from those of AI (Budinger *et al.*, 2000b; Morel *et al.*, 1993). Hence, latency differences in the activation of auditory cortical fields by a sound may reflect latency differences present already at the MGN level.

8.8 Functional differences of the multiple auditory fields

Our results indicate that the higher auditory fields seem not to represent frequency information of the sounds in distinct spatiotemporal activity patterns as found in AI. In part this may be explained by our finding that frequency integration in the belt fields, observed by overlapping response bands to different frequencies, is larger than that in the core fields. Multiple auditory fields were found in many species of mammals, but functional differences were found in only few. In the mustached bat auditory cortex, Suga and coworkers found multiple fields of clearly different functions in representation of echolocation calls. Higher auditory areas such as the CF/CF and FM–FM areas processed different biosonar information such as the speed and the range of the target (Suga, 1982). Neurons in these areas responded to combinations of frequency components in the emitted pulse and the echo that carried biologically important

8.8 Functional differences of the multiple auditory fields

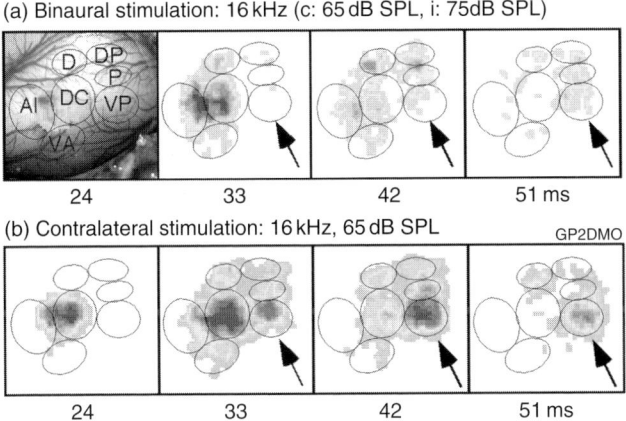

Figure 8.9. Optically recorded responses to binaural and contralateral 16 kHz stimulation. The binaural responses in AI, DC, D, DP and VA are similar to, in P and PV different from the contralateral responses. The stimulus is presented dichotically using two earphones connected to hollow ear bars. Numbers indicate time after the onset of the stimulus (16 kHz, contralateral 65 dB SPL, ipsilateral 55 dB SPL, 50 ms duration, 10 ms linear ramps). Arrows indicate the responses in P and PV. Abbreviations same as in Fig. 8.3.

(information bearing) parameters for the bat (Suga, 1982; Horikawa and Suga, 1991). In monkeys, belt areas responded differentially from the core fields to noise bands, vocalizations, and sound locations (Rauschecker and Tian, 2000; Tian et al., 2001). Neurons in the anterolateral (AL) of the lateral belt areas were most selective for monkey calls, while those in the caudolateral area (CL) were highly selective for spatial locations of sounds in a free field. Some neurons in the belt areas responded to temporal combinations of different vocalization components.

First studies on functional differences of the multiple fields of the guinea pig auditory cortex showed special sensitivity to broad-band stimuli in the fields DA to AI and around DC (Wallace et al., 2000) and a different binaural sensitivity compared to AI and DC in fields P and VP (Hosokawa et al., 2001, 2004). Figure 8.9 demonstrates the strong inhibition of responses in fields P and VP to binaural as compared with contralateral monaural stimulation. These fields were also sensitive to the location of the sound source (Hosokawa et al., 2001). Thus, fields P and VP seem to be related to sound location analysis. Studies of the posterior belt fields in the monkey auditory cortex indicated sensitivity to sound locations and suggested these fields to be part of the auditory "where" pathway (Rauscheker and Tian, 2000; Tian et al., 2001). This is encouraging to suggest a homology in the functional specifications of belt fields between monkeys and guinea pigs.

8.9 Principal component analysis applied to optical signals

As mentioned above, response characteristics such as representation of frequency, latency and duration differ among the fields in the guinea pig auditory cortex. To study differences in response characteristics other than the just mentioned ones, we applied the PCA to the optical signals. PCA has been used to find orthogonal coordinate systems in data sets where the coordinates represent independent variables of the data set. The mathematical procedures of PCA are described for example in the text book by Murtagh and Heck (1997).

PCA as we applied it to the optical data maximally decorrelates the temporal response functions of the different recording locations. The principal components were obtained as eigenvectors onto which the projection of the data indicated different degrees of variance. Back projections of the principal components into the image domain enabled the search for specific spatial distributions of cortical response properties and their correlations between the different auditory fields.

The analysis clearly showed field specific spatial response differences, especially between the core and their surrounding belt fields. Fig. 8.10a shows the eigenvectors (temporal response functions) of the first four principal components C1, C2, C3 and C4, for a 12 kHz pure tone stimulus. Time (ms) is plotted on the abscissa, variance on the ordinate. A high variance indicates an independent phenomenon occurring at that time. The peak of each component occurred at a different time. C4 was the earliest followed by C1, C2, and C3. Figure 8.10b shows the corresponding maps of C1, C2, C3 and C4 (12 kHz tone stimulus). The core fields AI and DC showed the highest variance in the data set, which strongly determined C1. The time course of the activity was principally the same in AI and DC. This means that the temporal response characteristics of AI and DC were very similar. Further, C2 was mainly determined by activity in the dorsal belt areas. This indicates that the response characteristics of the belt areas were different and independent of those of AI and DC. The variance of C3 was small in the ventral and caudal fields. This suggests that these fields were already well explained by C1 and C2, implying that the response characteristics in the dorsal belt were different from those in the ventral belt. C4 showed small spots of high variance in AI, DC and D. The early and transient response of C4 suggests that the inputs into the auditory cortex may be contained in this component.

8.9 *Principal component analysis applied to optical signals* 201

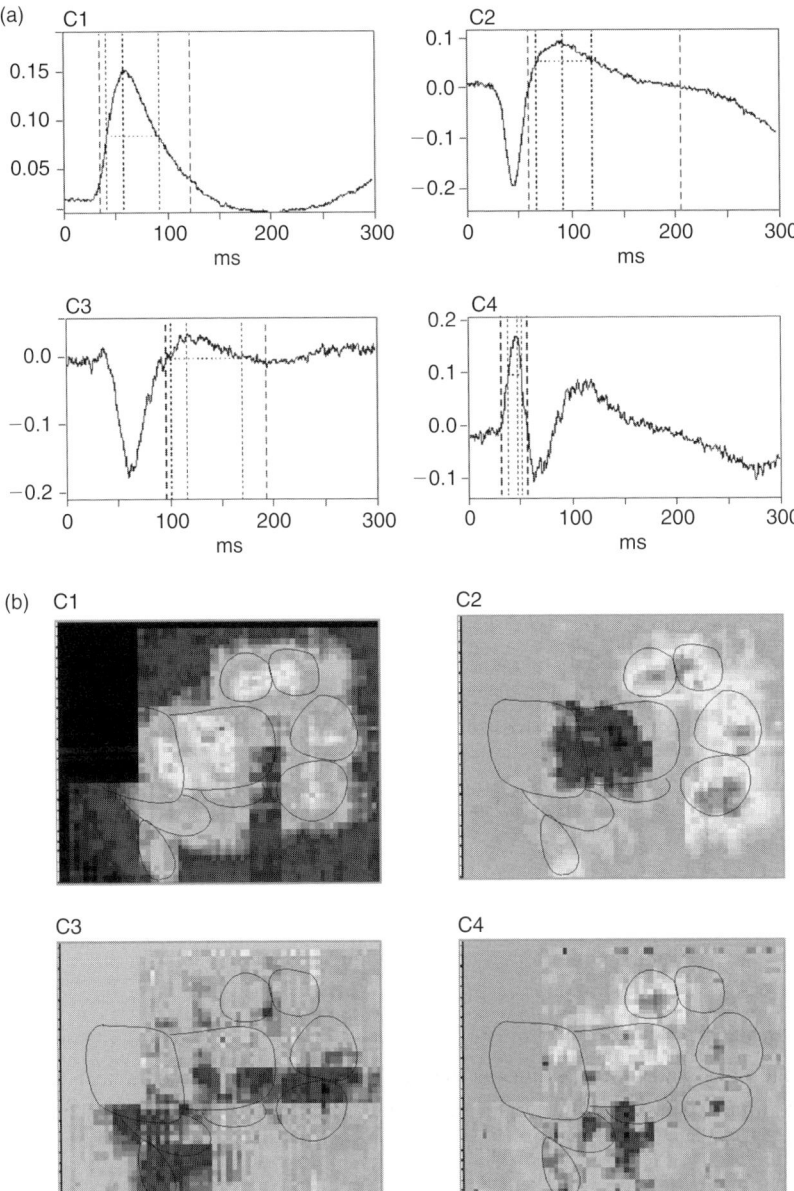

Figure 8.10. PCA of the response to 12 kHz stimulation (100 ms duration, 10 ms linear ramps, at 75 dB SPL). (a) The first four principal components C1, C2, C3 and C4. The abscissa indicates time in milliseconds and the ordinate the variance. (b) Maps of the first four principal components C1, C2, C3 and C4. The calculation was performed with the same data as in Fig. 8.7. The borders of the fields are superimposed on each map.

The advantages of the PCA analysis are the objective visualization of the differences in the temporal response properties for multiple fields, and the automatic separation of areas with different response characteristics. Although principal components are hard to interpret in terms of their physiological meaning, they may separate the inputs to the auditory cortex from other response properties that spread along the cortical fields, as has been shown and suggested above.

8.10 Conclusions

1. Optical recording clearly shows tonotopic organization in multiple fields of the guinea-pig auditory cortex.
2. AI may represent sounds as complex spatiotemporal (or spectrotemporal) response patterns.
3. Auditory information spreads from core fields AI and DC to the belt fields via three distinct (dorsocaudal, ventrorostral and caudal) pathways.
4. The belt fields, especially in the dorsal fields, show a larger frequency overlap, longer latency and longer duration of their activation.
5. Fields P and PV may be related to sound location analysis.
6. PCA analysis has shown differential temporal response characteristics between the core and the surrounding belt fields. This analysis also suggests a separation of the inputs to the auditory cortex from activities spreading along the cortex.

Acknowledgments

The work was supported by Grants-in-Aid for Scientific Research (13035019 and 13480096) and the 21st century COE Program "Intelligent Human Sensing" from the Ministry of Education, Science, Sports and Culture, Japan and The Mitsubishi Foundation.

REFERENCES

Blasdel GG, Salama G (1986) Voltage-sensitive dyes reveal a modular organization in monkey striate cortex. *Nature* 321: 579–585.

Bonhoeffer T, Grinvald A (1991) Iso-orientation domains in cat visual cortex are arranged in pinwheel-like patterns. *Nature* 353: 429–431.

Budinger E, Heil P, Scheich H (2000a) Functional organization of auditory cortex in the Mongolian gerbil (*Meriones unguiculatus*). III. Anatomical subdivisions and corticocortical connections. *Eur J Neurosci* 12: 2425–2451.

Budinger E, Heil P, Scheich H (2000b) Functional organization of auditory cortex in the Mongolian gerbil (*Meriones unguiculatus*). IV. Connections with anatomically characterized subcortical structures. *Eur J Neurosci* 12: 2452–2474.

Cohen LB, Lesher S (1986) Optical monitoring of membrane potential: methods of multisite optical measurement. In: de Weer P, Salzberg MM (eds), *Optical Methods in Cell Physiology*, pp. 71–99. John Wiley & Sons, New York.

Grinvald A, Segal M, Kuhnt U, Hildesheim R, Manker A, Anglister L, Freeman JA (1986) Real-time optical mapping of neuronal activity in vertebrate CNS *in vitro* and *in vivo*. In: de Weer P, Salzberg MM (eds), *Optical Methods in Cell Physiology*, pp. 165–197. John Wiley & Sons, New York.

Grinvald A, Lieke EE, Frostig RD, Hildesheim R (1994) Cortical point-spread function and long-range lateral interactions revealed by real-time optical imaging of macaque monkey primary visual cortex. *J Neurosci* 14: 2545–2568.

Hellweg FC, Koch R, Vollrath M (1977) Representation of the cochlea in the neocortex of guinea pigs. *Exp Brain Res* 29: 467–474.

Horikawa J, Suga N (1991) Neuroethology of auditory cortex. *Jpn J Physiol* 41: 671–691.

Horikawa J, Hosokawa Y, Kubota M, Nasu M, Taniguchi I (1996) Optical imaging of spatiotemporal patterns of glutamatergic excitation and GABAergic inhibition in the guinea-pig auditory cortex *in vivo*. *J Physiol Lond* 497: 629–638.

Horikawa J, Nasu M, Taniguchi I (1998) Optical recording of responses to frequency-modulated sounds in the auditory cortex. *NeuroReport* 9: 799–802.

Horikawa J, Hess A, Nasu M, Hosokawa Y, Scheich H, Taniguchi I (2001) Optical imaging of neural activity in multiple auditory cortical fields of guinea pigs. *NeuroReport* 15: 3335–3339.

Hosokawa Y, Horikawa J, Nasu M, Taniguchi I (2001) Optical imaging of azimuthal activities in multi-fields of the guinea pig auditory cortex. *Adv Ethol* 36: 180.

Hosokawa Y, Sugimoto S, Kubota M, Taniguchi I, Horikawa J (2004) Optical imaging of binaural interaction in multiple fields of the guinea pig auditory cortex. *NeuroReport* 15: 1093–1097.

Imig TJ, Reale RA (1980) Patterns of cortico-cortical connections related to tonotopic maps in cat auditory cortex. *J Comp Neurol* 192: 293–332.

Kamino K, Katoh Y, Komuro H, Sato K (1989) Multiple-site optical monitoring of neural activity evoked by vagus nerve stimulation in the embryonic chick brain stem. *J Physiol Lond* 409: 263–283.

Kanzaka Y, Yamaguchi Y, Horikawa J, Taniguchi I (1999) Analyses of feature binding on the auditory neural network based on wave sonagram hypothesis. *Trans Tech Comm Psychol Physiol Acoust* H-99-35: 1–8.

Morel A, Kaas JH (1992) Subdivisions and connections of auditory cortex in owl monkeys. *J Comp Neurol* 318: 27–63.

Morel A, Garraghty PE, Kaas JH (1993) Tonotopic organization, architectonic fields, and connections of auditory cortex in macaque monkeys. *J Comp Neurol* 335: 437–459.

Murtagh F, Heck A (1997) *Multivariate Data Analysis*. pp. 33–48. Kluwer, Dordrecht,

Ojima H, Honda CN, Jones EG (1991) Patterns of axon collateralization of identified supragranular pyramidal neurons in the cat auditory cortex. *Cereb Cortex* 1: 80–94.

Orbach HS, Cohen LB, Grinvald A (1985) Optical monitoring of neuron activity in the mammalian sensory cortex. *J Neurosci* 5: 1886–1895.

Rauschecker JP, Tian B (2000) Mechanisms and streams for processing of what and where in auditory cortex. *Proc Natl Acad Sci USA* 97: 11800–11806.

Redies H, Sieben U, Creutzfeldt OD (1989) Functional subdivisions in the auditory cortex of the guinea pig. *J Comp Neurol* 282: 473–488.

Suga N (1982) Functional organization of the auditory cortex, representation beyond tonotopy in the bat. In: Woolsey CN, Clifton, NJ (eds), *Cortical Sensory Organization*, pp. 157–218. Humana Press, Madison, WI.

Tanaka H, Komatsuzaki A, Taniguchi I (1994) Spatial distribution of response latency in the anterior field of the auditory cortex of the guinea pig. *Audiol Jpn* 37: 222–228.

Taniguchi I, Horikawa J, Moriyama T, Nasu M (1992) Spatio-temporal pattern of frequency representation in the auditory cortex of guinea pigs. *Neurosci Lett* 146: 37–40.

Tian B, Reser D, Durham A, Kustov A, Rauschecker JP (2001) Functional specialization in rhesus monkey auditory cortex. *Science* 292: 290–293.

Uno H, Murai N, Fukunishi K (1993) The tonotopic representation in the auditory cortex of the guinea pig with optical recording. *Neurosci Lett* 150: 179–182.

Wallace MN, Rutkowski RG, Palmer AR (1999) A ventrorostral belt is adjacent to the guinea pig primary auditory cortex. *NeuroReport* 10: 2095–2099.

Wallace MN, Rutkowski RG, Palmer AR (2000) Identification and localisation of auditory areas in guinea pig cortex. *Exp Brain Res* 132: 445–456.

Wallace MN, Rutkowski RG, Palmer AR (2002) Interconnections of auditory areas in the guinea pig neocortex. *Exp Brain Res* 143: 106–119.

Yamaguchi Y (1997) Synchronization in the brain. In: Nakamura R (ed.), *Unification of Information as the Emergence-Complexity and Diversity*, pp. 147–153. Springer-Verlag, Berlin.

Yamaguchi Y, Taniguchi I (1996) Neural activities of wave propagation in a neural network model of the auditory cortex. *IEICE Tech Rep NC95-141*: 197–204.

Yamaguchi Y, Horikawa J, Taniguchi I (2001) Neural dynamics of vocal processing in the auditory cortex. In: Poznanski RR (ed.), *Biophysical Neural Networks*, pp. 343–362. Mary Ann Liebert, New York.

9
Hierarchic Processing of Communication Sounds in Primates

Josef P. Rauschecker and Biao Tian

9.1 Auditory communication as a pattern recognition problem

Auditory communication involves the decoding of complex sounds and the assignment of specific sounds to behaviorally relevant meanings. While the latter problem has long been discussed in the realm of ethology, the former should be solvable more immediately by neurophysiology. But even the phonetic decoding problem is a difficult one, comparable in complexity with visual pattern recognition. Like in vision, it involves the combination of a multitude of features that occur simultaneously and in temporal sequence. The unique combination of features in the spectral and temporal domain, therefore, characterizes a specific communication sound. This alone, however, is not enough. The feature representation has to be robust and invariant against spurious changes and distortions caused by unpredictable influences.

The way such feature representations are generated in the brain is equally unclear. It could be that neurons in the auditory pathways become more and more specific for frequency, as one ascends the different levels of processing from the cochlea to auditory cortex and beyond. Such highly frequency-specific neurons could then be simultaneously active in concert when a complex sound, such as a communication sound, is present in the animal's environment. The opposite, however, is conceivable as well: neurons at higher levels of the auditory pathways could become more and more broadly tuned for frequency, as they combine more and more input across the frequency domain. As a consequence, they would become more and more specific for certain sounds and unresponsive to others.

It appears that both processes are in fact happening. While the tuning of neurons for frequency is still relatively broad at the level of the auditory nerve, it gets sharpened in the lemniscal pathways of brain stem and thalamus, with neurons in primary auditory cortex (A1) very narrowly tuned for frequency. As one travels beyond A1, however, the trend reverses dramatically. One of the discoveries of Merzenich and Brugge in their early mapping studies of rhesus monkey auditory cortex was that neurons in the lateral part of the supratemporal plane responded only poorly to pure tones and were, therefore, hard to characterize (Brugge and Merzenich, 1973; Merzenich and Brugge, 1973). Merzenich and Brugge's assumption was that the neurons in these lateral fields might respond to more complex sounds. As was discovered later, this is indeed the case (Rauschecker *et al.*, 1995): neurons in the lateral belt (LB) clearly prefer band-passed noise (BPN) bursts to tone pips and also respond well to other types of complex sounds, as we will review here.

It appears, therefore, that the coding problem of auditory pattern recognition does not end at the level of A1, it actually begins there. The realization of the existence of a multitude of auditory cortical representations (Merzenich and Brugge, 1973; Reale and Imig, 1980) on the one hand adds to the complexity of the problem but may at the same time be an important hint as to its solution. Could the various cortical fields be specialized for certain aspects of the auditory world and represent different levels of analysis in a hierarchic scheme? This is the answer suggested by research on visual pattern and object recognition over the last 25 years, in which it has become abundantly clear that different cortical areas do indeed specialize in certain aspects of the visual world. The specialization is, of course, not absolute – in fact, there exists substantial overlap – but there is clear evidence for processing streams that emphasize distinct aspects of our natural surroundings. In this sense, the ecologic adaptation of our brain during evolution is indeed a reality.

9.2 Early parallel processing in the auditory cortex

Parallel processing streams in the auditory cortex start as early as the core areas: areas A1 and R are both koniocortical areas with neurons sharply tuned for frequency and with tonotopic maps that are mirror-symmetric. Combined lesion and tracer studies (Rauschecker *et al.*, 1997) have shown that both cortical core areas receive input from the principal relay nucleus of the auditory thalamus, the ventral nucleus of the medial geniculate (MGv). By contrast, the other prominent area on the supratemporal plane, the caudomedial area (CM), does not receive input from MGv but only from the medial and dorsal subnuclei of the medial geniculate (MGd and MGm). As a consequence, lesions of A1 lead to unresponsiveness of neurons in CM to tonal stimulation, but not of neurons in area R, which receive independent input from MGv. To be sure, the parallel input to areas of the supratemporal plane may start even more peripherally than the thalamus. Studies of the auditory brain stem indicate that the ventral and dorsal cochlear nuclei have very different response characteristics and may subserve different functions of hearing, including auditory object and space processing, respectively (Yu and Young, 2000). Area CM, which contains large numbers of spatially tuned neurons (Rauschecker et al., 1997; Recanzone et al., 2000), could receive at least some of its input from the dorsal cochlear nucleus via the external nuclei of the inferior colliculus and the MGd (Rauschecker, 1997).

9.3 Processing of sounds with intermediate complexity

As has been recognized early on the basis of cytoarchitectonics (Pandya and Sanides, 1972), the auditory region in the superior temporal cortex consists of a "core" with a koniocortical appearance surrounded by a "belt". The cytoarchitecture is matched by distinct histochemical differences that make the core stand out by dark staining compared to the belt (with intermediate staining) and another zone termed "parabelt" (PB) (Morel et al., 1993; Hackett et al., 1998a) with very light staining (see schematic rendering in Fig. 9.1) (see also Jones et al., 1995). We will first deal with some processing characteristics of the belt and compare them with those of the core.

9.3.1 Selectivity for BPN

One fundamental finding that was secured in some initial studies (Rauschecker et al., 1995) and confirmed in detail later (Rauschecker and Tian, 2004), was the enhanced response of LB neurons to BPN compared to pure tones. This demonstrated the ability of LB neurons to integrate over a finite frequency spectrum in a facilitatory fashion (Fig. 9.2). By comparison, this integrative ability is largely absent in A1 neurons, a significant difference that we will return to later.

The finding of robust auditory responses in LB neurons to BPN stimuli also was of great practical value, however, because it permitted the systematic mapping of the LB. BPN bursts have a clearly defined center frequency as well as a defined bandwidth. Mapping of the LB along the rostro-caudal dimension reveals a smooth gradient for best center frequency with two reversals (Rauschecker et al., 1995; Rauschecker and Tian, 2004). This means that there exist at least three cochleotopically organized areas within the LB, which were termed antero-, middle-, and caudo-lateral (AL, ML, and CL, respectively; Fig. 9.1).

Not only do LB neurons integrate over frequency, they do so in a rather specific way, which produces the best response at a specific "best bandwidth" (BBW)

208   9 Hierarchic processing of communication sounds in primates

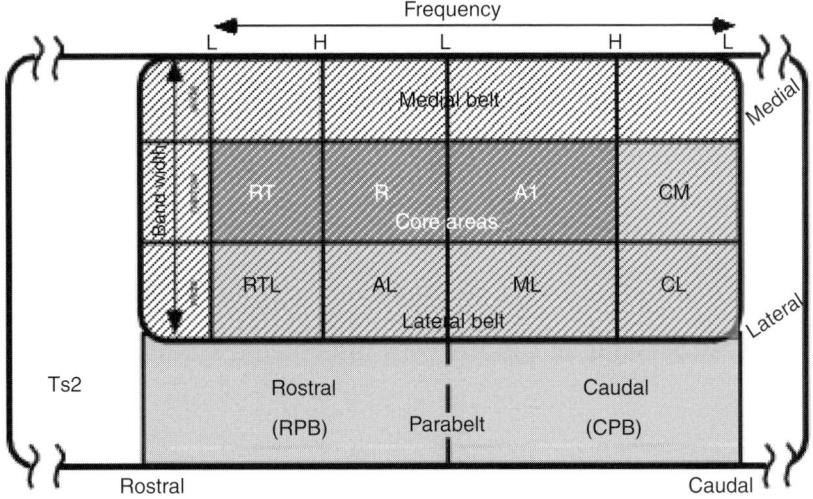

Figure 9.1. Schematic display of auditory cortical areas on the STG of the rhesus monkey with core areas (A1, R, and RT, respectively) in dark gray with lines, LB areas (AL, ML, CL, and RTL, respectively) in light gray with lines, and parabelt areas (RPB, CPB) in plain gray (modified from Rauschecker, 1998b) (data from Rauschecker et al., 1995; Hackett et al., 1998a). Ts2: anterior STG area (as cytoarchitectonically defined by Pandya and Sanides, 1972); L, H: low- and high-frequency representation, respectively; R: rostral area; RT: rostrotemporal area; RTL: rostrotemporal lateral area; RPB: rostral PB; CPB: caudal PB.

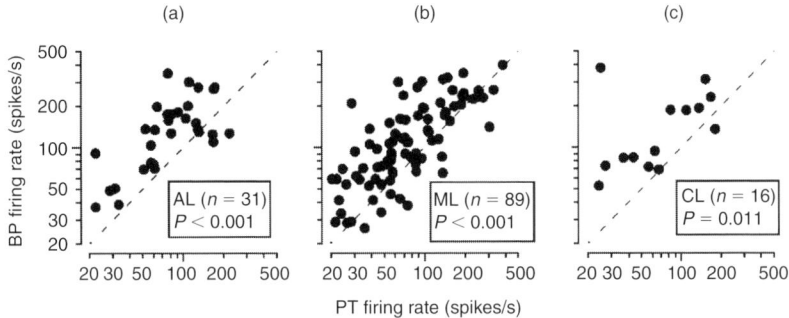

Figure 9.2. Scattergrams of neuronal responsiveness in the three LB areas (AL, ML, and CL, respectively) to BPN bursts versus tone pips. Almost invariably BP stimuli elicit a higher spike rate in the same neurons than PT stimuli (adapted from Rauschecker and Tian, 2004).

(Fig. 9.3). Presumably, this is the result of intricate interactions between excitatory and inhibitory inputs. BBWs in the LB are about equally distributed over the whole bandwidth spectrum, whereas A1 neurons clearly prefer pure tones to any BPN (Fig. 9.4). Thus, while BBW decreased in the medial direction towards the core regions, there was a clear trend for BBW to increase in the

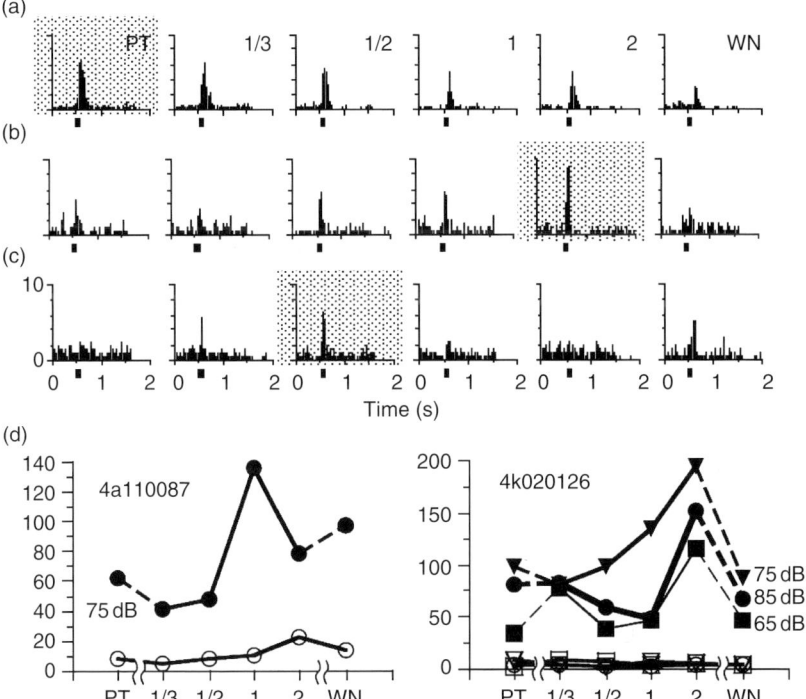

Figure 9.3. Responses (in spikes per second) of auditory cortical neurons to BPN bursts with different bandwidth. In (a) is shown a response typically found in core areas: the biggest response is found to pure tones (PT) with gradually declining responses as bandwidth increases (WN: white noise). (b) and (c) show typical responses of LB neurons: the response is most vigorous at intermediate bandwidths. In (d) two such responses are plotted diagrammatically as maximum firing rate against bandwidth. Filled symbols show net firing rate after subtracting spontaneous (base) firing rate (shown in open symbols) (data from Rauschecker and Tian, 2004).

lateral direction. It remains to be determined, however, whether the border between belt and PB is characterized by a sharp increase in BBW or whether the change is more gradual. This can only be determined by a precise correlation of histochemical staining and physiologic tuning in the same animals based on careful electrode track reconstructions.

It is tempting to conclude that neurons with selectivity for the center frequency and bandwidth of BPN bursts would be ideally suited to participate in the decoding of communication sounds. The latter contain many instances of BPN bursts (Fig. 9.5), not only in rhesus monkeys, but also in many other species (Wang, 2000), including humans. BPN detectors would, therefore, almost have to be included in the repertoire of feature detectors dealing with communication sounds. In order to perform such a task adequately, however, such feature

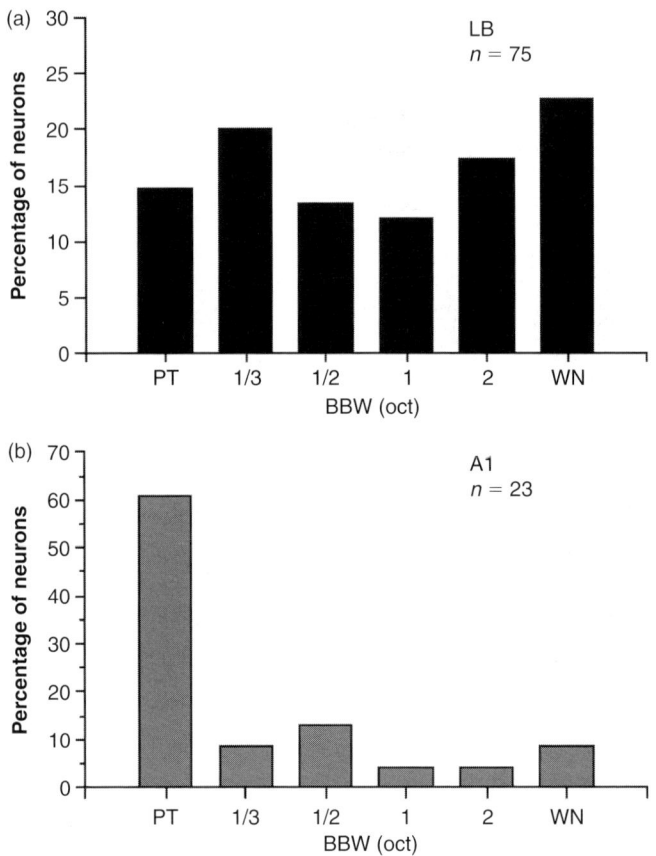

Figure 9.4. Distribution of BBWs in LB and A1 (from Rauschecker and Tian, 2004) WN: white noise; PT: pure tones.

detectors would have to preserve their selectivity regardless of sound intensity. Indeed, as the example in Fig. 9.3d shows, LB neurons generally do have the same best center frequency at different intensities.

9.3.2 Selectivity for parameters of frequency sweeps

Other features that are highly typical for communication sounds in most species are changes in frequency over time ("frequency-modulated (FM) sweeps"). FM sweeps are characterized by two parameters: FM rate and direction. Neurons in the LB are highly selective for both parameters (Tian and Rauschecker, 2004). First of all, 94% of LB neurons responded to FM stimuli in at least one direction. To characterize direction selectivity (DS) of FM quantitatively, a DS index was calculated. A neuron was considered direction-selective when the response in

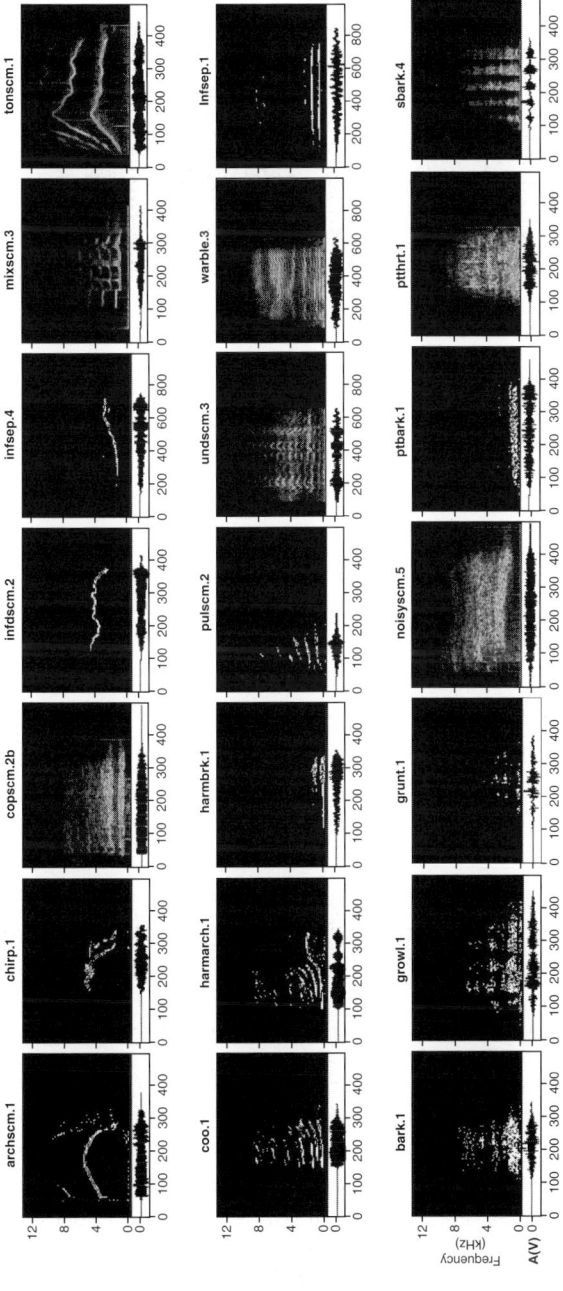

Figure 9.5. Twenty-one examples of rhesus monkey calls (MCs) in three phonetic–acoustic categories. Top row: tonal calls; middle row: harmonic calls; bottom row: noisy calls. Spectrograms and associated time signals are shown. Gray scale on bottom right shows relative energy content (amplitude) in different frequency bands (MCs recorded by Marc Hauser).

one FM direction for one or more FM rates was at least twice as large as that in the other direction (Mendelson and Cynader, 1985). An example is shown in Fig. 9.6. About 60% of LB neurons were classified as direction-selective on the basis of this criterion, with roughly equal proportions of neurons preferring upward and downward directions.

Even more striking was the selectivity of LB neurons for FM rate. The FM-direction-selective neuron in Fig. 9.6 also displays pronounced FM-rate selectivity. Various types of FM-rate tuning can be discerned in the LB, including

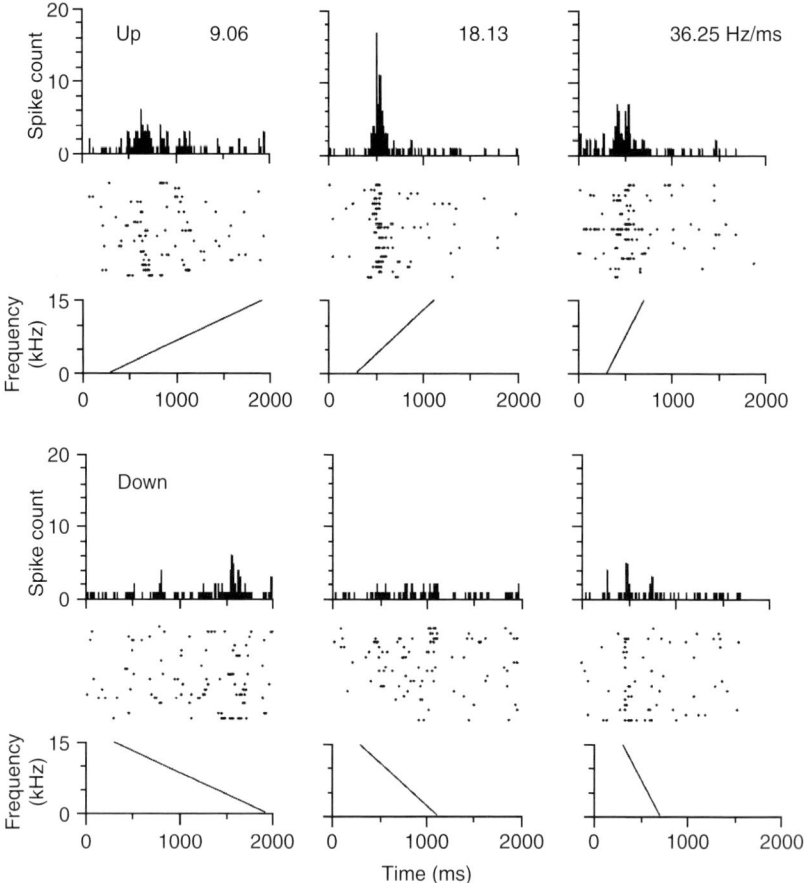

Figure 9.6. Response of a typical LB neuron to FM sweeps of different rate and direction. Peri-stimulus time histograms and raster dot displays are shown above a schematic display of the respective stimuli. Upward FM directions are shown in the top half, downward directions in the bottom half. FM rates are displayed on top of each column. A clear preference for a highly specific FM rate and for FM sweeps in the upward direction is found (from Tian and Rauschecker, 2004).

9.3 Processing of sounds with intermediate complexity 213

high-, low-pass tuning, and BP tuning. Neurons tuned to both FM direction and rate, like the neuron in Fig. 9.6, would be ideal candidates for the extraction of communication-sound features.

Preferred FM rate differed markedly between the three LB areas (Fig. 9.7). AL neurons preferred lower FM rates: More than half of AL neurons had their preferred FM rates below 64 Hz/ms, with medians of 25 and 50 Hz/ms for upward

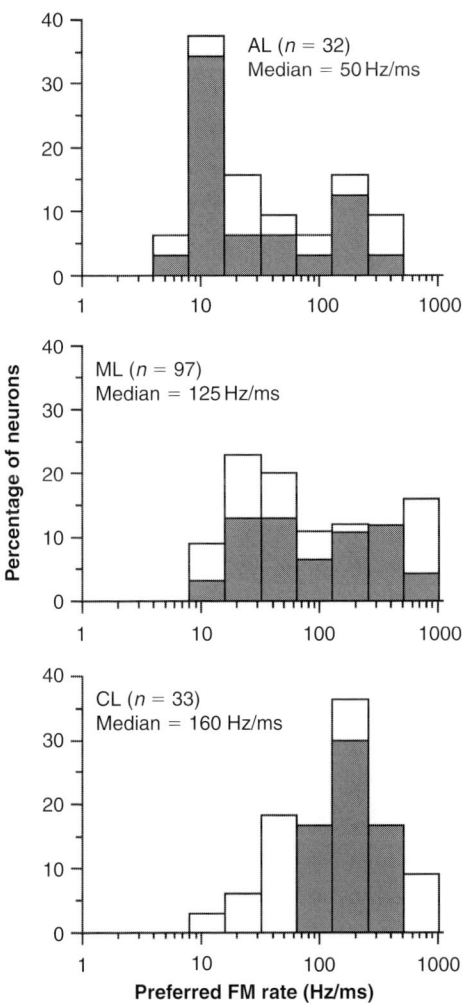

Figure 9.7. Distribution of preferred FM rates in different LB areas. Preferred FM rates in AL are about an order of magnitude lower than FM rates in CL, with preferred rates in ML in between. FM rate preferences in AL match the FM rates contained in species-specific vocalizations (from Tian and Rauschecker, 2004).

and downward sweeps, respectively. CL neurons, in contrast, preferred higher FM rates: About 70% of CL neurons preferred FM rates above 64 Hz/ms, with medians of 160 Hz/ms for both directions. ML neurons preferred FM rates in between.

According to these differences, AL neurons would be very well suited to participate in the decoding of species-specific vocalizations, which range mostly between 8 and 50 Hz/ms (Hauser, 1996; Rauschecker, 1998a). The various harmonics in the widely occurring "coo" calls fall between 10 and 40 Hz/ms. Only some of the "screams" contain FM rates above 100 Hz/ms (tonal scream: 103 Hz/ms; arch scream: 314 and 826 Hz/ms for the downward portion). Some of the neurons in AL do include responses to these faster sweeps. It is noteworthy that screams play an important role as alarm calls, which have to be well localizable by members of the same species. There was also a trend for neurons in AL to be more selective for FM direction, which would appear to be consistent with a role of theirs in the decoding of auditory patterns.

In its role, area AL can be likened to visual area V4, which contains neurons selective for the size of visual objects and plays a pivotal role in the ventral visual "what"-stream. Just as inferotemporal cortex, which receives input from V4, constitutes the later stages of visual object recognition, neurons in the rostral PB (and further anterior in the superior temporal gyrus, STG) are expected to rely on input from AL belt, compute invariances against distortions, and assure perceptual constancy.

9.4 Responses to species-specific calls

Neurons in the LB responded more vigorously to time-variant FM sweeps than to tones of constant frequency, and FM sweeps were also generally more effective than BPN bursts. LB neurons were also tested directly with whole monkey calls (MCs) or components thereof. Just as BPN bursts and FM sweeps, MC stimuli elicited more vigorous responses in LB than pure tones. MC stimuli were also generally more effective than BPN bursts but not necessarily more so than FM sweeps, which often remained the best stimuli.

9.4.1 Nonlinear integration mechanisms

LB neurons responded differentially to different types of MC. Although calls often had the same or comparable bandwidths, neuronal responses differed. Response selectivity, therefore, must be based on features contained in the phonetic fine structure of the calls. The conclusion is not far-fetched that it is the combination of features that causes a cell to respond to a specific type of call and not to others. Indeed, two fundamental mechanisms were identified as causing neuronal selectivity: nonlinear summation (a) in the spectral domain ("spectral facilitation", SFA) and (b) in the temporal domain ("temporal facilitation", TFA) (Fig. 9.8). This corresponds to spectral and temporal combination-sensitivity (CS), respectively, as it has been described previously in other species, such as bats, frogs, and songbirds (Suga *et al.*, 1978; Narins and Capranica, 1980; Margoliash and Fortune, 1992).

In spectral CS or SFA, inputs from lower-order neurons, such as BP-selective neurons, are combined in the frequency domain (Fig. 9.8a). In temporal CS or

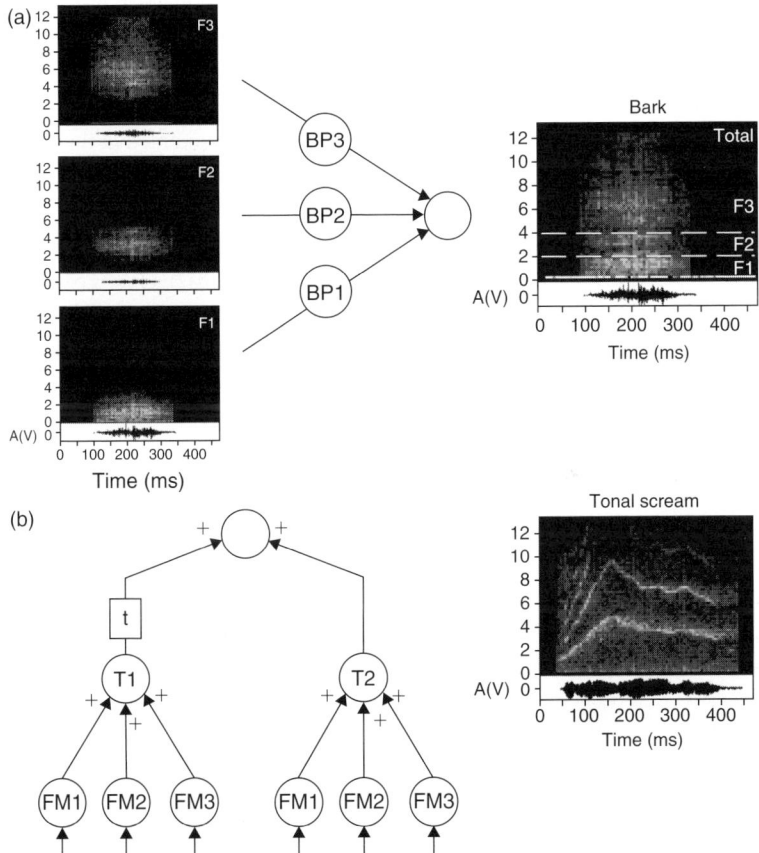

Figure 9.8. Models of nonlinear spectral (a) and temporal (b) integration ("combination-sensitivity", CS) in neurons of the LB (and presumably PB). Selectivity for specific communication calls is created by combining inputs from lower-order neurons that are BP-selective and/or FM-rate- and -direction-selective. Delay lines need to be implemented, as explained in the text, in order to create temporal CS (reprinted from Rauschecker 1998b, with permission of S. Karger AG, Basel).

TFA, inputs are combined in the time domain (Fig. 9.8b). However, both mechanisms are based on the same principle: coincidence detection by cell membranes with a relatively high threshold, that is, a logical AND-gate principle. Only with all inputs present simultaneously a response is evoked; with one input alone, no response follows. This explains why single components or syllables within a call usually are not sufficient to elicit a response. Temporal summation is accomplished by introducing staggered delays in the input pathways transmitting the early components; so all inputs eventually arrive simultaneously at the higher-order target neuron.

In some cases, however, the opposite is found: LB neurons respond decently to single components, but the response is suppressed by presenting the whole call. This is referred to as spectral or temporal suppression (SSU or TSU), respectively. We currently do not fully understand the significance of such units but assume that they are part of the logical alphabet implemented at that level.

One of the most striking differences between core and belt areas identified so far is the difference in their ability to nonlinearly integrate information both in the spectral and time domain (Fig. 9.9). While more than half of the neurons in LB show some form of nonlinear interaction (SFA, SSU; TFA, TSU), only ~10% (or less) of the neurons in A1 or R display the same form of behavior. This demonstrates a quantum leap in the processing characteristics of auditory cortex in nonhuman primates and is one of the strongest arguments for a hierarchic organization in auditory cortex.

9.4.2 MC and spatial selectivity

To quantify MC selectivity, a MC preference index (MCPI) can be calculated depending on the number of calls the neuron responds to. In most cases, a

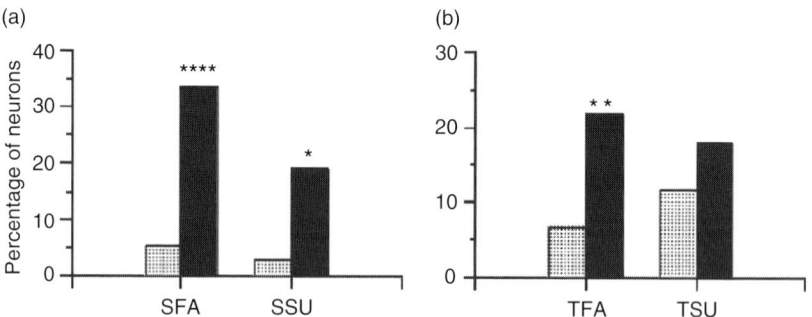

Figure 9.9. Percentage of neurons with nonlinear spectral and temporal integration in core and belt areas. Data from core areas are shown as stippled bars, data from belt in filled blocks. The amount of nonlinear integration was increased significantly in the belt for both spectral (a) and temporal (b) integration. $*P < 0.05$; $**P < 0.01$; $****P < 0.001$.

standard battery of seven of the most frequently occurring calls was used. An MCPI of seven, therefore, means that the cell responded to all the calls presented. An MCPI of three or less corresponds to a cell that responded to less than half of the calls and can be termed "MC-selective", whereas cells that responded to five or more of the calls are termed "MC-nonselective". The LB areas differed in their degree of MC selectivity, as quantified on this basis. Area AL had the greatest percentage of highly selective neurons (MCPI $\leq$ 2), followed by ML, whereas CL had the smallest percentage of highly selective neurons. Naturally, for the most non-selective neurons (MCPI $\geq$ 6), the opposite was found: CL had the greatest percentage of such non-selective neurons, AL the least, with ML somewhere between those two extremes.

The finding that spatial tuning in neurons of the LB shows the opposite areal distribution – highest selectivity is found in CL and lowest in AL – has led to the hypothesis that these two areas, which lie on opposite ends of the LB along its rostro-caudal extent, form the beginning of two processing streams for the processing of auditory object and space information (Rauschecker and Tian, 2000; Tian et al., 2001). The anterior "what"-stream may extend all the way to the temporal pole, which has recently been demonstrated unequivocally to be auditorily activated (Poremba et al., 2003) and to show a hemispheric difference for species-specific communication sounds (Poremba et al., 2004) using positron emission tomography (PET) and 2-deoxyglucose neuroimaging techniques in monkeys.

9.5 Auditory belt projections to prefrontal cortex

An anatomic study in rhesus monkeys has demonstrated the existence of largely separate pathways originating in the LB and projecting to different target regions in the prefrontal cortex (Romanski et al., 1999) (see Fig. 9.10). In this study, three different fluorescent tracers were injected into matched frequency regions of the three belt areas after these had been physiologically mapped. Injections into area AL produced label in ventrolateral and orbital regions of prefrontal cortex, whereas CL injections led to labeling of dorsolateral prefrontal cortex. The latter is known for its involvement in spatial working memory, whereas the former regions are assumed to participate in object working memory (Goldman-Rakic, 1996).

These projection patterns conform to the physiologic response properties found in the aforementioned study of Tian et al. (2001), which assigned superior selectivity for auditory patterns and space to areas AL and CL, respectively. The studies by Tian et al. (2001) and Romanski et al. (1999), therefore, form the cornerstones of a recent theory according to which dual processing streams in non primary auditory cortex underlie the perception of auditory objects and auditory space (Fig. 9.10; Rauschecker and Tian, 2000): One pathway projecting

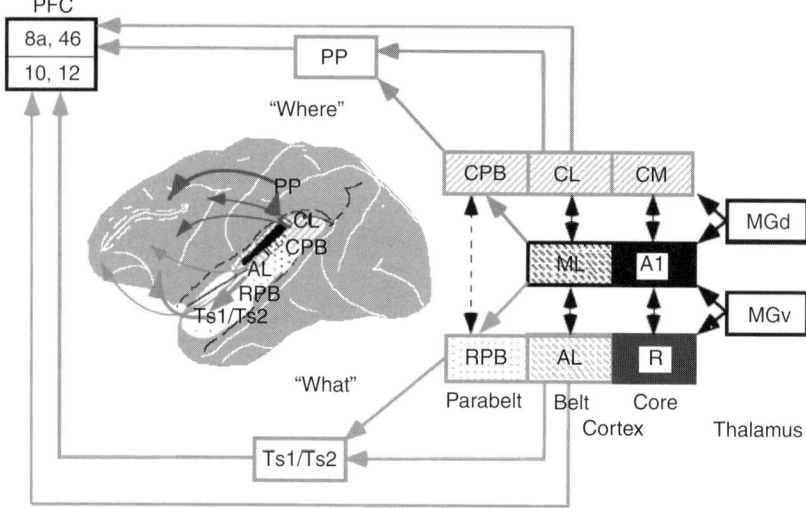

Figure 9.10. Schematic block diagram of processing streams in the auditory cortical system of primates (extended from Rauschecker, 1998b; Rauschecker and Tian, 2000). The projections are also shown, in the same colors, on a lateral view of the rhesus monkey brain (modified from Romanski et al., 1999). Shown in broken lines and dotted lines are areas in an anterior (ventral) stream for the identification of auditory objects ("what"), projecting directly (and indirectly, via rostral STG areas Ts1 and Ts2) to ventrolateral prefrontal cortex (PFC) (areas 10 and 12). Shown in diagonal lines are areas in a posterior (dorsal) stream for the processing of auditory space ("where") projecting directly (and indirectly, via posterior parietal (PP) cortex), to dorso-lateral PFC (areas 8a and 46). An analogous projection scheme has been proposed to hold for human auditory cortex (Rauschecker and Tian, 2000). Abbreviations of areas see text.

antero-ventrally from A1 through AL and the rostral STG and STS into orbitofrontal cortex forms the main substrate for auditory pattern recognition and object identification. Indeed, an auditory domain is found in ventrolateral prefrontal cortex, in which neurons show responses to complex sounds, including animal and human vocalizations (Romanski and Goldman-Rakic, 2002). Another pathway projecting caudo-dorsally into posterior parietal and dorsolateral prefrontal cortex is thought to be involved in auditory spatial processing.

9.6 Human imaging studies

Human neuroimaging studies have confirmed the organization of auditory cortex into core and belt areas by using the same types of stimuli as in the present study (Wessinger et al., 2001). Two core areas robustly activated by pure-tone stimuli and mirror-symmetric tonotopic organization were found along

Heschl's gyri. A third such area was sometimes seen more laterally. While the first two areas quite obviously correspond to areas A1 and R, the third area may be homologous to area ML, which (like the core areas A1 and R (Rauschecker et al., 1997)) has been shown to receive direct input from the MGv (Hackett et al., 1998b; Zhenochin et al., 1998). These three pure-tone responsive areas were surrounded by belt regions both medially and laterally, which were activated only by BPN bursts. An exploration of the medial belt region in the monkey with BPN bursts is therefore indicated.

Various findings from human neuroimaging strongly support the dual-stream hypothesis of auditory processing: AL areas of the superior temporal cortex are activated by intelligible speech or speech-like sounds (Binder et al., 2000, 2004; Scott et al., 2000; Alain et al., 2001; Maeder et al., 2001; Obleser et al., 2005), whereas caudal belt and PB areas (projecting up dorsally into posterior parietal cortex) are activated by auditory spatial discrimination tasks or tasks involving auditory motion (Maeder et al., 2001; Zatorre and Belin, 2001; Warren et al., 2002; Arnott et al., 2004; Krumbholz et al., 2005; Tata and Ward, 2005). Some of the areas in anterior human STG do indeed seem to represent species-specific sounds, because they light up only with speech but not with animal calls (Fecteau et al., 2004), whereas others may encode more general auditory object information (Zatorre et al., 2004). It becomes more and more obvious that behaviorally relevant auditory patterns are discriminated in an anterior auditory "what"-stream, although it had long been assumed that these processes are located posteriorly in a region called the planum temporale or Wernicke's area following human stroke studies of more than a century ago (Galaburda et al., 1978).

In conclusion, it appears that, like in the visual system, studies of nonhuman primates can serve as excellent models for future human studies. Conversely, human imaging studies can provide useful guidance for microelectrode studies in nonhuman primates, which permit analyses at much higher spatial and temporal resolution than would be possible in most human studies, with some exceptions (Howard et al., 2000).

9.7 Conclusions

Contrary to common belief, which places speech processing in posterior regions of the STG, three lines of evidence suggest that communication sounds are processed along an antero-ventral axis in the STG of nonhuman as well as human primates:

1. Neurophysiological single-unit studies in the rhesus monkey find an increased proportion of neurons with response selectivity for species-specific vocalizations in the AL area of the auditory belt cortex.

2. Area AL sends anatomic projections to the orbitofrontal cortex, which has previously been implicated in working memory for objects.
3. Neuroimaging studies in humans with functional magnetic resonance imaging (fMRI) and PET demonstrate that areas in the anterior STG are activated by species-specific sounds, that is, human voices, speech, and speech-related sounds.

From these and other studies emerges a picture of an anteriorly directed hierarchic processing stream dedicated to the identification and recognition of behaviorally relevant auditory patterns, which include those used for communication. Activity of neurons in this pathway signals "what" a complex sound represents. Areas in the LB, in particular area AL, are a relatively early station in this process, and the question arises what role exactly these intermediate areas play. We have proposed that neurons at the level of the belt participate in the extraction and integration of auditory features relevant for communication, as they are contained in conspecific calls as well as human speech. Examples are the selectivity of AL neurons for BPN bursts and FM sweeps. Neurons in the rostral PB (and further anterior in the STG) are expected to rely on input from AL belt, compute invariances against distortions, and assure perceptual constancy.

REFERENCES

Alain C, Arnott SR, Hevenor S, Graham S, Grady CL (2001) "What" and "where" in the human auditory system. *Proc Natl Acad Sci USA* 98: 12301–12306.

Arnott SR, Binns MA, Grady CL, Alain C (2004) Assessing the auditory dual-pathway model in humans. *Neuroimage* 22: 401–408.

Binder JR, Frost JA, Hammeke TA, Bellgowan PSF, Springer JA, Kaufman JN, Possing ET (2000) Human temporal lobe activation by speech and nonspeech sounds. *Cereb Cortex* 10: 512–528.

Binder JR, Liebenthal E, Possing ET, Medler DA, Ward BD (2004) Neural correlates of sensory and decision processes in auditory object identification. *Nat Neurosci* 7: 295–301.

Brugge JF, Merzenich MM (1973) Responses of neurons in auditory cortex of the macaque monkey to monaural and binaural stimulation. *J Neurophysiol* 36: 1138–1158.

Fecteau S, Armony JL, Joanette Y, Belin P (2004) Is voice processing species-specific in the human brain? An fMRI study. *Neuroimage* (23: 840–848).

Galaburda AM, Sanides F, Geschwind N (1978) Human brain. Cytoarchitectonic left–right asymmetries in the temporal speech region. *Arch Neurol* 35: 812–817.

Goldman-Rakic PS (1996) The prefrontal landscape: implications of functional architecture for understanding human mentation and the central executive. *Philos Trans R Soc Lond B Biol Sci* 351: 1445–1453.

Hackett TA, Stepniewska I, Kaas JH (1998a) Subdivisions of auditory cortex and ipsilateral cortical connections of the parabelt auditory cortex in macaque monkeys. *J Comp Neurol* 394: 475–495.

Hackett TA, Stepniewska I, Kaas JH (1998b) Thalamocortical connections of the parabelt auditory cortex in macaque monkeys. *J Comp Neurol* 400: 271–286.

Hauser MD (1996) *The Evolution of Communication*. MIT Press, Cambridge, MA.

Howard MA, Volkov IO, Mirsky R, Garell PC, Noh MD, Granner M, Damasio H, Steinschneider M, Reale RA, Hind JE, Brugge JF (2000) Auditory cortex on the human posterior superior temporal gyrus. *J Comp Neurol* 416: 79–92.

Jones EG, Dell'Anna ME, Molinari M, Rausell E, Hashikawa T (1995) Subdivisions of macaque monkey auditory cortex revealed by calcium-binding protein immunoreactivity. *J Comp Neurol* 362: 153–170.

Krumbholz K, Schönwiesner M, Cramon DYv, Rübsamen R, Shah NJ, Zilles K, Fink GR (2005) Representation of interaural temporal information from the left and right auditory space in the human planum temporale and inferior temporal lobe. *Cereb Cortex* 15: 324–324.

Maeder PP, Meuli RA, Adriani M, Bellmann A, Fornari E, Thiran JP, Pittet A, Clarke S (2001) Distinct pathways involved in sound recognition and localization: a human fMRI study. *Neuroimage* 14: 802–816.

Margoliash D, Fortune ES (1992) Temporal and harmonic combination-sensitive neurons in the zebra finch's HVc. *J Neurosci* 12: 4309–4326.

Mendelson JR, Cynader MS (1985) Sensitivity of cat primary auditory cortex (AI) neurons to the direction and rate of frequency modulation. *Brain Res* 327: 331–335.

Merzenich MM, Brugge JF (1973) Representation of the cochlear partition on the superior temporal plane of the macaque monkey. *Brain Res* 50: 275–296.

Morel A, Garraghty PE, Kaas JH (1993) Tonotopic organization, architectonic fields, and connections of auditory cortex in macaque monkeys. *J Comp Neurol* 335: 437–459.

Narins PM, Capranica RR (1980) Neural adaptations for processing the two-note call of the Puerto Rican treefrog, Eleutherodactylus coqui. *Brain Behav Evol* 17: 48–66.

Obleser J, Boecker H, Drzezga A, Haslinger B, Roettinger M, Eulitz C, Rauschecker JP (2005) Vowel sound extraction in anterior superior temporal cortex. *Human Brain Mapping* (in press).

Pandya DN, Sanides F (1972) Architectonic parcellation of the temporal operculum in rhesus monkey and its projection pattern. *Z Anat Entw-Gesch* 139: 127–161.

Poremba A, Saunders RC, Crane AM, Cook M, Sokoloff L, Mishkin M (2003) Functional mapping of the primate auditory system. *Science* 299: 568–572.

Poremba A, Malloy M, Saunders RC, Carson RE, Herscovitch P, Mishkin M (2004) Species-specific calls evoke asymmetric activity in the monkey's temporal poles. *Nature* 427: 448–451.

Rauschecker JP (1997) Processing of complex sounds in the auditory cortex of cat, monkey and man. *Acta Otolaryngol* 532: 34–38.

Rauschecker JP (1998a) Parallel processing in the auditory cortex of primates. *Audiol Neuro-otol* 3: 86–103.

Rauschecker JP (1998b) Cortical processing of complex sounds. *Curr Opin Neurobiol* 8: 516–521.

Rauschecker JP, Tian B (2000) Mechanisms and streams for processing of "what" and "where" in auditory cortex. *Proc Natl Acad Sci USA* 97: 11800–11806.

Rauschecker JP, Tian B (2004) Processing of band-passed noise in the lateral auditory belt cortex of the rhesus monkey. *J Neurophysiol* 91: 2578–2589.

Rauschecker JP, Tian B, Hauser M (1995) Processing of complex sounds in the macaque nonprimary auditory cortex. *Science* 268: 111–114.

Rauschecker JP, Tian B, Pons T, Mishkin M (1997) Serial and parallel processing in rhesus monkey auditory cortex. *J Comp Neurol* 382: 89–103.

Reale RA, Imig TJ (1980) Tonotopic organization in auditory cortex of the cat. *J Comp Neurol* 192: 265–291.

Recanzone GH, Guard DC, Phan ML, Su TK (2000) Correlation between the activity of single auditory cortical neurons and sound-localization behavior in the macaque monkey. *J Neurophysiol* 83: 2723–2739.

Romanski LM, Goldman-Rakic PS (2002) An auditory domain in primate prefrontal cortex. *Nat Neurosci* 5: 15–16.

Romanski LM, Tian B, Fritz J, Mishkin M, Goldman-Rakic PS, Rauschecker JP (1999) Dual streams of auditory afferents target multiple domains in the primate prefrontal cortex. *Nat Neurosci* 2: 1131–1136.

Scott SK, Blank CC, Rosen S, Wise RJS (2000) Identification of a pathway for intelligible speech in the left temporal lobe. *Brain* 123: 2400–2406.

Suga N, O'Neill WE, Manabe T (1978) Cortical neurons sensitive to combinations of information-bearing elements of biosonar signals in the mustache bat. *Science* 200: 778–781.

Tata MS, Ward LM (2005) Spatial attention modulates activity in a posterior "where" pathway. *Neuropsychologia* 43: 509–516.

Tian B, Rauschecker JP (2004) Processing of frequency-modulated sounds in the lateral auditory belt cortex of the rhesus monkey. *J Neurophysiol* 92: 2993–3013.

Tian B, Reser D, Durham A, Kustov A, Rauschecker JP (2001) Functional specialization in rhesus monkey auditory cortex. *Science* 292: 290–293.

Wang X (2000) On cortical coding of vocal communication sounds in primates. *Proc Natl Acad Sci USA* 97: 11843–11849.

Warren JD, Zielinski BA, Green GGR, Rauschecker JP, Griffiths TD (2002) Analysis of sound source motion by the human brain. *Neuron* 34: 1–20.

Wessinger CM, VanMeter J, Tian B, Van Lare J, Pekar J, Rauschecker JP (2001) Hierarchical organization of the human auditory cortex revealed by functional magnetic resonance imaging. *J Cogn Neurosci* 13: 1–7.

Yu JJ, Young ED (2000) Linear and nonlinear pathways of spectral information transmission in the cochlear nucleus. *Proc Natl Acad Sci USA* 97: 11780–11786.

Zatorre RJ, Belin P (2001) Spectral and temporal processing in human auditory cortex. *Cereb Cortex* 11: 946–953.

Zatorre RJ, Bouffard M, Belin P (2004) Sensitivity to auditory object features in human temporal neocortex. *J Neurosci* 24: 3637–3642.

Zhenochin S, Fritz J, Tian B, Ojima H, Rauschecker JP (1998) Thalamo-cortical projections underlying differential responsiveness in the lateral belt areas of rhesus monkey auditory cortex. *Assoc. Res. Otolaryngol. Abstr.* 21: 146.

10
Synaptic Mechanisms and Sensitive Periods for Song Learning

J. Matthew Kittelberger and Richard Mooney

10.1 Sensitive periods for learned behavior

The development of particular species-typical behaviors displays heightened sensitivity to experience during specific-time windows known as developmental sensitive periods.[1] During the sensitive period, specific learning abilities are greatly enhanced relative to earlier and later periods in life. After the sensitive period, the acquired behaviors typically display tremendous resistance to subsequent modification.

Behavioral sensitive periods have been described in a wide range of animals. Classic examples of developmental sensitive periods for learned behaviors include olfactory imprinting in salmon (Hasler and Scholz, 1983), visual and filial imprinting in young birds (Lorenz, 1937; Sluckin, 1972; Horn, 1985; Bolhuis and Honey, 1998), social imprinting in dogs (Scott, 1962, 1978) and possibly also monkeys (Harlow and Zimmerman, 1959; Harlow and Harlow, 1969), song learning in passerine songbirds (Thorpe, 1958; Marler and Tamura, 1964; Immelmann, 1969; Marler, 1970), and certain aspects of language acquisition in humans (Lenneberg, 1967; Snow, 1987; Doupe and Kuhl, 1999). Other examples include two forms of olfactory imprinting that arise during tightly proscribed sensitive periods in adults: maternal recognition in sheep (Poindron and Levy, 1990; Hudson, 1993), and the olfactory memories of female mice for urinary pheromones of recently mated males (Brennan and Keverne, 1997). Thus, sensitive periods for learning are largely but not exclusively restricted to juvenile life, and help shape the development of species-typical behaviors, especially those important to intraspecific communication, pair-bonding, and reproduction.

---

[1] When such periods are sharply delimited by innate biological processes, they have been referred to as "critical periods," while the less stringent term "sensitive period" encompasses, in addition, examples in which the opening and closing of the developmental window may also be relatively malleable and influenced by experience. For the most part, we will use the broader term "sensitive period" throughout this review.

Neurobiologists have long sought to understand the neural basis for sensitive periods for learned behaviors in two broad contexts. First, how does experience during the sensitive period change underlying neural circuitry to effect changes in behavior? Second, what special features of neural circuits endow them with such pronounced plasticity during sensitive periods? Answers to these two questions will provide insight into the specifics of the neurobiologic bases for sensitive periods for learned behavior, and are likely to have broader ramifications for our ideas about neural development and adult learning.

10.2 Neuronal and behavioral sensitive periods: common themes

Similar to sensitive period behavioral changes, neural circuit organization can also be especially sensitive to the influence of sensory input during defined periods in development. Indeed, one attractive idea is that behavioral sensitive periods are limited by sensitive periods for the experience-dependent modification of underlying neuronal circuitry. This idea has yet to be rigorously tested, because the best described neuronal sensitive periods (see below) are in primary sensory regions of the brain, complicating connections to specific behaviors. Nonetheless, the substantial understanding of neuronal sensitive periods, including growing insights into underlying mechanisms, provides an essential context for the discussion of neurobiologic mechanisms underlying behavioral sensitive periods.

10.3 Sensitive periods for neural circuit plasticity: search for mechanisms

Neuronal sensitive periods have been studied in a variety of systems. In the visual system of vertebrates, the topography of eye-specific inputs to the visual cortex (Hubel and Wiesel, 1970; Hubel et al., 1977; Wiesel, 1982) and to the optic tectum (Keating et al., 1975; Keating and Grant, 1992; Udin and Grant, 1999) can be driven by altered visual experience during developmentally defined sensitive periods. In the somatic sensory system of rodents, manipulating whiskers can induce dramatic changes in the representation of these whiskers in the somatosensory cortex, but only if whisker manipulation occurs during an early developmental period (Woolsey, 1990). In the midbrain of the barn owls, the alignment of visual and auditory representations of space is sensitive to manipulations of visual and auditory experience, and these manipulations are most effective during a particular juvenile-time window (Knudsen et al., 1984; Knudsen, 1985; Knudsen and Knudsen, 1990; Brainard and Knudsen, 1998).

10.3 Sensitive periods for neural circuit plasticity

Several common themes emerge from these diverse examples. First, patterned sensory experience, translated into patterned activity of sensory inputs to central brain structures, can drive changes in the spatial organization of the inputs themselves, and therefore in the post-synaptic response properties of target neurons (Hubel and Wiesel, 1965; Keating and Feldman, 1975; Udin and Keating, 1981; Knudsen, 1983, 1985; Reh and Constantine-Paton, 1985; Stryker and Harris, 1986; Shatz, 1990; Knudsen and Brainard, 1991; Antonini and Stryker, 1993; Weliky and Katz, 1997). Second, manipulations of sensory experience can effectively alter neuronal circuit topography only during particular sensitive periods (Hubel and Wiesel, 1970; LeVay et al., 1980; Knudsen, 1985; Knudsen and Knudsen, 1990; Keating and Grant, 1992; Brainard and Knudsen, 1998). Third, the activity-dependent changes appear to obey competitive rules, whereby active, or synchronously active, inputs to a particular post-synaptic cell or region are maintained, and inactive or asynchronous inputs are removed (Hubel and Wiesel, 1965, 1970; Wiesel and Hubel, 1965; Guillery, 1972; Weliky and Katz, 1997; Brickley et al., 1998; Kind et al., 2002). Together, these findings imply that during normal development, patterned sensory inputs act to help establish the structure and function of the neural circuits involved in processing these inputs (LeVay et al., 1978, 1980; Shatz, 1990; Antonini and Stryker, 1993; Knudsen, 1994; Katz and Shatz, 1996; Udin and Grant, 1999). The underlying mechanism is often asserted to be an initial overproduction of synaptic inputs, followed by a period of activity-dependent refinement of circuits through selective elimination of particular inputs/synapses (Hubel et al., 1977; Rakic, 1977; LeVay et al., 1978; Udin, 1989; Shatz, 1990; Antonini and Stryker, 1993; Katz and Shatz, 1996; Lichtman and Colman, 2000). It is worth noting that there is significant debate about the relationship between the effects of altered sensory experience on modifying established synaptic connections during sensitive periods and the role of sensory experience in shaping initial connectivity patterns during normal development. The initial formation of maps, and subsequent sensitive periods of experience-dependent map plasticity, may in fact be governed by separate neural mechanisms (Horton and Hocking, 1996; Crowley and Katz, 2000; Katz and Crowley, 2002). Nevertheless, it is clear that, for restricted periods of developmental time, sensory experience can exert profound effects on the structure and function of central neural circuits.

In these diverse systems, sensitive period experience- and activity-dependent neural changes have been well characterized. Furthermore, cellular and molecular mechanisms by which activity may induce the described neural plasticity also have been explored. N-methyl-D-aspartate (NMDA) receptor-dependent forms of long-lasting synaptic potentiation and depression (LTP and LTD) may be one means by which co-active, or more efficacious, inputs are selectively strengthened, and others weakened and eliminated (Cline *et al.*, 1987; Kleinschmidt *et al.*, 1987; Cline and Constantine-Paton, 1989, 1990; Scherer and Udin, 1989; Bear *et al.*, 1990; Cline *et al.*, 1990; Shatz, 1990; Schlaggar *et al.*, 1993; Rema *et al.*, 1998). Evidence also supports a role for the activity-dependent release of neurotrophins in sculpting neural circuit structure and function during developmental sensitive periods (Castren *et al.*, 1992; Cabelli *et al.*, 1995, 1996, 1997; Riddle *et al.*, 1995; Katz and Shatz, 1996; Bonhoeffer and Shatz, 1998).

The mechanisms regulating the onset and termination of sensitive periods of heightened neural circuit plasticity remain enigmatic, although several important candidates have emerged. One idea is that the maturational change in NMDA receptor subunit composition, which shortens the duration of Ca^{2+} currents passing through the receptors, contributes to closing the sensitive period. Shorter currents, and thus smaller rises in internal Ca^{2+}, could reduce the amount of LTP and/or LTD elicited at the relevant synapses, or could reduce the time constant for the integration of synchronous inputs, either of which would lessen the magnitude of activity-dependent plasticity (Carmignoto and Vicini, 1992; Crair and Malenka, 1995; Katz and Shatz, 1996; Flint *et al.*, 1997; Livingston and Mooney, 1997; Crair, 1999). The developmental transition in subunit composition and current duration of NMDA receptors can itself be regulated by activity and experience (Quinlan *et al.*, 1999a, b; Philpot *et al.*, 2001). However, blocking the transition to fast NMDA currents in mice fails to prevent the normal termination of the sensitive period for somatic sensory cortex plasticity (Lu *et al.*, 2001). Therefore, other mechanisms are likely to be more central to sensitive period regulation. One possibility is that a decline in the expression, or the activity dependence thereof, of neurotrophins or their receptors might trigger the end of sensitive period neural plasticity (Cohen-Cory and Fraser, 1994; Cabelli *et al.*, 1996; Bonhoeffer and Shatz, 1998). Another possibility is that the maturation of GABAergic inhibitory networks, which may in turn be regulated by neurotrophins, might terminate the sensitive period (Hensch *et al.*, 1998; Huang *et al.*, 1999; Fagliolini and Hensch, 2000), perhaps by limiting excitation-dependent forms of synaptic plasticity.

Despite insights into mechanisms underlying neuronal sensitive periods, our understanding of the neural mechanisms underlying heightened sensitive period behavioral plasticity remains more limited. Arguably, progress towards understanding neural regulation of behavioral sensitive periods will benefit from an analysis of those animals with well-delineated central circuits mediating behaviors acquired during a sensitive period. Songbirds are attractive organisms in these two regards: their courtship songs are acquired during a juvenile sensitive period and singing and song learning are mediated by a well-defined central circuit, referred to as the song system. Furthermore, songbirds are one of only a handful of animal groups, besides humans, that engage in vocal learning (Doupe and Kuhl, 1999). Therefore, elucidating the neurobiologic basis of sensitive periods for song learning may also help us understand the neural mechanisms underlying human vocal learning.

10.4 Birdsong development: sensitive periods for a learned behavior

In adult songbirds, courtship vocalizations are the primary means for mate choice and territorial defense and are necessary for reproductive success. The forms and functions of birdsong have long been a subject of fascination, in part because of the complexity and diversity of song, and because these features can be readily defined and quantified. Furthermore, in oscine perching birds (i.e. order: Passeriformes; suborder: Oscini), song is learned, thereby affording the opportunity to study the role of auditory experience in shaping neural circuits engaged in song production. Neurobiologists have used birdsong as a model system for examining:

1. auditory processing of species-typical communication sounds,
2. motor circuits involved in generating these sounds, and
3. interactions between auditory and motor domains of vocal communication.

A centuries-old observation is that young captive songbirds can copy the songs of adult tutors. In populations of wild songbirds, song dialects are culturally transmitted (Marler and Tamura, 1964; Kroodsma, 1974), while in the lab, studies of song development have led to the definition of particular sensitive periods during development when auditory experience shapes song learning. Isolation and cross-fostering experiments show that to properly develop their species-typical song, young birds must hear the song of an adult conspecific tutor during a particular time period, which typically ends before the young bird itself begins to sing (Thorpe, 1958; Marler and Tamura, 1964; Immelmann, 1969; Marler, 1970; Eales, 1985; Slater et al., 1988; Bohner, 1990). During this sensitive period, referred to as sensory acquisition, birds form a memory or template of the tutor's song to which they will eventually match their own song. The exact timing of this sensitive period varies from species to species. In many, including zebra finches, sensory acquisition is restricted from a few weeks to up to a few months after fledging (Immelmann, 1969; Marler and Peters, 1977, 1981; Eales, 1985; Bohner, 1990). In other species, it may extend into the second year of life when birds begin to establish and defend territories (Rice and

228 *10 Synaptic mechanisms and sensitive periods for song learning*

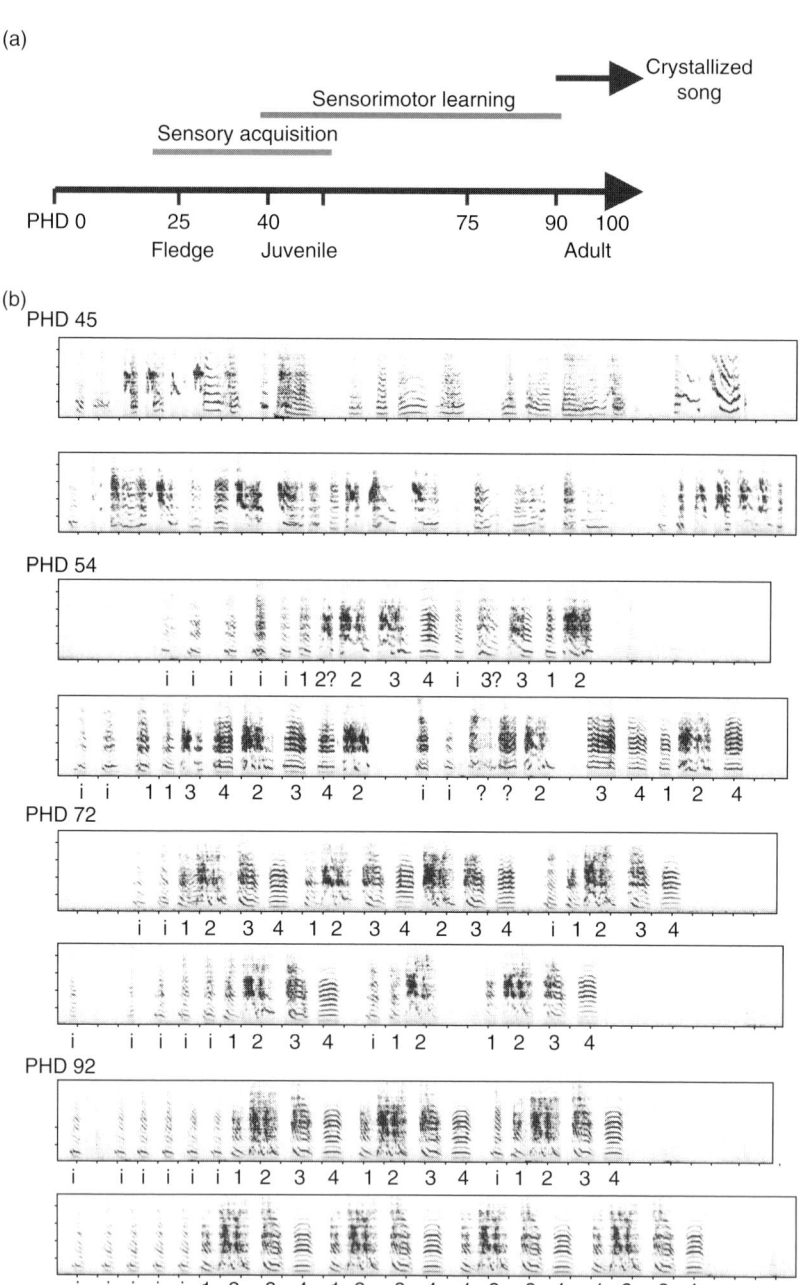

Figure 10.1. Zebra finch song development. A. Timeline of the major stages of song learning in the zebra finch. Sensory acquisition, the sensitive period when birds form an auditory memory of the song of an adult male tutor, normally lasts from post hatch day (PHD) 20 to PHD 60. Young males begin to sing at around PHD

10.4 Birdsong development: sensitive periods

Thompson, 1968; Kroodsma, 1974), and in others, the so-called "open-ended learners," acquisition of new song models may proceed throughout life (Nottebohm and Nottebohm, 1978; Mountjoy and Lemon, 1995). In zebra finches, normal sensory acquisition begins at around the 20th post hatch day (PHD 20) and lasts until about PHD 60 (see Fig. 10.1a) (Immelmann, 1969; Eales, 1985; Bohner, 1990). The timing of this sensitive period, however, is malleable to the influences of experience: zebra finch males isolated from social and/or auditory experience with other males before PHD 20 can learn new songs beyond PHD 120, well over 2 months past the normal closure of sensory acquisition (Eales, 1987; Morrison and Nottebohm, 1993; Jones *et al.*, 1996). In white-crowned sparrows, photoperiod and tutor exposure may both influence the closure of sensory acquisition (Whaling *et al.*, 1998).

Sensory acquisition is followed by a second discrete sensitive period typically referred to as sensorimotor learning, which lasts from the juvenile's earliest efforts to sing to the adult's stable song (Fig. 10.1a). During sensorimotor learning, birds use auditory feedback to progressively match their own song to the memorized model (Konishi, 1965). The first singing efforts (around PHD 35 in zebra finches) of young birds are known as subsongs: soft and rambling vocalizations that vary in the structure of individual acoustic elements (known as notes and syllables), and in the organization of these elements into the larger sequences (Thorpe and Pilcher, 1958; Immelmann, 1969; Arnold, 1975; Marler and Peters, 1982). Subsong has been likened in both form and function to the babbling of

Caption for fig. 10.1. (cont.)
35, first singing a highly variable and poorly defined subsong. During this sensitive period for sensorimotor learning, birds use auditory feedback to progressively match their own song to the memorized tutor song. Song evolves through a continuum of plastic song, gradually losing its variability as it comes to match the tutor song. Sensorimotor learning lasts until about PHD 90, when song crystallizes into a highly stable adult form. B. Sonograms illustrating the stages of song development in a single male zebra finch recorded from PHD 40 through adulthood. Three song bouts are shown for each of 4 ages; syllable identity is denoted beneath the sonograms ("i" = introductory note). At PHD 45 (late subsong), the song was highly variable in both syllable structure and order; syllables were not organized into repeating motifs. By PHD 54 (plastic song) a semblance of a repetitive structure had emerged, and syllables that eventually comprised the adult song were recognizable, though not unambiguously. There was significant variability in the order in which these syllables were sung. At PHD 72 (late plastic song), all syllables were sung in essentially the same order as in adulthood. Some continued variability in ordering was present: the bird occasionally dropped syllables or syllable combinations out of a motif. By PHD 92 song was mostly crystallized, with only minor variations (i.e., sometimes inserting an "i" between motifs). At PHD 162 (not shown), the bird always inserted an "i" between motifs, and the song was invariant from bout to bout. Y-axis: frequency, 2kHz/unit. X-axis: time, 0.1s/unit.

human infants (Fig. 10.1b) (Konishi, 1989; Doupe and Kuhl, 1999). Subsong is followed by plastic song, where notes and syllables with recognizable structures first emerge, and where imitations of the tutor song first become apparent. During plastic song, syllables are organized into repeating sequences, known as motifs (Fig. 10.1b), although the ordering of syllables into motifs remains variable (Marler, 1956; Arnold, 1975; Marler and Peters, 1982). Coincident with the onset of sexual maturity (around PHD 90 in zebra finches), song variability declines, resulting in a stable or "crystallized" form which is typically a close imitation of the tutor song (Fig. 10.1b) (Nice, 1943; Marler, 1956; Thorpe and Pilcher, 1958; Arnold, 1975; Marler and Peters, 1982). In many species, crystallization includes the selective attrition of vocal material; during plastic song, birds often overproduce syllables and songs that are not retained in the birds' crystallized adult repertoire (Marler and Peters, 1981, 1982; Nelson and Marler, 1994). In other species, including zebra finches, syllable variety, and number increase over song development in a generative, rather than a selective, process (Tchernichovski et al., 2001).

The two predominant hallmarks of song during sensorimotor learning are its variability and its plasticity. Song variability denotes the bout-to-bout variations in both the structure of individual song elements and in the sequencing of these elements. Song plasticity refers to the progressive, experience-dependent changes in song during this sensitive period. The variability and the plasticity of song appear to be linked. Both decline in parallel over song development (Immelmann, 1969; Arnold, 1975; Marler and Peters, 1982), and neural manipulations that block song variability also block song plasticity (Bottjer et al., 1984; Scharff and Nottebohm, 1991). It has been postulated that variability is, in fact, fundamental to sensorimotor plasticity, in both the selective (Marler and Peters, 1981; Nelson and Marler, 1994; Marler, 1997) and the generative (Tchernichovski et al., 2001) models of song learning. In the selective model, song variability provides the grist for young birds to select the motor patterns that best reproduce the memorized song; in the generative model, variability is essential for birds to develop multiple mature syllables from a single prototype sound.

Evidence that sensorimotor learning occurs during a distinct sensitive period derives from studies showing that song development is especially responsive to particular aspects of experience during this time window. Birds deafened after sensory acquisition but before or during plastic song show rapid and pronounced degradation of their songs and little evidence of successful imitation of tutor song (Konishi, 1965; Price, 1979). Thus, auditory feedback plays a critical role in the matching of each bird's own song to the memorized template. In contrast, after adult deafening, song typically degrades much more slowly and then sometimes only very mildly (Konishi, 1965; Price, 1979; Nordeen and

Nordeen, 1992). While there is significant variability between species in the degree to which auditory feedback is required for adult song maintenance (Nottebohm et al., 1976; Woolley and Rubel, 1997) (see below), it is clear that such feedback is especially important during sensorimotor learning. More recently, using a reversible technique for paralyzing the syrinx, Pytte and Suthers (2000) examined the effects of blocking vocal practice at different stages of song development in zebra finches. Interestingly, they found that the timing of the disruption was much more important than the duration: blocking motor practice even briefly across the time period when crystallization normally occurs (i.e. PHD 90) produced permanent defects in adult song, while much longer paralyses during subsong and plastic song had little effect, as long as the birds recovered from paralysis and were able to engage in even a limited amount of singing prior to PHD 90. In another recent study (Funabiki and Konishi, 2003), zebra finches were given access to a tutor during the normal-time period for sensory acquisition, but then were prevented from hearing their own songs until after the time when song normally crystallizes. Interestingly, after access to auditory feedback was restored, these birds were able to develop faithful copies of the syllables present in the tutor's song, but they were not able to replicate the correct ordering of these syllables unless hearing was restored prior to PHD 90. Thus, crystallization appears to be a discrete event marking the closure of a sensitive period for sensorimotor learning, particularly with regard to syllable sequence learning; aberrant songs cannot be corrected after this window closes. Importantly, both sets of results also indicate that the timing of crystallization is not entirely experience dependent, as birds allowed only a limited amount of vocal experience prior to PHD 90 nonetheless crystallized at the normal time.

Crystallization is not entirely independent of experience either, as deafening prevents the normal stabilization of song (Konishi, 1965), and both isolation and reversible blockade of auditory feedback reveal that sensorimotor learning of syllable structure can occur past the normal time of song crystallization (Morrison and Nottebohm, 1993; Funabiki and Konishi, 2003). High androgen levels may be required for song crystallization, as castrated males fail to crystallize normally (Arnold, 1975; Marler et al., 1988; Bottjer and Hewer, 1992), but can be rapidly induced to crystallize their song by administering testosterone (Marler et al., 1988). As in the vocal-paralysis study (Pytte and Suthers, 2000), these experiments show a dissociation between both the amount and the quality of song learning and song crystallization: adult castrates fail to crystallize despite extended singing experience and accurate sensory acquisition. In other studies, however, androgen expression is positively correlated with the amount of singing (Arnold, 1975; Doupe, 1993), suggesting that androgenic effects on crystallization may be indirectly mediated by the amount of vocal practice (Doupe, 1993). In canaries, which seasonally re-enter plastic song, androgen levels fluctuate annually such that low levels correlate with plastic song (Nottebohm *et al.*, 1987). In zebra finches, androgen levels increase over development (Adkins-Regan *et al.*, 1990; White *et al.*, 1999) and can be regulated by both auditory and social experience (Livingston *et al.*, 2000). Treatment of juvenile zebra finches with exogenous testosterone does disrupt normal song development, though not in a manner clearly equivalent to premature crystallization (Korsia and Bottjer, 1991), suggesting that high androgen levels may be necessary, though not sufficient, to induce crystallization in all species. Androgens are clearly involved in regulating song crystallization, though the precise role they play has not been entirely defined, and this role may vary from species to species.

Taken together, these data strongly support the idea that there are distinct mechanisms, both experience dependent and independent, regulating the degree of vocal plasticity across a sensitive period for sensorimotor learning. Indeed, the variability across species in the expression of vocal plasticity, from species that display plastic song only during a brief sensitive period in development (the closed-ended learners), to those re-expressing plastic song seasonally, to the open-ended learners that can produce new song types throughout life, implies species-specific mechanisms regulating the degree and timing of plasticity. Vocal plasticity is sustained in a variety of species and experimental conditions where extensive singing and song learning have clearly occurred already (Nottebohm and Nottebohm, 1978; Marler *et al.*, 1988; Mountjoy and Lemon, 1995). Conversely, in other species and conditions, crystallization can occur despite a paucity of vocal experience and learning (Pytte and Suthers, 2000; Funabiki and Konishi, 2003). The implication is that vocal plasticity is actively maintained by factors other than experience, and is not simply a reflection of insufficient learning. The neural correlates of vocal plasticity during sensorimotor learning, and the neural mechanisms involved in regulating this vocal plasticity, are the central topics of the remainder of this chapter.

10.5 The song system: a neural circuit for the learning and production of birdsong

An interconnected series of brain nuclei have been described that are critical for both the development and production of normal song (Fig. 10.2). Collectively known as the song system, these structures first drew the attention of neurobiologists because they are dramatically larger in singing males than in non-singing

10.5 The song system: a neural circuit for the learning 233

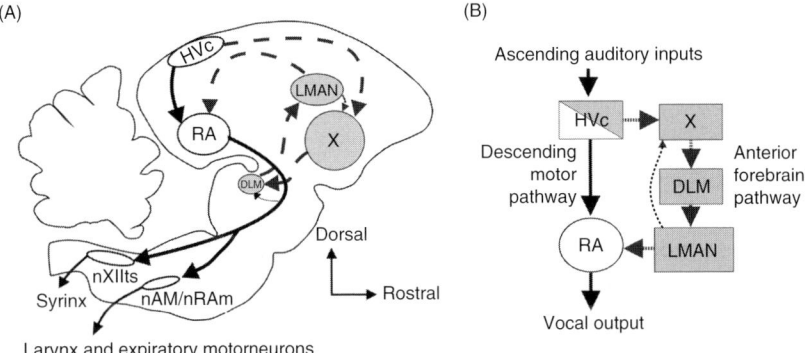

Figure 10.2. The song system. A. Simplified anatomical outline of the brain structures involved in avian song production and learning. The main components of the direct motor pathway critical for learned song production at all ages are outlined in white, with solid arrows denoting axonal connections. The basal-ganglia – like anterior forebrain pathway, which is necessary for feedback-dependent song plasticity, is shown in gray with dashed arrows. Abbreviations: HVC, nucleus HVC of the nidopallium; RA, robust nucleus of the arcopallium; nXIIts, tracheosyringeal portion of the hypoglossal nucleus; nAm, nucleus ambiguus; nRAm, nucleus retroambigualis; X, area X of the medial striatum; DLM, medial nucleus of the dorsolateral thalamus; LMAN, lateral magnocellular nucleus of the anterior nidopallium. Omitted structures include the ascending auditory inputs and two recursive feedback loops. B. A schematic of information flow within the song system.

females (Nottebohm and Arnold, 1976). A variety of tract tracing studies revealed a direct-motor pathway for the production of learned song including the telencephalic nucleus HVC of the nidopallium, the premotor robust nucleus of the arcopallium (RA) and brainstem vocal-respiratory areas that control the vocal (i.e. syringeal) and respiratory muscles (Fig. 10.2, white) (Nottebohm *et al.*, 1976, 1982; Vicario and Nottebohm, 1988; Vicario, 1991, 1993; Wild, 1993a, b) (updated nomenclature as per (Reiner *et al.*, 2004)). Several lines of evidence suggest that this direct pathway patterns song. First, lesions of nuclei in this pathway severely disrupt or abolish singing at all stages of development (Nottebohm *et al.*, 1976; Simpson and Vicario, 1990). Second, microstimulation of different nuclei in this direct pathway can elicit syringeal contractions and song-like vocalizations (Vicario and Simpson, 1995), and can disrupt song production when performed during singing (Vu *et al.*, 1994). Finally, chronic extracellular recordings in awake singing birds reveal bursts of patterned neuronal activity in these nuclei preceding and correlating with the temporal dynamics of song (McCasland, 1987; Yu and Margoliash, 1996; Chi and Margoliash, 2001; Hahnloser *et al.*, 2002).

A second, indirect pathway also connects HVC with RA via Area X of the medial striatum, the medial nucleus of the dorsolateral thalamus (DLM), and the lateral magnocellular nucleus of the anterior nidopallium (LMAN) (Fig. 10.2, gray) (Nottebohm *et al.*, 1982; Okuhata and Saito, 1987; Bottjer *et al.*, 1989). This pathway bears strong homologies to mammalian cortical-basal ganglia pathways (Reiner *et al.*, 1998; Perkel and Farries, 2000; Luo *et al.*, 2001). Lesions of this anterior forebrain pathway (AFP) in juvenile zebra finches dramatically derail normal song development (Bottjer *et al.*, 1984; Sohrabji *et al.*, 1990; Scharff and Nottebohm, 1991). In particular, juvenile LMAN lesions induce a rapid and premature stabilization of song, while preventing continued sensorimotor learning (Bottjer *et al.*, 1984; Scharff and Nottebohm, 1991). In contrast, lesions of the AFP in adult zebra finches have no noticeable effects on normal song production (Bottjer *et al.*, 1984; Sohrabji *et al.*, 1990; Scharff and Nottebohm, 1991; Nordeen and Nordeen, 1993). The effects of LMAN lesions on song development wane at around PHD 65 (Bottjer *et al.*, 1984), a time when song is becoming more adult-like, but well before crystallization. Finally, in isolated zebra finches, which normally display delayed learning, later LMAN lesions block the delayed learning (Morrison and Nottebohm, 1993). Taken together, these findings led to the hypothesis that the AFP is a specialized "learning pathway" critical for sensorimotor matching of the bird's own song to the acquired auditory template. It is worth noting, however, that juvenile deafening continues to disrupt song development past the time when LMAN lesions cease to exert song effects, implying some degree of dissociation between LMAN function and the role of auditory feedback in guiding sensorimotor learning.

More recent evidence shows that the AFP is also required for various forms of adult song plasticity, in addition to its essential role in sensorimotor learning. In adult zebra finches, auditory feedback is used to actively maintain the stable, crystallized song: when normal auditory feedback of song is blocked by deafening, song degrades; when auditory feedback is chronically distorted, the previously crystallized song destabilizes (Nordeen and Nordeen, 1992; Williams and McKibben, 1992; Okanoya and Yamaguchi, 1997; Woolley and Rubel, 1997; Leonardo and Konishi, 1999; Hough and Volman, 2002). Interestingly, song destabilization induced either by deafening or by vocal-nerve section is prevented by prior LMAN lesions, suggesting that an intact AFP is necessary for audition-dependent vocal plasticity in the adult (Williams and Mehta, 1999; Brainard and Doupe, 2000). In a similar vein, adult LMAN lesions perturb the seasonal re-expression of song plasticity in canaries and white-crowned sparrows (Nottebohm *et al.*, 1990; Benton *et al.*, 1998). Therefore, in several songbird species, LMAN is necessary for audition-dependent song plasticity in the adult, rather than functioning only to enable juvenile song learning. If so, one question that arises with regard to the sensitive period for sensorimotor learning is: what changes in the structure or function of LMAN, or its connections to RA, might influence the pronounced changes observed over development in the quality and quantity of vocal plasticity?

The AFP may play instructive and/or permissive roles in song plasticity (Mooney, 1999; Doupe et al., 2000). The instructive, or error-correction hypothesis asserts that the AFP provides information to the direct vocal motor pathway (i.e. RA) regarding the error between the memorized song model and the actual song, as reflected by auditory feedback. This error signal then adaptively modifies vocal motor circuitry to better match the desired output. The permissive model for AFP function postulates that LMAN activity enables plasticity in the direct pathway so that the circuit is able to respond to instructive, experience-dependent signals, generated either in the AFP, or elsewhere in the brain. Neither the juvenile nor the adult experiments examining the effects of LMAN lesions on song plasticity can distinguish between instructive or permissive roles, as both functions are presumably necessary for experience-dependent modification of song.

Support for an error-correction function of the AFP comes from several experimental directions. First, electrophysiologic recordings show that AFP neurons, like many other neurons in the song system, are selectively activated by presentation of the bird's own song (Doupe and Konishi, 1991; Doupe, 1997), and these auditory responses can be detected in alert as well as anesthetized zebra finches (Hessler and Doupe, 1999a, b). The selectivity of these responses develops gradually during sensory acquisition (Doupe, 1997; Solis and Doupe, 1997), can sometimes show preference for the tutor song (Solis and Doupe, 1999), and may be influenced by both auditory and vocal experience (Solis and Doupe, 1999, 2000). These properties are consistent with a role for AFP neurons in sensory calculations regarding the degree of match or mis-match between the actual and intended bird's own song. Second, LMAN and Area X neurons show song-related motor activity in singing adults, as expected of an error-correction circuit (Hessler and Doupe, 1999a, b). In the error-correction model, such activity in adults would actively maintain stable song, while in juveniles, it would provide adaptive signals that guide song learning. Social context, which is known to regulate aspects of song learning (Slater et al., 1988) and to have subtle influences on adult song production (Sossinka and Bohner, 1980; Jarvis et al., 1998), influences both this song-related neuronal activity (Hessler and Doupe, 1999a), and also singing-induced patterns of expression of the immediate early gene ZENK in the AFP (Jarvis et al., 1998). A final prediction of a presumptive error-correction circuit is that it should undergo activity-dependent plastic changes during sensory acquisition, as the template which will subsequently be used to compute motor errors is learned. In support of this prediction, particular synapses in LMAN display a developmentally limited form of activity- and NMDA-receptor dependent synaptic plasticity (Boettiger and Doupe, 2001). Furthermore, NMDA receptor blockade in LMAN during tutoring impedes sensory acquisition (Basham et al., 1996). Taken together, these data imply that experience-dependent modification of neuronal connectivity and activity in LMAN is a component of sensory acquisition and that, during sensorimotor learning, LMAN activity is part of an instructive error-correction signal guiding plasticity in the vocal motor pathway.

If LMAN activity conveys an online error-correction signal, either distorting auditory feedback or blocking auditory feedback altogether (via deafening) should produce significant changes in LMAN activity during singing. In two different experimental paradigms this is not the case: song-related LMAN activity patterns in adult zebra finches are essentially unaltered following deafening (Hessler and Doupe, 1999b) or when loud white noise is played every time the bird sings a particular syllable (Leonardo, 2004), even though both manipulations induce vocal plasticity. These data argue against a simple error-correction model for LMAN function, though LMAN activity may possibly convey an efference copy of song motor activity utilized as part of a presumptive instructional signal (Hessler and Doupe, 1999b; Troyer and Doupe, 2000a, b).

Although an instructional role for LMAN in sensorimotor learning is unclear, a permissive role is supported by the finding that LMAN exerts a trophic effect on RA early in development. Prior to the onset of singing (~PHD 20), lesions of LMAN induce significant cell death in RA (Akutagawa and Konishi, 1994; Johnson and Bottjer, 1994). LMAN lesions in older juveniles (at PHD 40–50) that disrupt sensorimotor learning do not induce RA cell death (Johnson and Bottjer, 1994), suggesting that gross regulation of RA cell number is unlikely to be the major mechanism for enabling vocal plasticity. However, LMAN input in the juvenile (and presumably the adult experiencing abnormal auditory feedback) could exert a more subtle effect on RA structure and function important for permitting vocal plasticity.

10.6 What are the neural correlates of song plasticity during sensorimotor learning?

A first step we undertook to better understand how LMAN might enable vocal plasticity is to describe properties of the juvenile direct vocal motor pathway associated with the sensitive period for sensorimotor learning. Our working hypothesis was that certain aspects of neuronal structure and function in the direct pathway allow the song to be plastic in juveniles prior to crystallization, and that these features are lost as vocal plasticity declines at the close of the sensitive period.

10.7 The nucleus RA is an important site to search for neuronal correlates of vocal plasticity

For a variety of reasons, we focused our attention on synaptic connectivity within the nucleus RA of the direct pathway. The nucleus RA is especially likely to be an important locus for changes in neuronal circuit structure and function that influence levels of sensitive period vocal plasticity. RA is a crucial site in the vocal motor pathway for controlling the production of learned song, because RA projection neurons (PNs) directly innervate the brainstem nuclei that control the activity of syringeal and respiratory muscles (Nottebohm *et al.*, 1976; Vicario, 1993; Wild, 1993a, b). Furthermore, individual RA PNs receive convergent excitatory inputs from LMAN (the AFP output) and from the descending premotor pathway, via the nucleus HVC (Canady *et al.*, 1988; Mooney and Konishi, 1991; Mooney, 1992). In fact, RA is the only nucleus in the direct pathway receiving AFP input (Fig. 10.2). Thus, instructive or permissive changes in song-motor circuitry mediated by the AFP are likely to occur at the level of RA.

10.8 Neuronal codes for song in the nucleus RA

Ultimately, an understanding of the neural mechanisms of vocal plasticity will depend on a thorough understanding of the neural mechanisms for learned vocal control. Fortunately, a variety of experiments have illuminated the electrophysiologic properties of RA neurons, including their activity patterns in singing birds. These studies give insight into neural codes for song and present clues as to which features of RA neuronal activity need to be modified to affect song quality. Individual RA PNs fire temporally precise bursts of action potentials during singing that are correlated with the production of specific syllables or notes (Yu and Margoliash, 1996; Chi and Margoliash, 2001). Recordings from multiple PNs in single birds suggest that distinct ensembles of RA PNs fire synchronously during the production of each syllable (Leonardo and Fee, 2005). Various models of song production based on these data postulate that the synchronous activation of specific ensembles of RA PNs drives the coordinated contraction of the different muscles of the syrinx and abdomen to produce particular vocal gestures. The precise sequential activation of different RA ensembles would thus produce sequences of notes and syllables (Yu and Margoliash, 1996; Troyer and Doupe, 2000a, b). During sensorimotor learning, changes in the composition of the different ensembles could influence which sounds were produced and when they were produced in a sequence of sounds (Doya and Sejnowski, 1994; Troyer et al., 1996; Troyer and Doupe, 2000a, b).

10.9 Extrinsic and intrinsic patterns of synaptic connectivity in the nucleus RA

If such a model correctly describes RA's role in song control, then both extrinsic and intrinsic patterns of synaptic connectivity onto RA PNs are likely to be the major factors for determining the composition of RA ensembles. As neurons in the song premotor nucleus HVC pattern song and directly innervate RA, the pattern of input from individual RA-projecting HVC neurons (HVC_{RA} neurons) to RA PNs poses an attractive means for determining ensemble composition. This projection appears to be both highly convergent and highly divergent. Evidence for synaptic convergence from HVC to RA comes from electrophysiologic recordings in singing birds, which show that HVC_{RA} neurons burst only once per motif (Hahnloser et al., 2002), in contrast to the multiple bursts per motif exhibited by RA PNs (Yu and Margoliash, 1996). Evidence for divergence comes from the reconstruction of single HVC_{RA} axons, which extend over vast regions of RA (Kittelberger and Mooney, 1997). Therefore, HVC neurons could coordinate the firing of RA neurons over a wide extent of the nucleus. The local RA circuit also provides a rich source of connections likely to be important to establishing which RA PNs participate in a given ensemble. Recordings in brain slices show that RA interneurons form widespread inhibitory connections that can synchronize the spiking of pairs of RA PNs, possibly providing an *in vitro* analog of the temporal coordination that these neurons display during singing (Spiro et al., 1999). Important to the consideration of how the AFP might influence vocal patterning, these interneurons also receive convergent excitatory inputs from HVC and LMAN (Spiro et al., 1999), and thus constitute an additional cellular locus where the AFP could affect song quality. A final source of local connections arises from the RA PNs themselves: they possess extensive axon collaterals that innervate other PNs as well as interneurons (Canady et al., 1988; Herrmann and Arnold, 1991a; Perkel, 1995).

10.10 The structure and function of RA subregions

The spatial organization of RA may also provide some insight into how modification of synaptic connections in RA could be important to song patterning. A series of detailed anatomical studies reveals that RA is topographically organized: neurons in the dorsal part of RA project to the respiratory premotor areas in the lateral medulla, and neurons in the more ventral parts project topographically onto the various motor-neuron pools innervating the different syringeal muscles (Vicario and Nottebohm, 1988; Vicario, 1991, 1993). Intriguingly, such topography is not maintained between HVC and RA, further underscoring the divergent and convergent nature of the HVC–RA pathway, and hinting that an important transformation in synaptic organization occurs at the level of the HVC–RA synapse. In contrast to the HVC–RA projection, LMAN inputs map topographically onto RA, preserving the spatial segregation of areas important to syringeal versus respiratory patterning (Johnson *et al.*, 1995; Vates and Nottebohm, 1995). Furthermore, this topography is maintained throughout the AFP: the nuclei of the AFP are topographically aligned so that the maps in all three AFP nuclei and in RA are in register (Luo *et al.*, 2001).

Functional connectivity among the different subregions of RA's vocal-respiratory map is likely to have important consequences for the coordination of downstream muscle groups to produce the complex acoustic elements of song. Specifically, the exact timing of individual syllables is regulated predominantly by expiratory–respiratory activity, while the spectral quality of individual notes is determined by (ventral) syringeal muscle tension. As both of these features are learned, and also highly synchronized with each other during singing, how neurons in different subregions of RA are recruited by HVC axons is likely to influence song structure.

The functional significance of topography in RA and the AFP is unknown, but one possibility is that it constitutes part of a motor map used in an instructive manner during song learning. Experience-dependent refinement of the topography of connections onto or within RA could play an important role in sensitive period vocal plasticity. The topography of the LMAN to RA projection is, in fact, refined developmentally, and this refinement is disrupted by early deafening, suggesting that it could be influenced during sensory acquisition (Iyengar *et al.*, 1999; Iyengar and Bottjer, 2002). The maturation of topography between LMAN and RA is completed quite early (by PHD 35), prior to the onset of singing. Although gross modifications to the map do not occur during sensorimotor learning, these results do not exclude the possibility that the establishment of the topographic projection is a necessary prelude to vocal learning.

10.11 A cellular analysis of the nucleus RA during sensorimotor learning

A variety of anatomical studies in zebra finches have cataloged numerous changes to RA's anatomical features over development. Early in development (i.e. ~PHD 15), RA is innervated by LMAN, but not by HVC (Mooney, 1992; Mooney and Rao, 1994), and there are no obvious sexual dimorphisms in RA volume, total cell number, or cell size (Konishi and Akutagawa, 1985, 1987, 1990). Just prior to the onset of singing in male zebra finches (around PHD 30), HVC axons enter RA and form synaptic connections (Konishi and Akutagawa, 1985; Mooney, 1992). This event is rapidly followed by a dramatic increase in RA's volume and cell soma area (Bottjer et al., 1985; Konishi and Akutagawa, 1985, 1987) and by the onset of singing (Immelmann, 1969; Arnold, 1975). In females, HVC axons fail to enter RA, which begins to atrophy as a result of selective apoptosis (Konishi and Akutagawa, 1987; Kirn and DeVoogd, 1989; Konishi and Akutagawa, 1990). In zebra finches, as in many other songbirds, sexual dimorphism in RA is paralleled by sex differences in the song behavior itself: only males sing.

The sexual differentiation of RA depends on the expression of estrogen receptors and innervation from HVC (Gurney and Konishi, 1980; Gurney, 1981; Konishi and Akutagawa, 1987; Herrmann and Arnold, 1991b; Akutagawa and Konishi, 1994; Burek et al., 1995), but not on auditory experience (Burek et al., 1991). In males, RA cell number remains constant (Konishi and Akutagawa, 1987; Kirn and DeVoogd, 1989), though RA continues to grow in both volume and cell size (and to decrease in cell density), reaching a peak at around PHD 45–60 days of age, and regressing slightly between this peak and song crystallization at PHD 90 (Konishi, 1985; Herrmann and Bischof, 1986; Konishi and Akutagawa, 1987). Electron microscopy studies show that the number and density of synapses in RA follow a similar time course of overproduction and regression (Herrmann and Arnold, 1991a). The density and number of synapses from LMAN axons onto RA neurons, in particular, decline over the course of sensorimotor learning (Herrmann and Arnold, 1991a), an interesting finding in light of the qualitative and quantitative changes in LMAN-mediated vocal plasticity over development.

The developmental trajectories of song plasticity and of these features of RA's anatomy are remarkably similar: the peak in synapse number, soma size, and RA volume coincides with the height of vocal plasticity and sensorimotor learning, and declines in parallel with vocal plasticity, stabilizing as song stabilizes. This time course is delayed relative to the maturation of surrounding non-song related brain structures (Herrmann and Bischof, 1986), strengthening the idea that these changes are functionally linked to song development. Developmental overproduction and winnowing of synapses is a common phenomenon in various neuronal systems (Purves and Lichtman, 1980; Rakic *et al.*, 1986; Lohof *et al.*, 1996; Huttenlocher and Dabholkar, 1997; Lichtman and Colman, 2000), and in at least some cases, synapse elimination contributes to experience-dependent neural circuit plasticity during the sensitive period (Hubel *et al.*, 1977; Wiesel, 1982; Udin, 1989; Antonini and Stryker, 1993). These findings raise the possibility that overproduction of synapses is an important prerequisite for neural and behavioral sensitive periods generally, and for vocal plasticity during sensorimotor learning in songbirds in particular (see below for further discussion of this point).

10.12 A functional synaptic analysis of the nucleus RA during sensorimotor learning

Interestingly, LMAN and HVC axons converge on the same RA neurons, where they activate different complements of post-synaptic receptors: LMAN axons primarily activate NMDA receptors, while HVC axons activate a mixture of alpha-amino-3-hydroxy-5-methyl-4-isoxazolepropionic acid (AMPA) and NMDA receptors in juvenile zebra finches during sensorimotor learning (PHD 35–90) (Mooney and Konishi, 1991; Mooney, 1992). The functional properties of these two sets of synapses change over the course of development, and some of these changes closely parallel the song learning process. At LMAN–RA synapses, the decay of the NMDA component of the post-synaptic response becomes faster over development (Stark and Perkel, 1999; White *et al.*, 1999). This form of synaptic maturation can be accelerated by androgens and delayed by deafening, treatments that respectively hasten or delay the onset of song crystallization (White *et al.*, 1999; Livingston *et al.*, 2000). Although these findings are consistent with the idea that the transition to fast currents plays a role in crystallization, the normal maturation of NMDA-EPSC decay times at the LMAN–RA synapse may be complete by PHD 50, well-before crystallization (Stark and Perkel, 1999), and is apparently not affected by isolation, which can delay crystallization (Livingston *et al.*, 2000). At the HVC–RA synapse, NMDA current decay times also decrease over development, and the relative contribution of NMDA versus AMPA currents declines. Furthermore, this latter feature may decrease late enough in sensorimotor learning to contribute to crystallization (Stark and Perkel, 1999). Thus, there are changes in synaptic connectivity, measured both anatomically and physiologically, which could be functionally relevant for developmental declines in vocal plasticity.

242   10 *Synaptic mechanisms and sensitive periods for song learning*

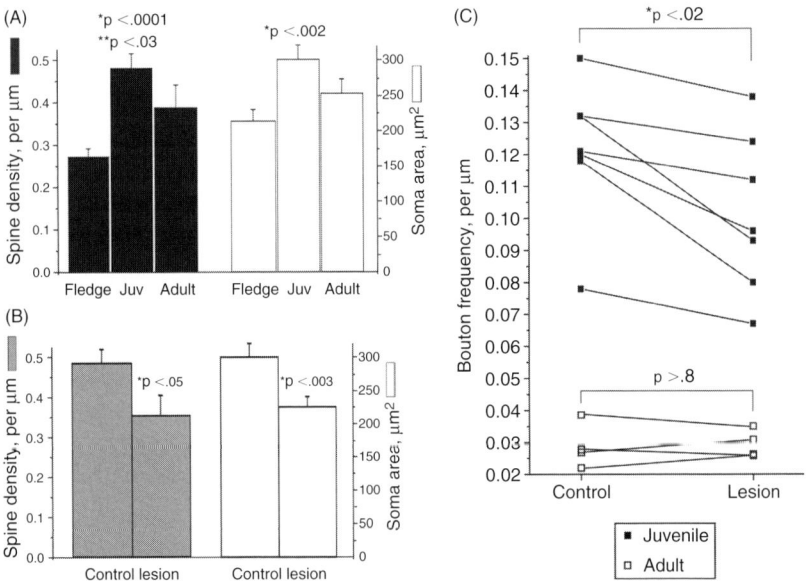

Figure 10.3. The morphology of RA projection neurons and of HVC-RA axonal arbors change over development and following juvenile LMAN lesions. A. Both dendritic spine density and soma size of RA projection neurons peak developmentally in juvenile birds (~PHD 45) during early sensorimotor learning, relative both to fledglings (*) (~PHD 28) who have not yet begun to sing, and in whom HVC has yet to innervate RA, and to adults (**) (>PHD 120) who have crystallized their songs. B. Juvenile LMAN lesions, which precipitate the rapid stabilization of song and block further progressive matching to the memorized song template, also have significant effects on the morphology of RA projection neurons. Within 4–7 days post lesion, RA projection neurons had significantly lower spine densities and soma sizes than in age-matched controls. Note that the lesion-induced morphological changes are of similar magnitude as the normal developmental changes, illustrated in A, observed over the course of sensorimotor learning. C. The frequency of presumptive synaptic boutons along HVC axons in RA was significantly reduced by juvenile LMAN lesions, and was lower in adults than in juveniles. Linear bouton frequency was measured after unilateral lesion of LMAN. In juveniles, between-hemisphere, within-bird comparisons revealed consistently lower bouton frequency in RA on the side receiving the LMAN lesion. In adults, on the other hand, in whom LMAN lesions have no obvious effects on normal song production, lesions did not affect HVC bouton frequency in RA, which was higher on the lesioned side just as often as it was lower. From Kittelberger and Mooney, 1999.

Studies we have conducted (Kittelberger and Mooney, 1999) extend these anatomical and electrophysiologic findings, revealing significant developmental changes in a variety of different structural and functional measures of synaptic connectivity within RA. These changes include decreases in RA PN spine density (Fig. 10.3a), in the length and complexity of RA PN local axon

10.12 A functional synaptic analysis of the nucleus RA 243

collateral arbors, and in the frequency of pre-synaptic boutons along HVC axons innervating RA (Fig. 10.3c), as well as an increase in the strength of the HVC–RA synapse (Fig. 10.4). All of these changes occur between PHD 45 and adulthood, in parallel with the developmental decline of vocal plasticity. These findings, together with the studies of others described above, support the hypothesis that synapses are overproduced in the juvenile RA, and then eliminated as sensorimotor learning proceeds.

While elevated synapse number in RA correlates with juvenile periods of song plasticity, these additional synapses may be functionally unrelated to behavioral plasticity. One way we probed this correlation further was to ask whether lesions of LMAN in juvenile birds, which cause a rapid premature crystallization of song, also influence synaptic connectivity in RA (Kittelberger and Mooney, 1999). In addition to testing whether higher numbers of synapses correlate with song plasticity, these experiments also provide insights into the function of LMAN with regard to RA circuit development. If LMAN's role is primarily instructive, then lesions of LMAN should block developmental decreases in RA synapses, as they block the progressive development of song. If, on the other hand, LMAN's role is primarily permissive, supporting a state of connectivity in RA that enables vocal plasticity, then juvenile LMAN lesions should trigger premature synapse elimination, resulting in juvenile birds with adult-like synaptic connectivity in RA.

Our findings support the idea that LMAN plays a permissive role in maintaining high numbers of synapses in the juvenile RA: within several days after juvenile LMAN lesions, synapse number in RA declines significantly, as measured by RA PN spine density (Fig. 10.3b), HVC–RA axonal bouton frequency (Fig. 10.3c), and the complexity of RA PN local axon collateral arbors. Importantly, these structural changes coincide with increases in the strength of excitatory synaptic transmission in RA by two separate measures, suggesting that synapse elimination is paralleled by synaptic consolidation, as occurs more slowly over normal development (Fig. 10.4) (Kittelberger and Mooney, 1999). In contrast, adult LMAN lesions, which do not affect normal adult song production or stereotypy, had no effect on at least one of these parameters, namely, HVC–RA bouton frequency (Fig. 10.3c).

The impact of LMAN lesions on the synaptic connectivity of the juvenile RA supports two conclusions. First, synaptic connectivity in RA covaries with the level of vocal plasticity rather than with age, supporting the hypothesis that the high number of relatively weak synapses in RA in juveniles could enable song plasticity. Second, LMAN input to RA in juveniles is needed to maintain the high number of synapses found in RA at this time, consistent with a trophic role for LMAN. One consequence of such a trophic effect could be to create a synaptic state in RA permissive for vocal plasticity.

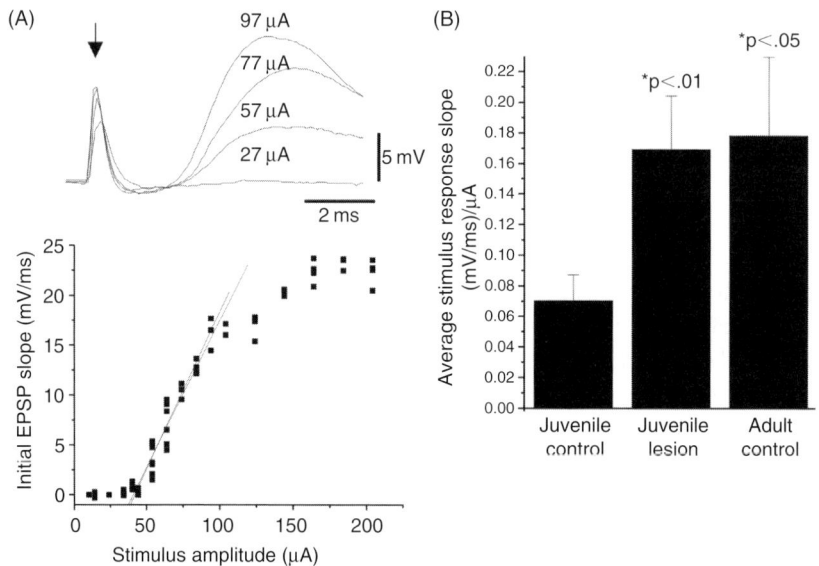

Figure 10.4. The strength of the HVC-RA synapse increases during sensorimotor learning and rapidly after LMAN lesions. The stimulus–response relationship of EPSPs evoked in RA by electrical stimulation of HVC fibers in brain slices increased both over normal development and after LMAN lesions. A. The onset slope and amplitude of EPSPs evoked in an adult RA projection neuron (*top traces*) by HVC fiber stimulation (at time marked by *arrow*) increased as a function of current amplitude (stimulus levels shown above each trace). Currents above 40mA evoked an EPSP from this RA projection neuron; higher currents elicited increasingly larger EPSPs, with steeper onset slopes. At even higher stimulus intensities, these EPSPs triggered action potentials, distinguishing them as excitatory (not shown). Measurements of the initial slope of the EPSP at different stimulus currents were used to plot the stimulus–response relationship for this cell (*bottom graph*), for which the two best linear fits of the stimulus–response data are superimposed. B. The average stimulus–response relationship (i.e., the mean of the slopes of the two best linear fits of the stimulus–response data) is shown for HVC EPSPs recorded from RA projection neurons in normal juveniles, age-matched juveniles 4–7 days after ipsilateral LMAN lesions, and adults. The stimulus–response relationship was significantly steeper both in LMAN-lesioned juveniles and in adults relative to control juveniles. From Kittelberger and Mooney, 1999.

10.13 Molecular effectors of synaptic and vocal plasticity

What are the molecular signals involved in actively maintaining circuit structure and function in RA in juveniles to enable vocal plasticity? A wide variety of findings support the idea that neurotrophins in general, and brain-derived neurotrophic factor (BDNF) in particular, can influence synapse structure and

10.13 Molecular effectors of synaptic and vocal plasticity

function, and that such influence could be important to the regulation of neural and behavioral sensitive periods. In the mammalian and avian brain, BDNF can exert effects on neuronal morphology, synapse number, and synaptic strength (Lohof *et al.*, 1993; Cohen-Cory and Fraser, 1995; Kang, H and Schuman, E 1995; Kang, HJ and Schuman, EM 1995; McAllister *et al.*, 1995; Murphy and Segal, 1997; Shimada *et al.*, 1998; Poo, 2001). These effects can be both cell-type specific and activity dependent (see McAllister *et al.*, 1999; Poo, 2001, for reviews). Neurotrophins have been shown to control total synapse number and density at several specific synapses both *in vivo* and *in vitro* (Nja and Purves, 1978; Purves *et al.*, 1988; Causing *et al.*, 1997; Murphy *et al.*, 1998; Shimada *et al.*, 1998). The synaptic effects of BDNF may be involved in regulating structural and functional plasticity underlying the sensitive period for ocular dominance plasticity in the visual cortex (Cabelli *et al.*, 1995, 1996, 1997; Bonhoeffer and Shatz, 1998; Huang *et al.*, 1999). Both adding exogenous BDNF and blocking the endogenous BDNF receptor tyrosine kinase B (trkB) results in unsegregated LGN inputs and a lack of ocular dominance columns (Cabelli *et al.*, 1995, 1997). One interpretation is that LGN axons compete for limited amounts of BDNF, whose release from layer four neurons is induced by synchronous patterns of pre-synaptic activity, and which in turn strengthens existing synapses or induces the formation of new synapses from the active inputs (Bonhoeffer and Shatz, 1998; McAllister *et al.*, 1999). Permissive models are also plausible: one possibility is that developmental regulation of BDNF expression or function could regulate the timing of the sensitive period itself, enabling plasticity only during this particular time window. Consistent with such a model, it has been shown that changes in cortical inhibition can alter sensitive period timing (Hensch *et al.*, 1998; Fagliolini and Hensch, 2000), and that BDNF can induce the maturation of these inhibitory interneurons and thus indirectly induce the opening of the sensitive period (Huang *et al.*, 1999). Finally, BDNF may also be involved in some forms of adult learning (Minichiello *et al.*, 1999; Hall *et al.*, 2000; Tokuyama *et al.*, 2000). In short, BDNF could provide a link between synaptic modification and sensitive periods for neural and behavioral plasticity.

Numerous findings suggest that BDNF plays an important role in the song system, and could mediate synaptic and cellular effects important to song learning and maintenance. First, BDNF is expressed in the right place: BDNF and its receptor trkB are expressed in RA (Johnson *et al.*, 1997; Akutagawa and Konishi, 1998; Dittrich *et al.*, 1999), and BDNF mRNA is expressed in both HVC and LMAN, pointing to both of these areas as potential sources of the BDNF protein seen in RA (Dittrich *et al.*, 1999; Li and Jarvis, 2001). Second, BDNF levels increase following singing (Li and Jarvis, 2001) and after steroid hormone treatment (Dittrich *et al.*, 1999). Third, BDNF expression in RA may

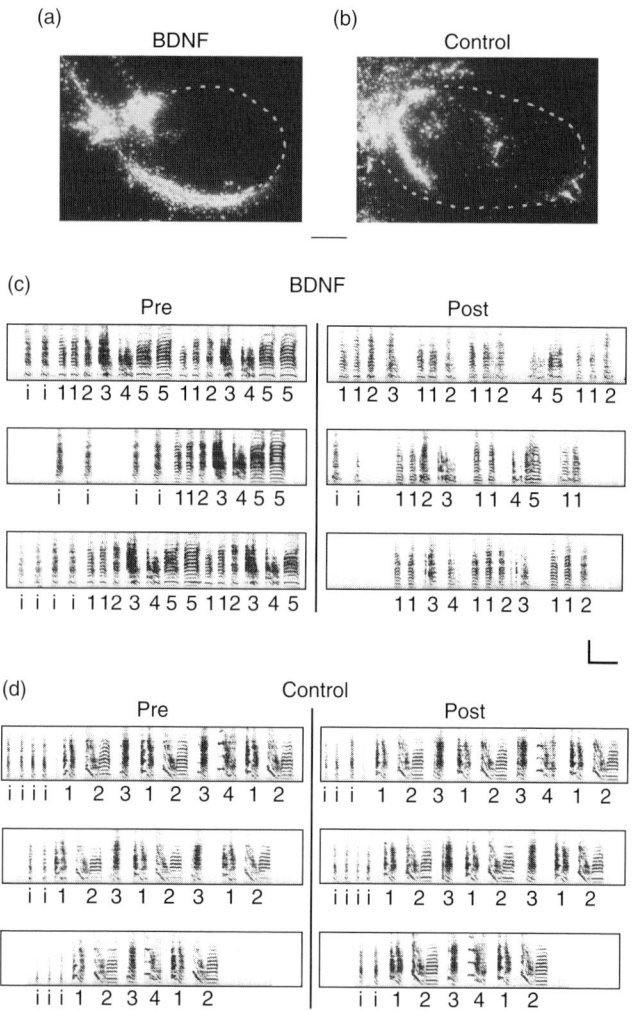

Figure 10.5. BDNF, but not control, injections to RA induced variable deletion of syllables in adult zebra finches previously singing stable songs. (A) and (B) Examples of size-matched BDNF and heat-inactivated BDNF injections to RA. Fluorescent latex microspheres, from which bound BDNF does not diffuse (Riddle et al., 1997), were injected into RA bilaterally. RA's borders, as determined from adjacent sections stained with cresyl violet, are outlined with dashed white lines. Scale bar: 200 mm. (C) and (D) Sonograms of three pre- and three post-injection songs of the two birds whose injections are shown in A and B. Note that, prior to injection, both birds' songs are highly stereotyped in syllable order. As early as the first day of song post-injection, the BDNF-injected bird (C, right) began to drop syllables, varying which syllables were present in each motif. Both qualitatively and quantitatively, the song variability following BDNF injections to

10.13 Molecular effectors of synaptic and vocal plasticity

decline between PHD 45 and adulthood (Akutagawa and Konishi, 1998), supporting the hypothesis that BDNF contributes to vocal plasticity, although it is important to note that these data have been disputed (Johnson et al., 2000). Fourth, RA cell death normally triggered by LMAN lesions prior to PHD 20 can be prevented by the infusion of BDNF into RA (Johnson and Bottjer, 1994; Johnson et al., 1997). Finally, BDNF can be anterogradely transported from LMAN to RA (Johnson et al., 1997), suggesting that LMAN could be an endogenous source of BDNF protein in RA. Indeed, anterograde transport and release of BDNF have been previously described in other systems (Altar et al., 1997; Altar and DiStefano, 1998; Kohara et al., 2001). One idea is that LMAN-mediated release of BDNF into the RA of the juvenile maintains the exuberant synaptic connectivity necessary for vocal plasticity, and developmental declines in BDNF signaling trigger synapse elimination and song crystallization.

If developmental declines in BDNF expression contribute to song crystallization, then experimentally elevating BDNF levels in adult birds should induce juvenile-like RA circuit properties and destabilize song. To test this idea, we made focal injections of BDNF-coated and control fluorescent latex microspheres into RA in adult male zebra finches (Kittelberger and Mooney, 2005). Notably, BDNF injections into RA were accompanied by a transient increase in the variability of song sequencing (Fig. 10.5). This BDNF-induced variability shows qualitative and quantitative similarities to the sequence variability expressed by juvenile birds prior to the closure of the sensitive period for sensorimotor learning (as in Fig. 10.1b). Furthermore, the transient destabilization of song stereotypy by BDNF led to permanent changes in song structure: when BDNF-injected birds' songs re-stabilized (typically around 1 week post injection), the majority of these birds had permanently deleted one or more of their pre-injection syllables (Kittelberger and Mooney, 2005). BDNF treatment also resulted in structural changes to the RA neuropil reminiscent of the juvenile: HVC–RA bouton frequency increased over the same time scale as the behavioral effects, strengthening the link between BDNF, synaptic exuberance, and behavioral plasticity. Thus, in both song phenotype and one measure of RA synaptic connectivity, BDNF-injected adults resembled normal juveniles. The most direct conclusion from these experiments is that low levels of BDNF in the adult are necessary for song stability. These data also lend indirect support

Caption for fig. 10.5. (cont.)
RA resembled that observed in normal juveniles during sensorimotor learning (compare C, right, with Figure 2B). In contrast, the control bird (D, right) sang a song essentially unchanged in syllable order (and the variability thereof) from before injection. Scale bar: y: 5 kHz, x: 0.2s. From Kittelberger and Mooney, 2005.

248   *10 Synaptic mechanisms and sensitive periods for song learning*

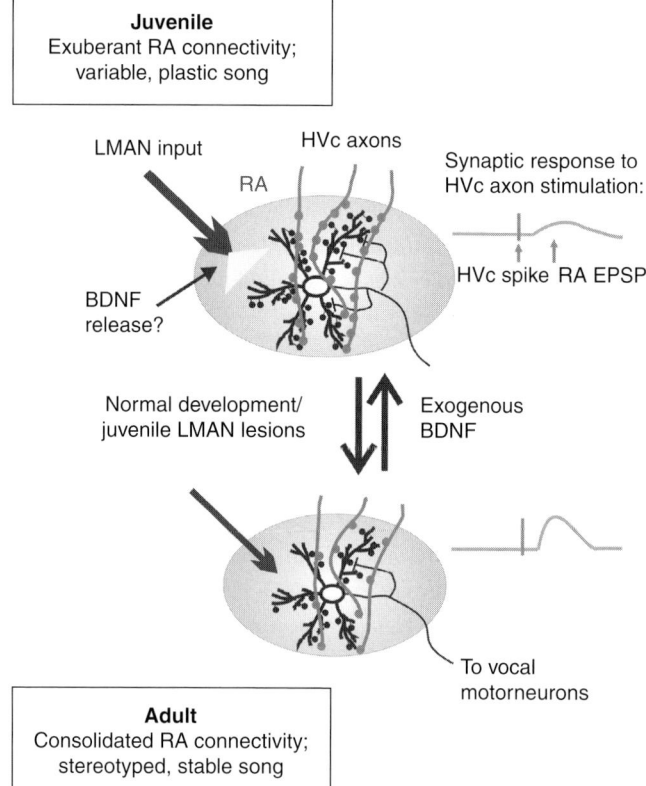

Figure 10.6. Model: Neural circuit correlates of the sensitive period for sensorimotor learning. This model outlines in brief our experimental findings, emphasizing the developmental, BDNF- and LMAN-dependent overproduction of synapses in RA. In juvenile birds singing plastic song, RA is characterized by an excess of relatively weak synaptic connections, including large numbers of RA projection neuron (PN) dendritic spines (black circles) and HVC axon synaptic boutons (which synapse onto these spines (Canady et al., 1988), (gray circles) and elaborate RA PN local axon collaterals. Over development, synapses are eliminated and vocal plasticity declines: the adult RA has fewer RA PN spines, fewer HVC axonal boutons, and less extensive RA PN collateral arbors. At the same time, synaptic responses to HVC stimulation increase. Juvenile LMAN lesions cause, simultaneously, both dramatically decreased vocal plasticity and an RA circuit state that resembles the normal adult phenotype in all of these structural and functional parameters. Exogenous BDNF injections to the adult RA cause reversible increases in song variability, and an increase in HVC-RA boutons, suggesting that BDNF is pushing both circuit and behavior back towards a juvenile-like phenotype. These results are consistent with a model in which an LMAN- and BDNF-dependent state of synaptic connectivity in the juvenile RA enables song plasticity, and raise the possibility that developmental decreases in BDNF signaling trigger the observed synaptic consolidation in RA and song crystallization.

to the hypotheses that higher levels of BDNF in the juvenile RA promote a state of synaptic connectivity in RA permissive for vocal plasticity, and that maturational declines in BDNF expression trigger song crystallization, terminating the sensitive period for sensorimotor learning.

Together, our own studies and those of others lend support to a model whereby the developmental, BDNF-dependent overproduction of synaptic connections in RA is an important contributor to juvenile vocal plasticity during sensorimotor learning (Fig. 10.6). Computational models of song learning suggest that high levels of uniformly weak HVC–RA synapses may fail to drive well-correlated ensemble activity in RA, resulting in song variability (Troyer and Doupe, 2000a, b). Song variability itself may be critical for sensorimotor learning, allowing the young bird to fully explore the relevant vocal-acoustic space, and thus providing the grist for subsequent selective instruction (Marler, 1997), much as genetic variation provides the substrate for natural selection. With maturation, song variability and plasticity wane as synapses are consolidated and those that remain are strengthened. We believe that BDNF is a key factor that maintains the individually weak but numerically exuberant HVC–RA synapses essential to this model.

10.14 Alternative possibilities and further tests of the model

A variety of additional experiments are needed to directly test the role of *endogenous* BDNF in regulating vocal plasticity and synaptic connectivity. A crucial test is to block endogenous BDNF signaling (using, e.g. soluble trkB-IgG fusion proteins (Shelton *et al.*, 1995)) in juvenile birds. If endogenous BDNF acts as proposed here, such treatment should reduce juvenile vocal plasticity, potentially triggering premature crystallization similar to that seen following LMAN lesions. In addition, the number and strength of HVC–RA synapses in trkB-IgG treated juveniles should resemble those found in normal adults and LMAN-lesioned juveniles. If BDNF acts more generally throughout life to enable vocal plasticity, then other LMAN-dependent forms of vocal plasticity, such as the degradation of song following adult deafening (Brainard and Doupe, 2000), should be blocked by preventing BDNF signaling in RA. Further characterization of the details of BDNF and trkB expression in RA will also contribute to our understanding of the function of endogenous BDNF. When and in which cell types are BDNF and trkB expressed? How do periods of behavioral plasticity correlate with patterns of BDNF expression in the song system? Is BDNF expression LMAN- and/or activity dependent? In HVC, BDNF expression can be induced by singing, and enhances the survival of new HVC–RA neurons in adult canaries (Rasika *et al.*, 1999; Li *et al.*, 2000), but neither the functional ramifications of these effects within HVC, nor the corresponding details regarding BDNF expression in RA have as yet been fully elucidated.

Correlations between HVC–RA synaptic connectivity and vocal plasticity can be further tested through a variety of behavioral manipulations. If the developmental overproduction of synapses in RA is crucial for sensitive period vocal plasticity, then manipulations affecting vocal plasticity, or the timing of crystallization, should have predictable effects on synaptic connectivity. For example, juvenile deafening, castration, and acoustic isolation can all, to some degree, extend the sensitive period for sensorimotor learning and/or delay song crystallization (Konishi, 1965; Arnold, 1975; Price, 1979; Marler *et al.*, 1988; Morrison and Nottebohm, 1993). If declines in synapse number or increases in synapse strength are functionally related to song crystallization, then each of these manipulations should delay the developmental maturation of RA synaptic connectivity. Conversely, manipulations that induce vocal plasticity and variability in adult birds, such as deafening (Nordeen and Nordeen, 1992; Lombardino and Nottebohm, 2000), delayed auditory feedback (Leonardo and Konishi, 1999), or distorted auditory feedback due to tracheosyringeal nerve cuts (Williams and McKibben, 1992), or implanting beads in the syrinx (Hough and Volman, 2002), should cause the re-expression of juvenile patterns of synaptic connectivity in RA. One assumption of such an experimental approach is that these adult forms of vocal plasticity rely on the same mechanism that enables sensorimotor learning in the juvenile. A common mechanism for juvenile and adult vocal plasticity is likely because LMAN lesions block vocal plasticity in deafened and nerve-sectioned adults, and also prevent extended learning in adult isolates (Morrison and Nottebohm, 1993; Williams and Mehta, 1999; Brainard and Doupe, 2000).

A more extensive survey of synapse number and strength in RA over normal development could help confirm or reject the hypothesis that these features underlie aspects of song plasticity. We have only sparsely sampled the developmental period when song learning occurs: in fledglings at PHD 25–30, in juveniles singing highly plastic song at PHD 45, and in adults singing stable crystallized song. What is the temporal profile of the consolidation in synaptic connectivity that occurs between PHD 45 and adulthood? Is this decrease gradual, or delayed and sudden, with a sharp drop at song crystallization? A third possibility is that synapse number declines rapidly after PHD 45, so that by PHD 60, when song is still highly variable, the structural and functional properties of the HVC–RA synapse are already adult like. The first two patterns would be consistent with the idea that synapse number and strength contribute to vocal plasticity,

10.14 Alternative possibilities and further tests of the model 251

but the latter would not be. Thus, assaying synapse number and strength in RA at PHD 60 will be an important test of the validity of the proposed model.

Finally, further characterization of the effects of exogenous BDNF injections in adults, and the mechanisms by which these injections influence vocal plasticity, will be important. The results of our developmental and LMAN-lesion experiments would predict that BDNF overexpression in the adult RA should cause increases in other measures of synapse number, such as RA PN spine density, and decreases in excitatory strength at HVC–RA synapses. Another possibility is that BDNF may influence inhibition in RA. In other systems, BDNF signaling can influence the maturation of inhibitory interneurons and levels of functional inhibition (Rutherford *et al.*, 1997, 1998; Huang *et al.*, 1999; Rico *et al.*, 2002), and developmental changes in inhibition in the visual cortex can, in turn, influence the timing of the critical period for ocular dominance plasticity (Hensch *et al.*, 1998; Huang *et al.*, 1999; Fagliolini and Hensch, 2000). In RA, GABAergic interneurons can synchronize the activity of RA PNs (Spiro *et al.*, 1999), and BDNF-mediated changes in inhibition would likely have important consequences for vocal output. Examining whether BDNF can influence inhibition in RA to modulate vocal plasticity could be an important test of the general applicability of this new model for the regulation of sensitive period plasticity. In this light, it is worth noting an important distinction between our results and those of Hensch *et al.* (1998) and Fagliolini and Hensch (2000). We show that re-elevating BDNF levels in adults can re-induce aspects of juvenile sensitive period plasticity, whereas re-expressing high levels of inhibition pharmacologically did not re-open the sensitive period for sensitivity to monocular deprivation once this window had been closed (Fagliolini and Hensch, 2000). The significance of this difference will be an important area to explore in future experiments.

A remaining question is whether developmental regulation of HVC–RA connectivity reflects an instructive or merely permissive process. The various studies we have already discussed are consistent with a permissive role for LMAN in vocal learning. One possible extension of the experimental data presented here is that the developmental elimination of synapses within RA contains an instructive component directly related to sensorimotor learning. Perhaps, as LMAN-dependent trophic support wanes (reflected in the developmental decline in LMAN input (Herrmann and Arnold, 1991a), and BDNF expression in RA possibly decreases (Akutagawa and Konishi, 1998), in RA), activity-dependent selective mechanisms determine which synapses are maintained and strengthened. Such instructive selection could strengthen those synapses contributing to motor patterns that produce song matching the memorized song template. Thus, selective maintenance of RA synapses should be dependent on auditory feedback.

One caveat is that demonstrating experience-dependent synaptic *selection* in RA is important for sensorimotor learning could be challenging. If synapse elimination *per se* is not experience dependent, as suggested by the results of the LMAN-lesion experiments, it will be necessary to show that the pattern, or identity, of the maintained synapses is guided by auditory feedback. While RA is roughly topographically mapped relative to the muscles engaged in singing (Vicario and Nottebohm, 1988; Vicario, 1991, 1993), there is no obvious mapping of HVC inputs to RA (Kittelberger and Mooney, 1997), or of RA's intrinsic connections. Thus, there is no known spatial patterning of synaptic connectivity in RA, as there is, for example, in the visual cortex, somatosensory barrel cortex, or barn owl midbrain, on which to base an assay for the instructive effects of experience. While our own experiments do not speak directly to the relationship of either synapse elimination or BDNF function to auditory feedback dependent instruction of vocal output, such relationships represent a potentially exciting area of future study.

10.15 Conclusions

In summary, our studies show:

1. A peak density of synaptic connections in RA at a developmental time corresponding with the height of song plasticity. While HVC–RA synaptic connectivity is numerically exuberant, individual synapses are functionally weak. These synapses undergo consolidation over the course of song learning, as synapse number declines and the strength of the remaining synapses increases.
2. The maintenance of high numbers of relatively weak synapses in the juvenile RA depends on the presence of LMAN inputs.
3. Exogenous BDNF injections to RA destabilize particular aspects of song phenotype, transiently increasing the variability of syllable sequencing and causing permanent changes in song structure. At the same time, BDNF injections cause increased HVC–RA bouton frequency.
4. In three separate experimental paradigms (normal development, juvenile LMAN lesions and adult BDNF injections) aspects of synaptic connectivity vary consistently in parallel with levels of song plasticity and variability.

These observations lead us to suggest a model wherein endogenous, LMAN-dependent signals, most likely including BDNF, actively enable vocal plasticity during the sensitive period for sensorimotor learning by maintaining high numbers of weak excitatory synaptic connections within RA, particularly between HVC and RA neurons (Fig. 10.6). Developmental changes in these permissive signals, which may include the retraction of LMAN inputs and/or a decrease in BDNF signaling, trigger synaptic consolidation in RA, resulting in diminished vocal plasticity and song crystallization. Although the mechanisms underlying sensitive period behavioral plasticity are likely to be both multifactorial and interdependent, the various results discussed here support the idea that neurotrophin-dependent synaptic overproduction is an important regulator of sensitive periods for vocal plasticity in songbirds.

REFERENCES

Adkins-Regan E, Abdelnabi M, Mobarak M, Ottinger MA (1990) Sex steroid levels in developing and adult male and female zebra finches (*Poephila guttata*). *Gen Comp Endocrinol* 78: 93–109.

Akutagawa E, Konishi M (1994) Two separate areas of the brain differentially guide the development of a song control nucleus in the zebra finch. *Proc Natl Acad Sci USA* 91: 12413–12417.

Akutagawa E, Konishi M (1998) Transient expression and transport of brain-derived neurotrophic factor in the male zebra finch's song system during vocal development. *Proc Natl Acad Sci USA* 95: 11429–11434.

Altar CA, DiStefano PS (1998) Neurotrophin trafficking by anterograde transport. *Trends Neurosci* 21: 433–437.

Altar CA, Cai N, Bliven T, Juhasz M, Conner JM, Acheson AL, Lindsay RM, Wiegand SJ (1997) Anterograde transport of brain-derived neurotrophic factor and its role in the brain. *Nature* 389: 856–860.

Antonini A, Stryker MP (1993) Development of individual geniculocortical arbors in cat striate cortex and effects of binocular impulse blockade. *J Neurosci* 13: 3549–3573.

Arnold AP (1975) The effects of castration on song development in zebra finches (*Poephila guttata*). *J Exp Zool* 191: 261–278.

Basham ME, Nordeen EJ, Nordeen KW (1996) Blockade of NMDA receptors in the anterior forebrain impairs sensory acquisition in the zebra finch. *Neurobiol Learn Mem* 66: 295–304.

Bear MF, Kleinschmidt A, Gu QA, Singer W (1990) Disruption of experience-dependent syanptic modifications in striate cortex by infusion of an NMDA receptor antagonist. *J Neurosci* 10: 909–925.

Benton S, Nelson DA, Marler P, DeVoogd TJ (1998) Anterior forebrain pathway is needed for stable song expression in adult male white-crowned sparrows (*Zonotrichia leucophrys*). *Behav Brain Res* 96: 135–150.

Boettiger CA, Doupe AJ (2001) Developmentally restricted synaptic plasticity in a songbird nucleus required for song learning. *Neuron* 31: 809–818.

Bohner J (1990) Early acquisition of song in the zebra finch, *Taeniopygia guttata*. *Anim Behav* 39: 369–374.

Bolhuis JJ, Honey RC (1998) Imprinting, learning and development: from behaviour to brain and back. *Trends Neurosci* 21: 306–311.

Bonhoeffer T, Shatz CJ (1998) Neurotrophins and visual system plasticity. In: Carew T, Menzel R, Shatz CJ (eds), Mechanistic Relationships between Development and Learning, pp. 93–112. Wiley, Chichester, England.

Bottjer SW, Hewer SJ (1992) Castration and antisteroid treatment impair vocal learning in male zebra finches. *J Neurobiol* 23: 337–353.

Bottjer SW, Miesner EA, Arnold AP (1984) Forebrain lesions disrupt development but not maintenance of song in passerine birds. *Science* 224: 901–903.

Bottjer SW, Glaessner SL, Arnold AP (1985) Ontogeny of brain nuclei controlling song learning and behavior in zebra finches. *J Neurosci* 5: 1556–1562.

Bottjer SW, Halsema KA, Brown SA, Miesner EA (1989) Axonal connections of a forebrain nucleus involved with vocal learning in zebra finches. *J Comp Neurol* 279: 312–326.

Brainard MS, Knudsen EI (1998) Sensitive periods for visual calibration of the auditory space map in the barn owl optic tectum. *J Neurosci* 18: 3929–3942.

Brainard MS, Doupe AJ (2000) Interruption of a basal ganglia-forebrain circuit prevents plasticity of learned vocalizations. *Nature* 404: 762–766.

Brennan PA, Keverne EB (1997) Neural mechanisms of mammalian olfactory learning. *ProgNeurobiol* 51: 457–481.

Brickley SG, Dawes EA, Keating MJ, Grant S (1998) Synchronizing retinal activity in both eyes disrupts binocular map development in the optic tectum. *J Neurosci* 18: 1491–1504.

Burek MJ, Nordeen KW, Nordeen EJ (1991) Neuron loss and addition in developing zebra finch song nuclei are independent of auditory experience during song learning. *J Neurobiol* 22: 215–223.

Burek MJ, Nordeen KW, Nordeen EJ (1995) Initial sex differences in neuron growth and survival within an avian song nucleus develop in the absence of afferent input. *J Neurobiol* 27: 85–96.

Cabelli RJ, Hohn A, Shatz CJ (1995) Inhibition of ocular dominance column formation by infusion of NT-4/5 or BDNF. *Science* 267: 1662–1666.

Cabelli RJ, Allendoerfer KL, Radeke MJ, Welcher AA, Feinstein SC, Shatz CJ (1996) Changing patterns of expression and subcellular localization of TrkB in the developing visual system. *J Neurosci* 16: 7965–7980.

Cabelli RJ, Shelton DL, Segal RA, Shatz CJ (1997) Blockade of endogenous ligands of TrkB inhibits formation of ocular dominance columns. *Neuron* 19: 63–76.

Canady RA, Burd GD, DeVoogd TJ, Nottebohm F (1988) Effect of testosterone on input received by an identified neuron type of the canary song system: a golgi/electron microscopy/degeneration study. *J Neurosci* 8: 3770–3784.

Carmignoto G, Vicini S (1992) Activity-dependent decrease in NMDA receptor responses during development of the visual cortex. *Science* 258: 1007–1011.

Castren E, Zafra F, Thoenen H, Lindholm D (1992) Light regulates expression of brain-derived neurotrophic factor mRNA in rat visual cortex. *Proc Natl Acad Sci USA* 89: 9444–9448.

Causing CG, Gloster A, Aloyz R, Bamji SX, Chang E, Fawcett J, Kuchel G, Miller FD (1997) Synaptic innervation density is regulated by neuron-derived BDNF. *Neuron* 18: 257–267.

Chi Z, Margoliash D (2001) Temporal precision and temporal drift in brain and behavior of zebra finch song. *Neuron* 32: 899–910.

Cline HT, Constantine-Paton M (1989) NMDA receptor antagonists disrupt the retinotectal topographic map. *Neuron* 3: 413–426.
Cline HT, Constantine-Paton M (1990) NMDA receptor agonist and antagonists alter retinal ganglion cell arbor structure in the developing frog retinotectal projection. *J Neurosci* 10: 1197–1216.
Cline HT, Debski EA, Constantine-Paton M (1987) N-methyl-D-aspartate receptor antagonist desegregates eye-specific stripes. *Proc Natl Acad Sci USA* 84: 4342–4345.
Cline HT, Debski EA, Constantine-Paton M (1990) The role of the NMDA receptor in the development of the frog visual system. *Adv Exp Med Biol* 268: 197–203.
Cohen-Cory S, Fraser SE (1994) BDNF in the development of the visual system of Xenopus. *Neuron* 12: 747–761.
Cohen-Cory S, Fraser S (1995) Effects of brain-derived neurotrophic factor on optic axon branching and remodelling *in vivo*. *Nature* 378: 192–196.
Crair MC (1999) Neuronal activity during development: permissive or instructive? *Curr Opin Neurobiol* 9: 88–93.
Crair MC, Malenka RC (1995) A critical period for long-term potentiation at thalamocortical synapses. *Nature* 375: 325–328.
Crowley JC, Katz LC (2000) Early development of ocular dominance columns. *Science* 290: 1321–1324.
Dittrich F, Feng Y, Metzdorf R, Gahr M (1999) Estrogen-inducible, sex-specific expression of brain-derived neurotrophic factor mRNA in a forebrain song control nucleus of the juvenile zebra finch. *Proc Natl Acad Sci USA* 96: 8241–8246.
Doupe AJ (1993) A neural circuit specialized for vocal learning. *Curr Opin Neurobiol* 3: 104–111.
Doupe AJ (1997) Song- and order-selective neurons in the songbird anterior forebrain and their emergence during vocal development. *J Neurosci* 17: 1147–1167.
Doupe AJ, Konishi M (1991) Song-selective auditory circuits in the vocal control system of the zebra finch. *Proc Natl Acad Sci USA* 88: 11339–11343.
Doupe AJ, Kuhl PK (1999) Birdsong and human speech: common themes and mechanisms. *Annu Rev Neurosci* 22: 567–631.
Doupe AJ, Brainard MS, Hessler NA (2000) The song system: neural circuits essential throughout life for vocal behavior and plasticity. In: Gazzaniga M (ed.), *The New Cognitive Neurosciences*, pp. 451–467. MIT Press, Boston.
Doya K, Sejnowski TJ (1998) A computational model of birdsong learning by auditory experience and auditory feedback. In: *Poon PWF, Brugge JF (eds)*, Central Auditory Processing and Neural modeling, pp. 77–88 Plenum, New York.
Eales LA (1985) Song learning in zebra finches: some effects of song model availability on what is learnt and when. *Anim Behav* 33: 1293–1300.
Eales LA (1987) Song learning in female-raised zebra finches: another look at the sensitive phase. *Anim Behav* 37: 1356–1365.
Fagliolini M, Hensch TK (2000) Inhibitory threshold for critical-period activation in primary visual cortex. *Nature* 404: 183–186.
Flint AC, Maisch US, Weishaupt JH, Kriegstein AR, Monyer H (1997) NR2A subunit expression shortens NMDA receptor synaptic currents in developing neocortex. *J Neurosci* 17: 2469–2476.
Funabiki Y, Konishi M (2003) Long memory in song learning by zebra finches. *J Neurosci* 23: 6928–6935.

Guillery RW (1972) Binocular competition in the control of geniculate cell growth. *J Comp Neurol* 144: 117–130.

Gurney ME (1981) Hormonal control of cell form and number in the zebra finch song system. *J Neurosci* 1: 658–673.

Gurney ME, Konishi M (1980) Hormone-induced sexual differentiation of brain and behavior in zebra finches. *Science* 208: 1380–1383.

Hahnloser RHR, Kozhevnikov A, Fee MS (2002) An ultra-sparse code underlies the generation of neural sequences in a songbird. *Nature* 419: 65–70.

Hall J, Thomas KL, Everitt BJ (2000) Rapid and selective induction of BDNF expression in the hippocampus during contextual learning. *Nat Neurosci* 3: 533–535.

Harlow HF, Harlow MK (1969) Effects of various mother–infant relationships on rhesus monkey behaviors. In: Foss BM (ed.), *Determinants of Infant Behaviour*, pp. 15–36. Methuen, London.

Harlow HF, Zimmerman RR (1959) Affectional responses in the infant monkey. *Science* 130: 421–432.

Hasler AD, Scholz AT (1983) *Olfactory Imprinting and Homing in Salmon: Investigations into the Mechanism of the Imprinting Process.* Springer-Verlag, New York.

Hensch TK, Fagiolini M, Mataga N, Stryker MP, Baekkeskov S, Kash SF (1998) Local GABA circuit control of experience-dependent plasticity in developing visual cortex. *Science* 282: 1504–1508.

Herrmann K, Arnold A (1991a) The development of afferent projections to the robust archistriatal nucleus in male zebra finches: a quantitative electron microscopic study. *J Neurosci* 11: 2063–2074.

Herrmann K, Arnold AP (1991b) Lesions of HVc block the developmental masculinizing effects of estradiol in the female zebra finch song system. *J Neurobiol* 22: 29–39.

Herrmann K, Bischof H-J (1986) Delayed development of song control nuclei in the zebra finch is related to behavioral development. *J Comp Neurol* 245: 167–175.

Hessler NA, Doupe AJ (1999a) Social context modulates singing-related neural activity in the songbird forebrain. *Nat Neurosci* 2: 209–211.

Hessler NA, Doupe AJ (1999b) Singing-related neural activity in a dorsal forebrain-basal ganglia circuit of adult zebra finches. *J Neurosci* 19: 10461–10481.

Horn G (1985) *Memory, Imprinting and the Brain: An Inquiry into Mechanism.* Clarendon Press, Oxford.

Horton JC, Hocking DR (1996) An adult-like pattern of ocular dominance columns in striate cortex of newborn monkeys prior to visual experience. *J Neurosci* 16: 1791–1807.

Hough GE, Volman SF (2002) Short-term and long-term effects of vocal distortion on song maintenance in zebra finches. *J Neurosci* 22: 1177–1186.

Huang ZJ, Kirkwood A, Pizzorusso T, Porciatti V, Morales B, Bear MF, Maffei L, Tonegawa S (1999) BDNF regulates the maturation of inhibition and the critical period of plasticity in mouse visual cortex. *Cell* 98: 739–755.

Hubel DH, Wiesel TN (1965) Binocular interaction in striate cortex of kittens reared with artificial squint. *J Neurophysiol* 28: 1041–1059.

Hubel DH, Wiesel TN (1970) The period of susceptibility to the physiological effects of unilateral eye closure in kittens. *J Physiol* 206: 419–436.

Hubel DH, Wiesel TN, LeVay S (1977) Plasticity of ocular dominance columns in monkey striate cortex. *Philos Trans Roy Soc Lond B* 278: 377–409.

Hudson R (1993) Olfactory imprinting. *Curr Opin Neurobiol* 3: 548–552.

Huttenlocher PR, Dabholkar AS (1997) Regional differences in synaptogenesis in human cerebral cortex. *J Comp Neurol* 387: 167–178.

Immelmann K (1969) Song development in the zebra finch and other estrildid finches. In: Hinde RA (ed.), *Bird Vocalizations*, pp. 61–74. Cambridge University Press, Cambridge.

Iyengar S, Bottjer SW (2002) The role of auditory experience in the formation of neural circuits underlying vocal learning in zebra finches. *J Neurosci* 22: 946–958.

Iyengar S, Viswanathan SS, Bottjer SW (1999) Development of topography within song control circuitry of zebra finches during the sensitive period for song learning. *J Neurosci* 19: 6037–6057.

Jarvis ED, Scharff C, Grossman MR, Ramos JA, Nottebohm F (1998) For whom the bird sings: context-dependent gene expression. *Neuron* 21: 775–788.

Johnson F, Bottjer SW (1994) Afferent influences on cell death and birth during development of a cortical nucleus necessary for learned vocal behavior in zebra finches. *Development* 120: 13–24.

Johnson F, Sablan MM, Bottjer SW (1995) Topographic organization of a forebrain pathway involved with vocal learning in zebra finches. *J Comp Neurol* 358: 260–278.

Johnson F, Hohmann SE, DiStefano PS, Bottjer SW (1997) Neurotrophins suppress apoptosis induced by deafferentiation of an avian motor-cortical region. *J Neurosci* 17: 2101–2111.

Johnson F, Norstrom E, Soderstrom K (2000) Increased expression of endogenous biotin, but not BDNF, in telencephalic song regions during zebra finch vocal learning. *Dev Brain Res* 120: 113–123.

Jones AE, Ten Cate C, Slater PJB (1996) Early experience and plasticity of song in adult male zebra finches. *J Comp Psychol* 110: 354–369.

Kang H, Schuman E (1995) Long-lasting neurotrophin-induced enhancement of synaptic transmission in the adult hippocampus. *Science* 267: 1658–1662.

Kang HJ, Schuman EM (1995) Neurotrophin-induced modulation of synaptic transmission in the adult hippocampus. *J Physiol Paris* 89: 11–22.

Katz LC, Shatz CJ (1996) Synaptic activity and the construction of cortical circuits. *Science* 274: 1133–1138.

Katz LC, Crowley JC (2002) Development of cortical circuits: lessons from ocular dominance columns. *Nat Rev Neurosci* 3: 34–42.

Keating MJ, Feldman JD (1975) Visual deprivation and intertectal neuronal connexions in *Xenopus laevis*. *Proc Roy Soc Lond B* 191: 467–474.

Keating MJ, Grant S (1992) The critical period for experience-dependent plasticity in a system of binocular visual connections in *Xenopus laevis*: its temporal profile and relation to normal developmental requirements. *Eur J Neurosci* 4: 27–36.

Keating MJ, Beazley L, Feldman JD, Gaze RM (1975) Binocular interaction and intertectal neuronal connexions: dependence upon developmental stage. *Proc R Soc Lond B* 191: 445–466.

Kind PC, Mitchell DE, Ahmed B, Blakemore C, Bonhoeffer T, Sengpiel F (2002) Correlated binocular activity guides recovery from monocular deprivation. *Nature* 416: 430–433.

Kirn JR, DeVoogd TJ (1989) Genesis and death of vocal control neurons during sexual differentiation in the zebra finch. *J Neurosci* 9: 3176–3187.

Kittelberger JM, Mooney R (1997) Individual HVc axons innervate RA subdomains that control temporal and spectral elements of learned song. *Soc Neurosci Abstr* 23: 100–115.

Kittelberger JM, Mooney R (1999) Lesions of an avian forebrain nucleus that disrupt song development alter synaptic connectivity and transmission in the vocal premotor pathway. *J Neurosci* 19: 9385–9398.

Kittelberger JM, Mooney R (2005) Acute injections of brain-derived neurotrophic factor in a vocal premotor nucleus reversibly disrupt adult birdsong and trigger syllable deletion. *J Neurobiol* 62: 406–424.

Kleinschmidt A, Bear MF, Singer W (1987) Blockade of "NMDA" receptors disrupts experience-dependent plasticity of kitten striate cortex. *Science* 238: 355–358.

Knudsen EI (1983) Early auditory experience aligns the auditory map of space in the optic tectum of the barn owl. *Science* 222: 939–942.

Knudsen EI (1985) Experience alters the spatial tuning of auditory units in the optic tectum during a sensitive period in the barn owl. *J Neurosci* 5: 3094–3109.

Knudsen EI (1994) Supervised learning in the brain. *J Neurosci* 14: 3985–3997.

Knudsen EI, Brainard MS (1991) Visual instruction of the neural map of auditory space in the developing optic tectum. *Science* 253: 85–87.

Knudsen EI, Knudsen PF (1990) Sensitive and critical periods for visual calibration of sound localization by barn owls. *J Neurosci* 10: 222–232.

Knudsen EI, Esterly SD, Knudsen PF (1984) Monaural occlusion alters sound localization during a sensitive period in the barn owl. *J Neurosci* 4: 1001–1011.

Kohara K, Kitamura A, Morishima M, Tsumoto T (2001) Activity-dependent transfer of brain-derived neurotrophic factor to postsynaptic neurons. *Science* 291: 2419–2423.

Konishi M (1965) The role of auditory feedback in the control of vocalization in the white-crowned sparrow. *Z Tierpsychol* 22: 770–783.

Konishi M (1985) Birdsong: from behavior to neuron. *Annu Rev Neurosci* 8: 125–170.

Konishi M (1989) Birdsong for neurobiologists. *Neuron* 3: 541–549.

Konishi M, Akutagawa E (1985) Neuronal growth, atrophy and death in a sexually dimorphic song nucleus in the zebra finch brain. *Nature* 315: 145–147.

Konishi M, Akutagawa E (1987) Hormonal control of cell death in a sexually dimorphic song nucleus in the zebra finch. *Ciba Found Symp: Select Neuronal Death* 126: 173–185.

Konishi M, Akutagawa E (1990) Growth and atrophy of neurons labeled at their birth in a song nucleus of the zebra finch. *Proc Natl Acad Sci USA* 87: 3538–3541.

Korsia S, Bottjer SW (1991) Chronic testosterone treatment impairs vocal learning in male zebra finches during a restricted period of development. *J Neurosci* 11: 2362–2371.

Kroodsma DE (1974) Song learning, dialects, and dispersal in the Bewick's wren. *Z Tierpsychol* 35: 352–380.

Lenneberg EH (1967) *Biological Foundations of Language*. Wiley, New York.

Leonardo A, Fee MS (2005) Ensemble coding of vocal control in birdsong. *J Neurosci* 25: 652–661.

Leonardo A (2004) Experimental test of the birdsong error-correction model. *Proc Natl Acad Sci USA* 101:16935–16940.

Leonardo A, Konishi M (1999) Decrystallization of adult birdsong by perturbation of auditory feedback. *Nature* 399: 466–470.

LeVay S, Stryker MP, Shatz CJ (1978) Ocular dominance columns and their development in layer IV of the cat's visual cortex. *J Comp Neurol* 179: 223–244.
LeVay S, Wiesel TN, Hubel DH (1980) The development of ocular dominance columns in normal and visually deprived monkeys. *J Comp Neurol* 191: 1–51.
Li X-C, Jarvis ED (2001) Sensory- and motor-driven BDNF expression in a vocal communication system. *Soc Neurosci Abstr* 27: 538.
Li X-C, Jarvis ED, Alvarez-Borda B, Lim DA, Nottebohm F (2000) A relationship between behavior, neurotrophin expression, and new neuron survival. *Proc Natl Acad Sci USA* 97: 8584–8589.
Lichtman JW, Colman H (2000) Synapse elimination and indelible memory. *Neuron* 25: 269–278.
Livingston FS, Mooney R (1997) Development of intrinsic and synaptic properties in a forebrain nucleus essential to avian song learning. *J Neurosci* 17: 8997–9009.
Livingston FS, White SA, Mooney R (2000) Slow NMDA-EPSCs at synapses critical for song development are not required for song learning in zebra finches. *Nat Neurosci* 3: 482–488.
Lohof AM, Ip NY, Poo M-M (1993) Potentiation of developing neuromuscular synapses by the neurotrophins NT-3 and BDNF. *Nature* 363: 350–353.
Lohof AM, Delhaye-Bouchaud N, Mariani J (1996) Synapse elimination in the central nervous system: functional significance and cellular mechanisms. *Rev Neurosci* 7: 85–101.
Lombardino AJ, Nottebohm F (2000) Age at deafening affects the stability of learned song in adult male zebra finches. *J Neurosci* 20: 5054–5064.
Lorenz K (1937) The companion in the bird's world. *The Auk* 54: 245–273.
Lu H-C, Gonzalez E, Crair MC (2001) Barrel cortex critical period plasticity is independent of changes in NMDA receptor subunit composition. *Neuron* 32: 619–634.
Luo M, Ding L, Perkel DJ (2001) An avian basal ganglia pathway essential for vocal learning forms a closed topographic loop. *J Neurosci* 21: 6836–6845.
Marler P (1956) Behaviour of the chaffinch. *Behav Suppl* 6: 1–186.
Marler P (1970) A comparative approach to vocal learning: song development in white-crowned sparrows. *J Comp Physiol Psych Monogr* 71: 1–25.
Marler P (1997) Three models of song learning: evidence from behavior. *J Neurobiol* 33: 501–516.
Marler P, Tamura M (1964) Culturally transmitted patterns of vocal behavior in sparrows. *Science* 146: 1483–1486.
Marler P, Peters S (1977) Selective vocal learning in a sparrow. *Science* 198: 519–521.
Marler P, Peters S (1981) Sparrows learn adult song and more from memory. *Science* 213: 780–782.
Marler P, Peters S (1982) Structural changes in song ontogeny in the swamp sparrow *Melospiza georgiana*. *The Auk* 99: 446–458.
Marler P, Peters S, Ball GF, Dufty AMJ, Wingfield JC (1988) The role of sex steroids in the acquisition and production of birdsong. *Nature* 336: 770–772.
McAllister AK, Lo DC, Katz LC (1995) Neurotrophins regulate dendritic growth in developing visual cortex. *Neuron* 15: 791–803.
McAllister AK, Katz LC, Lo DC (1999) Neurotrophins and synaptic plasticity. *Annu Rev Neurosci* 22: 295–318.
McCasland JS (1987) Neuronal control of bird song production. *J Neurosci* 7: 23–39.

Minichiello L, Korte M, Wolfer D, Kuhn R, Unsicker K, Cestari V, Rossi-Arnaud C, Lipp H-P, Bonhoeffer T, Klein R (1999) Essential role for trkB receptors in hippocampus-mediated learning. *Neuron* 24: 401–414.

Mooney R (1992) Synaptic basis for developmental plasticity in a birdsong nucleus. *J Neurosci* 12: 2464–2477.

Mooney R (1999) Sensitive periods and circuits for learned birdsong. *Curr Opin Neurobiol* 9: 121–127.

Mooney R, Konishi M (1991) Two distinct inputs to an avian song nucleus activate different glutamate receptor subtypes on individual neurons. *Proc Natl Acad Sci USA* 88: 4075–4079.

Mooney R, Rao M (1994) Waiting periods versus early innervation: the development of axonal connections in the zebra finch song system. *J Neurosci* 14: 6532–6543.

Morrison RG, Nottebohm F (1993) Role of a telencephalic nucleus in the delayed song learning of socially isolated zebra finches. *J Neurobiol* 24: 1045–1064.

Mountjoy DJ, Lemon RE (1995) Extended song learning in wild European starlings. *Anim Behav* 49: 357 366.

Murphy DD, Segal M (1997) Morphological plasticity of dendritic spines in central neurons is mediated by activation of cAMP response element binding protein. *Proc Natl Acad Sci USA* 94: 1482–1487.

Murphy DD, Cole NB, Segal M (1998) Brain-derived neurotrophic factor mediates estradiol-induced dendritic spine formation in hippocampal neurons. *Proc Natl Acad Sci USA* 95: 11412–11417.

Nelson DA, Marler P (1994) Selection-based learning in bird song development. *Proc Natl Acad Sci USA* 91: 10498–10501.

Nice MM (1943) Studies in the life history of the song sparrow (Part II). *Trans Linn Soc NY* 6: 1–328.

Nja A, Purves D (1978) The effects of nerve growth factor and its antiserum on synapses in the superior cervical ganglion of the guinea-pig. *J Physiol* 277: 53–75.

Nordeen KW, Nordeen EJ (1992) Auditory feedback is necessary for the maintenance of stereotyped song in adult zebra finches. *Behav Neural Biol* 57: 58–66.

Nordeen KW, Nordeen EJ (1993) Long-term maintenance of song in adult zebra finches is not affected by lesions of a forebrain region involved in song learning. *Behav Neural Biol* 59: 79–82.

Nottebohm F, Arnold AP (1976) Sexual dimorphism in vocal control areas of the songbird brain. *Science* 194: 211–213.

Nottebohm F, Nottebohm ME (1978) Relationship between song repertoire and age in the canary *Serinus canaria*. *Z Tierpsychol* 46: 298–305.

Nottebohm F, Stokes TM, Leonard CM (1976) Central control of song in the canary, *Serinus canarius*. *J Comp Neurol* 165: 457–486.

Nottebohm F, Kelley DB, Paton JA (1982) Connections of vocal control nuclei in the canary telencephalon. *J Comp Neurol* 207: 344–357.

Nottebohm F, Nottebohm ME, Crane L, Wingfield JC (1987) Seasonal change in gonadal hormone levels of adult male canaries and their relation to song. *Behav Neural Biol* 47: 197–211.

Nottebohm F, Alvarez-Buylla A, Cynx J, Kirn J, Ling C-Y, Nottebohm ME, Suter R, Tolles A, Williams H (1990) Song learning in birds: the relation between perception and production. *Philos Trans Roy Soc Lond B* 329: 115–124.

Okanoya K, Yamaguchi A (1997) Adult bengalese finches (*Lonchura striata var. domestica*) require real-time auditory feedback to produce normal song syntax. *J Neurobiol* 33: 343–356.

Okuhata S, Saito N (1987) Synaptic connections of a forebrain nucleus involved with vocal learning in zebra finches. *Brain Res Bull* 18: 35–44.

Perkel DJ (1995) Effects of neuromodulators on excitatory synaptic transmission in nucleus RA of the zebra finch. *Soc Neurosci Abstr* 21: 960.

Perkel DJ, Farries MA (2000) Complementary "bottom-up" and "top-down" approaches to basal ganglia function. *Curr Opin Neurobiol* 10: 725–731.

Philpot BD, Sekhar AK, Xhouval HZ, Bear MF (2001) Visual experience and deprivation bidirectionally modify the composition and function of NMDA receptors in visual cortex. *Neuron* 29: 157–169.

Poindron P, Levy F (1990) Physiological, sensory and experiential determinants of maternal behaviour in sheep. In: Krasnegor NA, Bridges RS (eds), *Mammalian Parenting*, pp. 133–156. Oxford University Press, New York.

Poo M-M (2001) Neurotrophins as synaptic modulators. *Nat Rev Neurosci* 2: 24–32.

Price PH (1979) Developmental determinants of structure in zebra finch song. *J Comp Physiol A* 93: 260–277.

Purves D, Lichtman JW (1980) Elimination of synapses in the developing nervous system. *Science* 210: 153–157.

Purves D, Snider WD, Voyvodic JT (1988) Trophic regulation of nerve cell morphology and innervation in the autonomic nervous system. *Nature* 336: 123–128.

Pytte CL, Suthers RA (2000) Sensitive period for sensorimotor integration during vocal motor learning. *J Neurobiol* 142: 172–189.

Quinlan EM, Olstein DH, Bear MF (1999a) Bidirectional, experience-dependent regulation of N-methyl-D-aspartate receptor subunit composition in the rat visual cortex during postnatal development. *Proc Natl Acad Sci USA* 96: 12876–12880.

Quinlan EM, Philpot BD, Huganir RL, Bear MF (1999b) Rapid, experience-dependent expression of synaptic NMDA receptors in visual cortex *in vivo*. *Nat Neurosci* 2: 352–357.

Rakic P (1977) Prenatal development of the visual system in rhesus monkey. *Philos Trans Roy Soc Lond B* 278: 245–260.

Rakic P, Bourgeois J-P, Eckenhoff MF, Zecevic N, Goldman-Rakic PS (1986) Concurrent overproduction of synapses in diverse regions of the primate cerebral cortex. *Science* 232: 232–235.

Rasika S, Alvarez-Buylla A, Nottebohm F (1999) BDNF mediates the effects of testosterone on the survival of new neurons in an adult brain. *Neuron* 22: 53–62.

Reh TA, Constantine-Paton M (1985) Eye-specific segregation requires neural activity in three-eyed *Rana pipiens*. *J Neurosci* 5: 1132–1143.

Reiner A, Medina L, Veenman CL (1998) Structural and functional evolution of the basal ganglia in vertebrates. *Brain Res Rev* 28: 235–285.

Reiner A, Perkel DJ, Bruce LL, Butler AB, Csillag A, Kuenzel W, Medina L, Paxinos G, Shimizu T, Striedter G, Wild JM, Ball GF, Durand S, Guturkun O, Lee DW, Mello CV, Powers A, White SA, Hough GE, Kubikova L, Smulders TV, Wada K, Dugas-Ford J, Husband S, Yamamoto K, Yu J, Siang C, Jarvis ED (2004) Revised nomenclature for avian telencephalon and some related brainstem nuclei. *J Comp Neurol* 473: 377–414.

Rema V, Armstrong-James M, Ebner FF (1998) Experience dependent plasticity of adult rat S1 cortex requires local NMDA receptor activation. *J Neurosci* 18: 10196–10206.

Rice JO, Thompson WL (1968) Song development in the indigo bunting. *Anim Behav* 16: 462–469.

Rico B, Xu B, Reichardt LF (2002) TrkB receptor signaling is required for establishment of GABAergic synapses in the cerebellum. *Nat Neurosci* 5: 225–233.

Riddle DR, Lo DC, Katz LC (1995) NT-4-mediated rescue of lateral geniculate neurons from effects of monocular deprivation. *Nature* 378: 189–191.

Rutherford L, DeWan A, Lauer H, Turrigiano G (1997) Brain-derived neurotrophic factor mediates the activity-dependent regulation of inhibition in neocortical cultures. *J Neurosci* 17: 4527–4535.

Rutherford LC, Nelson SB, Turrigiano GG (1998) BDNF has opposite effects on the quantal amplitude of pyramidal neuron and interneuron excitatory synapses. *Neuron* 21: 521–530.

Scharff C, Nottebohm F (1991) A comparative study of the behavioral deficits following lesions of various parts of the zebra finch song system: implications for vocal learning. *J Neurosci* 11: 2896–2913.

Scherer WJ, Udin SB (1989) N-methyl-D-aspartate antagonists prevent interaction of binocular maps in Xenopus tectum. *J Neurosci* 9: 3837–3843.

Schlaggar BL, Fox K, O'Leary DDM (1993) Postsynaptic control of plasticity in developing somatosensory cortex. *Nature* 364: 623–626.

Scott JP (1962) Critical periods in behavioral development. *Science* 138: 949–958.

Scott JP (ed.) (1978) *Critical Periods*. Dowden, Hutchinson and Ross, Stroudsburg, PA.

Shatz CJ (1990) Impulse activity and the patterning of connections during CNS development. *Neuron* 5: 745–756.

Shelton DL, Sutherland J, Gripp J *et al.* (1995) Human trks: molecular cloning, tissue distribution, and expression of extracellular domain immunoadhesins. *J Neurosci* 15: 477–491.

Shimada A, Mason CA, Morrison ME (1998) TrkB signaling modulates spine density and morphology independent of dendrite structure in cultured neonatal Purkinje cells. *J Neurosci* 18: 8559–8570.

Simpson HB, Vicario DS (1990) Brain pathways for learned and unlearned vocalizations differ in zebra finches. *J Neurosci* 10: 1541–1556.

Slater PJB, Eales LA, Clayton NS (1988) Song learning in zebra finches (*Taeniopygia guttata*): progress and prospects. *Adv Stud Behav* 18: 1–34.

Sluckin W (1972) *Imprinting and Early Learning*. Methuen, London.

Snow C (1987) Relevance of the notion of a critical period to language acquisition. In: Bornstein MH (ed.), *Sensitive Periods in Development: Interdisciplinary Perspectives*, pp. 183–209. Erlbaum, Hillsdale, NJ.

Sohrabji F, Nordeen EJ, Nordeen KW (1990) Selective impairment of song learning following lesions of a forebrain nucleus in the juvenile zebra finch. *Behav Neural Biol* 53: 51–63.

Solis MM, Doupe AJ (1997) Anterior forebrain neurons develop selectivity by an intermediate stage of birdsong learning. *J Neurosci* 17: 6447–6462.

Solis MM, Doupe AJ (1999) Contributions of tutor and bird's own song experience to neural selectivity in the songbird anterior forebrain. *J Neurosci* 19: 4559–4584.

Solis MM, Doupe AJ (2000) Compromised neural selectivity for song in birds with impaired sensorimotor learning. *Neuron* 25: 109–121.

Sossinka R, Bohner J (1980) Song types in the zebra finch (*Poephila guttata castanotis*). *Z Tierpsychol* 53: 123–132.

Spiro JE, Dalva MB, Mooney R (1999) Long-range inhibitory circuits in a forebrain nucleus critical for production of learned birdsong. *J Neurophysiol* 81: 3007–3020.

Stark LL, Perkel DJ (1999) Two-stage, input-specific synaptic maturation in a nucleus essential for vocal production in the zebra finch. *J Neurosci* 19: 9107–9116.

Stryker MP, Harris W (1986) Binocular impulse blockade prevents the formation of ocular dominance columns in cat visual cortex. *J Neurosci* 6: 2117–2130.

Tchernichovski O, Mitra PP, Lints T, Nottebohm F (2001) Dynamics of the vocal imitation process: how a zebra finch learns its song. *Science* 291: 2564–2569.

Thorpe WH (1958) The learning of song patterns by birds, with especial reference to the song of the Chaffinch *Fringilla coelebs*. *Ibis* 100: 535–570.

Thorpe WH, Pilcher PM (1958) The nature and characteristics of subsong. *Br Birds* 51: 509–514.

Tokuyama W, Okuno H, Hashimoto T, Li YX, Miyashita Y (2000) BDNF upregulation during declarative memory formation in monkey inferior temporal cortex. *Nat Neurosci* 3: 1134–1142.

Troyer TW, Doupe AJ (2000a) An associational model of birdsong sensorimotor learning. I. Efference copy and the learning of song syllables. *J Neurophysiol* 84: 1204–1223.

Troyer TW, Doupe AJ (2000b) An associational model of birdsong sensorimotor learning. II. Temporal hierarchies and the learning of song sequence. *J Neurophysiol* 84: 1224–1239.

Troyer TW, Doupe AJ, Miller KD (1996) An associational hypothesis for sensorimotor learning of birdsong. In: Bower JM (ed.), *Computational Neuroscience, Trends in Research, 1995*, pp. 409–414. Academic Press San Diego, CA.

Udin SB (1989) Development of the nucleus isthmi in Xenopus. II. Branching patterns of contralaterally projecting isthmotectal axons during maturation of binocular maps. *Visual Neurosci* 2: 153–163.

Udin SB, Grant S (1999) Plasticity in the tectum of *Xenopus laevis*: binocular maps. *Prog Neurobiol* 59: 81–106.

Udin SB, Keating MJ (1981) Plasticity in a central nervous pathway in xenopus: anatomical changes in the isthmotectal projection after larval eye rotation. *J Comp Neurol* 203: 575–594.

Vates GE, Nottebohm F (1995) Feedback circuitry within a song-learning pathway. *Proc Natl Acad Sci USA* 92: 5139–5143.

Vicario DS (1991) Organization of the zebra finch song control system. II. Functional organization of outputs from nucleus robustus archistriatalis. *J Comp Neurol* 309: 486–494.

Vicario DS (1993) A new brain stem pathway for vocal control in the Zebra finch song system. *NeuroReport* 4: 983–986.

Vicario DS, Nottebohm F (1988) Organization of the zebral finch song control system. I. Representation of syringeal muscles in the hypoglossal nucleus. *J Comp Neurol* 271: 346–354.

Vicario DS, Simpson HB (1995) Electrical stimulation in forebrain nuclei elicits learned vocal patterns in songbirds. *J Neurophysiol* 73: 2602–2607.

Vu ET, Mazurek ME, Kuo YC (1994) Identification of a forebrain motor programming network for the learned song of zebra finches. *J Neurosci* 13: 6924–6934.

Weliky M, Katz LC (1997) Disruption of orientation tuning in visual cortex by artificially correlated neuronal activity. *Nature* 386: 680–685.

Whaling CS, Soha JA, Nelson DA, Lasley B, Marler P (1998) Photoperiod and tutor access affect the process of vocal learning. *Anim Behav* 56: 1075–1082.

White SA, Livingston FS, Mooney R (1999) Androgens modulate NMDA receptor-mediated EPSCs in the zebra finch song system. *J Neurophysiol* 82: 2221–2234.

Wiesel TN (1982) Postnatal development of the visual cortex and the influence of environment. *Nature* 299: 583–591.

Wiesel TN, Hubel DH (1965) Comparison of the effects of unilateral and bilateral eye closure on cortical unit responses in kittens. *J Neurophysiol* 28: 1029–1040.

Wild JM (1993a) Descending projections of the songbird nucleus robustus archistriatalis. *J Comp Neurol* 338: 225–241.

Wild JM (1993b) The avian nucleus retroambigualis: a nucleus for breathing, singing and calling. *Brain Res* 606: 119–124.

Williams H, McKibben JR (1992) Changes in stereotyped central motor patterns controlling vocalization are induced by peripheral nerve injury. *Behav Neural Biol* 57: 67–78.

Williams H, Mehta N (1999) Changes in adult zebra finch song require a forebrain nucleus that is not necessary for song production. *J Neurobiol* 39: 14–28.

Woolley SMN, Rubel EW (1997) Bengalese finches *Lonchura striata domestica* depend upon auditory feedback for the maintenance of adult song. *J Neurosci* 17: 6380–6390.

Woolsey TA (1990) Peripheral alteration and somatosensory development. In: Coleman EJ (ed.), *Development of Sensory Systems in Mammals*, pp. 461–516. Wiley, New York.

Yu AC, Margoliash D (1996) Temporal hierarchical control of singing in birds. *Science* 273: 1871–1875.

11
Neuronal Substrates of Sensory Processing for Song Perception and Learning in Songbirds: Lessons from the Mormyrid Electric Fish

Claudio V. Mello and Patrick D. Roberts

11.1 Introduction

The sensory processing and decoding of behaviorally relevant information is a basic problem that animals must resolve in order to succeed in an ever-changing environment. Sensory systems of various modalities have specialized in detecting key features conveying information for purposes as diverse as feeding, navigation, identification of conspecifics, defense from predators, and reproduction. A growing body of evidence indicates that sensory systems often compare incoming external patterns of sensory stimulation with predictions or expectations about these patterns. These predictions can be based on previous knowledge of self-generated sensory patterns, as occurs in some electric fish, or on sensory memories acquired through experience and learning, as is the case in songbirds.

Sensory systems of different modalities sometimes solve similar computational problems of information processing. By studying the mechanisms of sensory processing in one system, we can develop new hypotheses of mechanisms in another system.

The electrosensory system of mormyrid electric fish provides a good example of a well-characterized sensory system that has developed mechanisms to distinguish self-generated sensory stimuli from stimuli that arise from external sources. Songbirds, on the other hand, need to compare the sounds of the songs they and their neighbors produce with previously formed song auditory memories in order to effectively perform the perceptual discriminations and vocal learning required for reproductive success. This chapter explores aspects of the mormyrid electrosensory system that may help us to understand how the songbird brain is organized to extract behaviorally relevant information from auditory signals, either to identify other individual birds or to generate and maintain the bird's own song (BOS).

In mormyrid electric fish, a weak electric field is generated by a modified muscle in the fish's tail and the electric field is detected by electrosensory receptors in the fish's skin (Bullock and Heiligenberg, 1986). Objects in the vicinity of the fish cast an electrical shadow that contains information about the objects' shape and composition. Mormyrid fish use this active sensory system to identify objects, but fish must distinguish between self-generated electric fields and electric fields generated by other organisms that may indicate the proximity of predators or prey. Recent characterizations of the adaptive mechanisms in the mormyrid electrosensory system (Bell, 1990; Bell *et al.*, 1997) yield a model for temporal pattern learning (Bell and Szabo, 1986; Roberts and Bell, 2000) that depends on synaptic plasticity in the sensory-processing circuits.

In songbirds, vocal communication is very prominent, playing a significant role in basic behaviors such as territoriality, mate attraction and mate selection (see reviews in Catchpole and Slater, 1995; Kroodsma and Miller, 1996). To express these behaviors, songbirds need to perceive, discriminate, and remember the songs they encounter throughout life. A bird's exposure to song can impact a broad range of behaviors, both vocal and non-vocal ones, and auditory memories resulting from such exposure are required for at least two general purposes, namely: (1) the identification and discrimination of different conspecific individuals based on acoustic features of song and (2) the learning and maintenance of the BOS. Behaviorally, these two contexts are quite distinct and involve different brain pathways and mechanisms. In principle, however, the auditory processing and memorization of song, whether it subserves perceptual discrimination or vocal learning and maintenance, is fundamentally a sensory/perceptual process. Thus, it seems parsimonious to postulate that the same basic neuronal pathways and mechanisms may be used for the sensory processing and memorization of song required in the two different contexts.

We first describe some general aspects of perceptual processing and vocal learning and maintenance of song in songbirds to establish the key characteristics we must consider for our comparison with adaptation and learning in the electrosensory system. We follow this with a section that provides details of the mormyrid electrosensory system and discusses the known mechanisms for learning and adaptation. We then discuss neural substrates of the songbird auditory system that might contribute to the generation of song templates for perceptual discrimination and vocal learning. Finally, we compare some critical features between the two systems and conclude with the proposition that the neural pathways and mechanisms that the bird uses for storing memories of other conspecifics could also contribute towards a template for generating and maintaining the BOS.

11.2 Song perceptual processing and song learning

Male songbirds from each of several species are able to identify conspecifics based on their songs and use this ability to help establish long-lasting relationships with birds in neighboring breeding territories. Females frequently select males based on acoustic features of their songs, often showing preferences for specific song characteristics. These preferences can relate to general variables such as rates of singing, song length and repertoire size, or to the presence of syllables or phrases with specific spectro-temporal features. Such preferences can potentially shape the properties of song over generations, and thus exert a strong effect on population differences in vocal production patterns (for reviews, see Catchpole and Slater, 1995; Muller and Kroodsma, 1996). Thus, both male and female songbirds must have an elaborate song recognition system. As a consequence of their use of vocal communication, songbirds of both sexes must rely heavily on their auditory systems to process and discriminate the songs they encounter, as well as to form and retrieve long-lasting auditory memories of these songs.

The need for long-lasting auditory memories is also supported by the fact that song is a learned behavior in songbirds (Thorpe, 1958; Marler and Peters, 1977), as well as in two other avian orders, parrots and hummingbirds (Nottebohm, 1972). In many species that have been investigated, the development of species-specific song appears to be fundamentally dependent on vocal imitation (see discussions in Catchpole and Slater, 1995; Marler, 1997). This imitative vocal learning process consists of two phases, both of which require intact hearing. The first, or sensory acquisition phase of song learning consists of the young bird hearing and memorizing the adult song (Konishi, 1965). In some species, this occurs simply by the juvenile bird's exposure to the sound of song (e.g. tape-recorded song, as in Marler and Peters, 1977). Even though other variables may affect this sensory acquisition process, such as social interactions through visual/tactile stimulation, an absolute requirement is that the bird be exposed to the appropriate song auditory model. This sensory phase occurs, or in some cases starts, at an age when the vocal control circuitry has not yet matured and the juvenile bird is not yet able to actively produce its own song (Marler and Peters, 1977, 1981; Eales, 1985; Bohner, 1990). Since some species show innate recognition preferences for song elements of their own species over those of other species, the process of song memorization during this sensory phase appears to involve the modification of an intrinsic or innate template (Marler, 1970; Marler and Peters, 1987, 1988; Marler, 1997). The end product of the sensory phase of song learning (i.e. the acquired song template) is arguably an acquired auditory representation of song. In other words, it is a neural representation of what the song to be imitated sounds like.

Consistent with this notion is the observation that song in Bengalese finches deteriorates after deafening in adults, but normal song production can then be reinstated upon recovery of hearing function in the absence of further external auditory input (Woolley and Rubel, 2002). Thus, a representation of the acoustic properties of the song to be imitated is formed early in life, but this acquired song auditory memory is long lasting and has an extended influence on the bird's vocal behavior. It is likely that the song auditory memories for vocal imitation contain information on both the spectral structure of individual song syllables as well as on the temporal or syntactic aspects of song, since both aspects of song are disrupted by isolation or deafening.

The second, or sensorimotor phase of song learning consists of the bird's active attempts to modify its vocal motor program(s) so that the sound of its own immature song can eventually match the acoustic properties of the internalized song model (Konishi, 1965; Marler and Peters, 1982; Marler, 1997; Tchernichovski et al., 2001; Tchernichovski and Mitra, 2002). In the previous sensory acquisition phase, the male songbird cannot have been instructed on how to execute specific song motor gestures and can only have acquired an auditory memory of the song to be imitated. In consequence, the sensory processing that occurs during the sensorimotor phase most likely consists of a comparison of the acoustic properties of the BOS with those of the song to be imitated. Intact hearing is needed during this phase of learning, as shown by the effect of cochlear lesions on song crystallization (Konishi, 1965). Through auditory feedback, the brain can be informed about the properties of the BOS, and use that information to guide vocal development. Importantly, once song has been learned, the brain's vocal motor program for producing the BOS is not sufficient, by itself, to maintain song structure. Rather, the brain requires continuous information that can be obtained through auditory feedback about the vocal output (Nordeen and Nordeen, 1992; Leonardo and Konishi, 1999; Woolley and Rubel, 2002).

Arguably, the comparison the brain needs to perform between the BOS and the song template is primarily sensory/perceptual in nature and consists of the comparison between the sound of the BOS and the internalized representation of auditory features contained in the song template (Fig. 11.1a). As the bird vocalizes, the resulting auditory feedback activates the cochlea and is then processed at various stations of the auditory pathways, as represented schematically in Fig. 11.1a left (BOS to auditory input). At some point, this incoming auditory feedback input must be compared with the internalized song auditory template, which contains information about how the BOS should sound like (Fig. 11.1a left, comparator element). The song template could be seen as a previously acquired auditory representation of song that sets an expectation for

11.2 Song perceptual processing and song learning

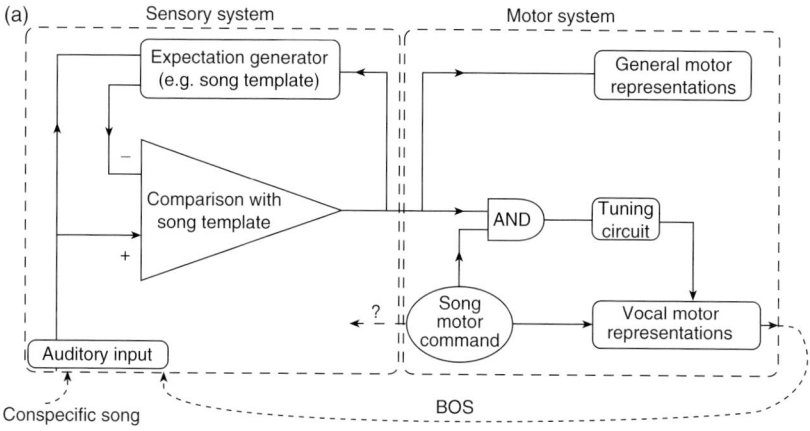

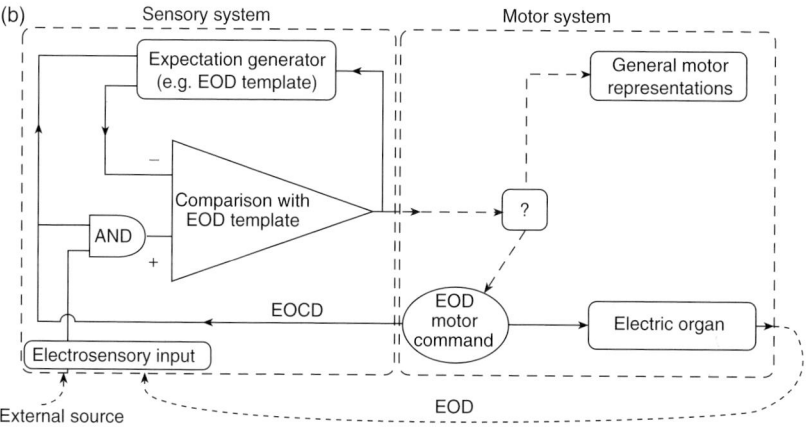

Figure 11.1. Comparison of two sensorimotor processing systems. (a) *Songbird model*: auditory input is compared with a song template by the subtraction of the expected spectral-temporal pattern of auditory stimuli from the direct input. The result of the comparison is used to update the song template by a mechanism that is not yet known. The output of the sensory system influences the song motor learning system by adjusting the motor program only if the bird is actively singing. This condition of active singing is enforced by the AND-gate. Deviations of the song from the expected song are used as an error signal to tune song production. (b) *Mormyrid electrosensory system*: electrosensory input is gated by a centrally generated electric organ corollary discharge (EOCD) signal to limit processing to only the fish's own self-generated electric field. The electrosensory pattern is compared with the expected electric organ discharge (EOD) template. The difference between the sensory pattern caused by the EOD and the expected pattern is passed for further processing and represents novel sensory information. This difference is also used to update the EOD template through a mechanism based on spike-timing dependent synaptic plasticity.

the incoming auditory input resulting from the act of singing (Fig. 11.1a left, expectation generator). In case a significant difference, or mismatch, is detected between this auditory feedback input and the template, an error signal is passed on to centers where vocal motor representations (i.e. the set of commands that lead to the generation of the song) are encoded, informing on the need to modify these representations (Fig. 11.1a right). Importantly, a message informing about an error or mismatch and the need to modify the current vocal motor representations should only be passed on to motor representation centers if such a mismatch occurs during singing. This could be accomplished by a filter mechanism (Fig. 11.1a right, AND-filter) that allows an error signal to be passed on to motor representation centers only if it is concurrent with a signal from motor command centers indicating that the bird is vocalizing. Changes in vocal motor representation would then be accomplished by the activation of tuning circuit(s) (Fig. 11.1a right), resulting in a modified vocal motor output.

The comparison process diagrammed in Fig. 11.1a left, is a particular case of comparing an external auditory input to an internal expectation. Such a process is analogous to the comparison the brain performs in the context of perceptual discrimination, for example between the song a bird hears at any given moment (Fig. 11.1a left, conspecific song) and the previously internalized auditory representations, or memories, of familiar songs that play the same role as the song template. In this situation, the brain is comparing the acoustic properties of a song heard with an internal representation of how a familiar song sounds like. Similarly to the song template for vocal learning, a previous auditory representation of a familiar song would set an expectation for the auditory system to analyze incoming input patterns from other songs (Fig. 11.1a left, expectation generator). A "mismatch" in this case signals an unfamiliar song (e.g. a potential intruder or rival) and informs motor control centers that mediate the appropriate behavioral responses (Fig. 11.1a right, general motor representations). As a song motor command is absent due to the lack of active singing behavior in this situation, the AND-filter prevents the song tuning circuits from being activated, so that vocal motor representations are not modified.

Based on the above discussion, the nature of the comparison process is fundamentally similar in the contexts of both perceptual discrimination and auditory feedback evaluation for vocal learning. In both cases, the task of comparing demands an analysis of the degree of similarity between the auditory stimulus and an internal auditory representation. The result of this analysis can then be used to inform brain centers that control behavioral programs or actions. Importantly, these comparisons differ markedly in terms of their behavioral consequences. The behavioral responses to the recognition of a familiar song or to the identification of an intruder's song (Catchpole and Slater, 1995) require the coordinated recruitment and action of the motor programs involved in the control of these responses: for example, aggressive behavior while confronting an intruder, or copulatory behavior triggered upon recognition of a familiar mate. In contrast, in the case of vocal learning, it is vocal motor centers that need to be informed on whether vocal motor representations require modification. Thus, the motor targets of the processed sensory information differ markedly in the two situations above, depending on the behavioral context in which the song percept is generated. Nonetheless, the same sensory-processing machinery, that is the auditory system, could, in principle, subserve the basic auditory task in both situations.

It can be argued that the sensory information provided by auditory feedback during singing can only be effectively compared with the internal song auditory representation if the neuronal circuits involved in this comparison are informed that a motor command for singing has just been given by the song motor centers. In other words, the auditory system would need to be aware that the bird has just sung in order to process the resulting sound in a timely manner and to be able to compare it with the internalized song template. It is possible that the motor act of singing itself sets the auditory system into a mode whereby the auditory input that reaches the brain after a given time delay is interpreted as resulting from that motor act. This possibility requires a central pathway and mechanisms through which the song motor centers can influence the auditory system (Fig. 11.1a, question mark), but their existence has not been demonstrated.

11.3 Electrosensory processing in the mormyrid electric fish

Mormyrid electric fish are a fresh-water family of fish native to African rivers and lakes (Bullock and Heiligenberg, 1986). Mormyrids generate weak electric pulses using a modified muscle in their tail called an electric organ. The resultant weak electric field caused by the electric organ discharge (EOD) surrounds the fish and is detected by an array of electroreceptors on the body surface. Distortions of this self-generated electric field are then used by the fish to identify objects in its surrounding environment, allowing the fish to effectively navigate in the dark. This form of navigation that uses a self-generated electric field is called active electrolocation.

The signal to initiate an EOD originates in a motor command nucleus (Bell and Emde, 1995). In addition to initiating an electrical discharge, the command nucleus of mormyrids generates a so-called corollary discharge that informs the electrosensory circuits of *when* the electric pulse occurred (Zipser and Bennet, 1976; Bell, 1989). The timing information provided by the corollary discharge signal from the motor control center allows the electrosensory system to generate an internal representation, or expectation, of the reafferent electrosensory image that is generated as a result of an EOD (Bell, 1981). By comparing this internal expectation with the actual electrosensory image the fish can determine whether any distortions are present in its electric field caused by objects in the environment.

Figure 11.1b shows a schematic diagram of the active electrosensory processing system (Bell and Szabo, 1986). The EOD motor command (Fig. 11.1b right) causes the electric organ to discharge. Primary afferents from electrosensory receptors converge with the electric organ corollary discharge (EOCD) signal in a functional AND-gate (Fig. 11.1b left). The AND-gate eliminates signals that originate from external sources such as the EODs of other mormyrids. A branch of the EOCD pathway provides a timing signal to the "expectation generator" (Bell *et al.*, 1997) that contains a template of the reafferent EOD sensory image. This predicted image is compared with the reafferent electrosensory image and is effectively subtracted so that the output of the initial electrosensory processing circuit passes only novel stimuli. The output of the comparison proceeds to further processing, but also informs the expectation generator of any deviations of the sensory image from the predicted image to update the EOD template.

The motor system of mormyrid electric fish (Fig. 11.1b right) is less well characterized than the sensory system, particularly with regards to any direct influence that the sensory image cancellation system has on the motor output. Of course, the general motor representation of an EOD is quite simple when compared to vocalization in songbirds. The motor command initiates a single pulse from the electric organ during each cycle. However, the "novelty response" (Hall *et al.*, 1995) behavior of mormyrid electric fish suggests that some direct link (Fig. 11.1b, question mark) between the sensory system and the motor system exists. These fish respond to novel stimuli with a brief increase in the rate of their EOD. One potential trigger for the novelty response would be a deviation of the sensory image from the EOD template, but the exact neural pathways have yet to be determined. Our lack of knowledge of the motor pathways in the mormyrid EOD system is in contrast to songbirds where much of the research has concentrated on the song generation system. However, much of the neuronal circuitry involved in the electrosensory processing, corollary discharge, and expectation generation in mormyrids is quite well

11.3 Electrosensory processing in the mormyrid electric fish

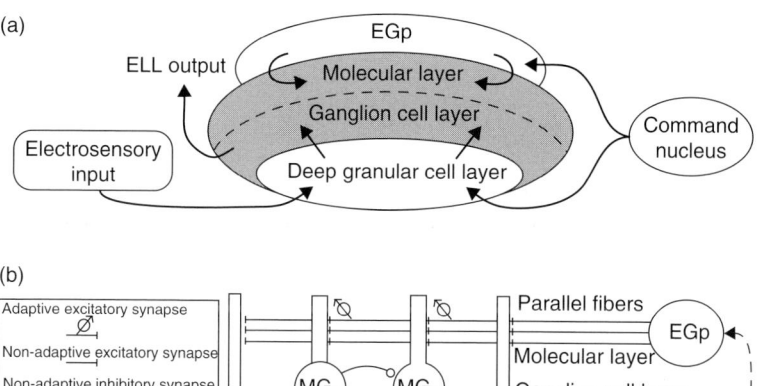

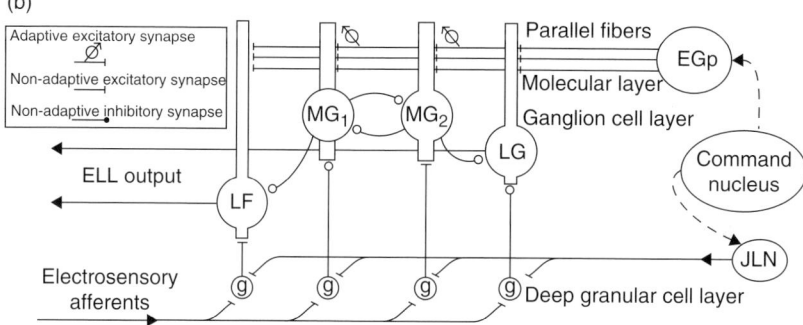

Figure 11.2. Adaptive sensory processing in the electrosensory lateral line (ELL) lobe. (a) Primary electrosensory input enters the deep granular cell layer and combines with information from the command nucleus about *when* there was an electric pulse. This combination is passed to the neurons of the ganglion cell layer. Neurons with cell bodies in the ganglion cell layer have dendrites in the molecular layer that receive command information. The summation of the sensory and motor pathways is transmitted out of the ELL for further processing. (b) The cell types and connectivity of the ELL (Han *et al.*, 1999; Meek *et al.*, 1999) shows how electrosensory signals are combined with command timing-information. Electrosensory afferents converge onto granular cells (g) with inputs from the juxtalobar nucleus (JLN) that carry a spike following each EOD command. The granular cells are the AND-gates of Fig. 11.1b, and transfer the self-generated electrosensory signal to the medium ganglion cells (MG_1 and MG_2) and the efferent cells, large ganglion (LG) and large fusiform (LF) cells. The MG cells exhibit the strongest synaptic plasticity at their apical dendrites that are contacted by parallel fibers from the eminentia granularis posterior (EGp) that receives EOD command timing-information.

characterized (Han *et al.*, 1999; Meek *et al.*, 1999) as shown in Fig. 11.2. Adaptive electrosensory processing initially takes place in the electrosensory lateral line (ELL) lobe, a region of the mormyrid brain that receives primary electrosensory afferents. The ELL is a member of a class of neural structures called "cerebellum-like" structures because of their laminar similarity to the cerebellum. This class includes the gymnotid ELL (Bastian, 1995), the octavolateral nucleus of sharks and rays (Montgomery and Bodznick, 1994), and the dorsal cochlear nucleus of mammals (Oertel and Young, 2004).

In the active electrosensory regions of the ELL, electrosensory information enters the ELL via electrosensory primary afferents in the deep granule cell layer (Fig. 11.2a). The afferents encode electric field strength in a latency code where the time delay between the EOD and the first spike of the afferent response to the EOD provides a precise representation of the reafferent field strength in the center of the receptive field. The AND-gate appears to be located in the deep granule cell layer and the information about self-generated electric fields is delivered to the ganglion cell layer, where a comparison is made with the EOD template. The output of the ELL (Fig. 11.2b) represents the electrosensory image that is filtered by a neural expectation of that image.

The EOCD from the command nucleus arrives via two separate pathways. One pathway is through the EGp, a granule cell layer that projects to the ELL by way of parallel fibers into the molecular layer. The EGp also receives sensory inputs such as proprioceptive afferents (Bell and Grant, 1992) that aid in generating an EOD template (Fig. 11.2a). The second pathway by which the command signals arrive in the ELL is to the deep granule cell layer. The final nucleus on this pathway before the ELL is the juxtalobar nucleus (JLN) where the neurons generate a single spike during each EOD (Fig. 11.2b). Between the command nucleus and the JLN there are five synaptic junctions, yet the variability of the JLN spikes is less than a millisecond with respect to the motor command (Bell and Emde, 1995). This remarkable precision of the spike arrival time suggests that the timing of the EOD is critical for the efficient operation of active electrosensory processing.

The AND-gate in Fig. 11.1b is critical for the fish to filter externally generated electrical fields so that the active electrosensory system is focused on the fish's own field during electrolocation tasks. In addition to the mechanism of the AND-gate, the convergence of the EOCD signal from the JLN with the primary afferents onto granular cells provides a decoding mechanism for the afferent spikes. The present hypothesis for the decoding of the afferent latency is that the primary sensory afferents converge with JLN afferents onto deep granular cells (Bell, 1990), and the overlap of excitatory postsynaptic potentials from these two synaptic inputs determines the response of the deep granular cells. Recordings of granular cells show a burst response that varies in number of spikes during their burst depending on the electrosensory stimulus delay with respect to the EOD command (Bell and Grant, 1992).

The number of spikes in each burst of the granular cells provides electrosensory information to the principal cells of the ELL (i.e. the medium ganglion cells, MG_1 and MG_2), and to the efferent cells of the ELL (i.e. the large fusiform (LF) and the large ganglion (LG) cells). The MG cells are GABAergic, Purkinje-like cells that inhibit each other and the efferent cells. All of the principal cells

11.3 Electrosensory processing in the mormyrid electric fish

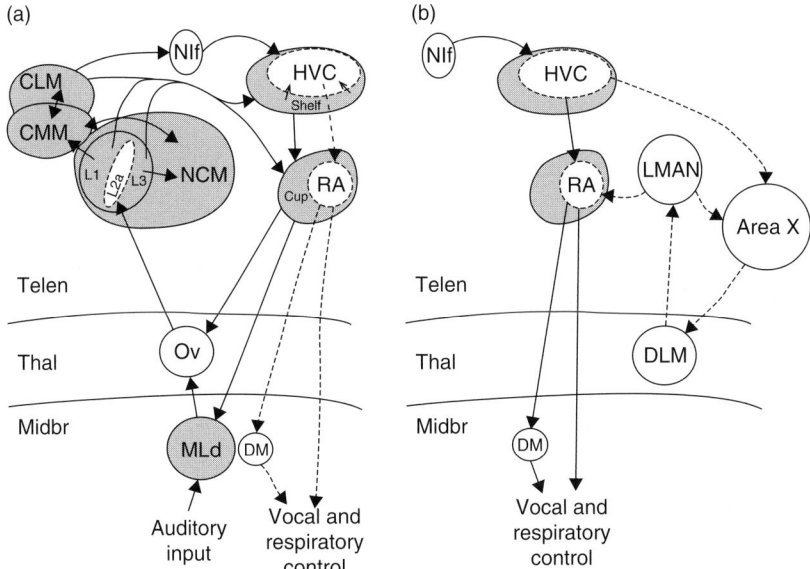

Figure 11.3. Organization of auditory and motor control pathways in songbirds (modified from Mello, 2002). (a) The auditory system consists of an ascending pathway that includes midbrain and thalamic nuclei, intratelencephalic projections among the thalamo-recipient area field L and its targets, and a descending projection from the cup to nuclei of the ascending auditory pathway. Note the close apposition of nuclei of the direct song motor pathway (HVC and RA) to auditory structures. (b) The song control system consists of the direct motor pathway (solid lines) and the anterior forebrain pathway (dashed lines). Areas at the interface between the auditory and motor control systems (e.g. NIf, shelf) are likely candidate mediators of sensorimotor integration. Abbreviations: CLM, caudolateral mesopallium; CMM, caudomedial mesopallium; DLM, medial nucleus of the dorsolateral thalamus; DM, dorsomedial intercollicular nucleus; L1/L2/L3, field L subdivisions; LMAN, lateral magnocellular nucleus of the anterior nidopallium; midbr, midbrain; MLd, dorsal part of the lateral mesencephalic nucleus; NCM, caudomedial nidopallium; NIf, interfacial nucleus of the nidopallium; Ov, nucleus ovoidalis; telen, telencephalon; thal, thalamus.

of ELL receive parallel fiber inputs, but the MG cells show the strongest adaptive properties.

A comparison of the electrosensory input with the EOD template appears to take place in the MG cells. The basilar dendrites of MG cells receive inputs from deep granular cells providing an EOD-command-gated sensory input, and the apical dendrites receive an EOD template. The mechanism for generating the EOD template is based on spike-timing dependent plasticity (STDP) at the synapse from parallel fibers onto the MG cells (Bell *et al.*, 1997). A special

STDP learning rule causes the MG cells to sculpt a negative image of the electrosensory image from parallel fiber input (Roberts and Bell, 2000). The parallel fibers carry all of the timing information necessary to generate a negative image of the sensory response to the EOD so that, when combined with the predicted sensory inputs, the output of MG cells is constant. This learning mechanism requires many EODs, so the slow adaptive process generates a prediction of the sensory image based on the recent history of sensory stimulation, and novel sensory information is emphasized.

Thus, the neurons that process the comparison between an internal expectation and the electrical field that is actually sensed by the fish belong to sensory-processing circuits and constitute an eminently sensory processing system. Importantly, even though the ELL receives a motor-related input and is modulated by a motor command, it is still primarily involved in the sensory processing of the electric field.

The electrosensory system described above is used for active electrolocation. However, a parallel electrosensory system is used for passive detection of electrical signals that originate in the environment. Electroreceptors specialized for passive electrolocation, the ampullary receptors, modulate their spike rate in response to changing electric field strength across the skin (Bell and Szabo, 1986). Electrosensory information from ampullary afferents are also initially processed in the ELL, but in a separate region (Bell, 1990) with similar structure to the regions that process the active electrolocation signals. An important difference between the two regions of ELL is the absence of an AND-gate in the passive electrolocation region so that all electrosensory information is passed into the ELL. The passive electrolocation region also receives strong EOCD that is used in an expectation generator to eliminate predictable electrosensory patterns. Most notably, the passive electrolocation system eliminates the response of the ampullary afferents to the fish's own EOD in the ELL (Bell, 1981; Bell *et al.*, 1993). The mechanism is the same as described above, where synaptic plasticity generates an EOD template of the predictable electrosensory response that cancels the afferent input. A similar template of predictable sensory patterns is presumed to exist in the songbird sensory system, but the mechanisms for the generation of the song template, and for making comparisons with auditory stimuli, are still unknown.

11.4 The perceptual processing and memorization of song: in search of a neuronal substrate

Remarkable progress has been made in identifying the brain areas and circuits involved in the production and learning of birdsong. Some of the song control

11.4 The perceptual processing and memorization of song

areas have also been implicated in the perceptual aspects of song. We will start this section by a brief review of the set of discrete brain nuclei known as the "song control system" (Nottebohm et al., 1976). The song control system is involved in vocal motor control of song production and in song learning, based on the combined evidence from lesions, anatomic tract-tracing, electrophysiologic recordings, and the mapping of activity-dependent gene expression (see review articles in Brenowitz et al., 1997; Zeigler and Marler, 2004; see also Mello, 2002). This system consists of a series of interconnected forebrain nuclei whose main output is to brainstem areas involved in vocal and respiratory control. These nuclei are: the HVC of the nidopallium, the robust nucleus of the arcopallium (RA), area X of the medial striatum, the medial nucleus of the dorsolateral thalamus (DLM), and the lateral magnocellular nucleus of the anterior nidopallium (LMAN). We follow here the revised avian brain nomenclature as detailed in Reiner et al. (2004).

As shown in Fig. 11.3b, two main pathways have been described in the song control system. The direct motor pathway consists of the projections from HVC to RA onto the dorsomedial intercollicular nucleus (DM) and the tracheosyringeal portion of the hypoglossal nucleus (nXIIts), the latter containing motorneurons that innervate the syrinx, the avian vocal organ (Nottebohm et al., 1982; Vicario and Nottebohm, 1988; Vicario 1991; Wild, 1993, 1997). This direct motor pathway is required for the production of song, its maturation during development correlates with the emergence of song vocal behavior in juveniles (Nottebohm et al., 1976; Konishi and Akutagawa, 1985), and its component nuclei are activated coordinately with singing behavior (Yu and Margoliash, 1996; Hahnloser et al., 2002). The anterior forebrain pathway consists of the serial projections from LMAN to area X to DLM and back to LMAN; area X also receives pallial input from HVC (Okuhata and Saito, 1987; Bottjer et al., 1989; Luo et al., 2001). The main output of this anterior pathway is to song motor nucleus RA, through the LMAN to RA projection. In many respects, this anterior pathway is analogous to cortical-basal ganglia-thalamic-cortical loops found in motor control systems in mammals (Farries and Perkel, 2002). The pallial nucleus interface (NIf) and the thalamic nucleus uvaeformis (Uva) provide major inputs to HVC (Nottebohm et al., 1982). Telencephalic nuclei that resemble those in both the direct motor and anterior forebrain pathways have also been identified in the brains of parrots and hummingbirds (Paton et al., 1981; Striedter, 1994; Brauth et al., 1997; Durand et al., 1997; Jarvis and Mello, 2000; Jarvis et al., 2000), two other avian orders that exhibit vocal learning (Nottebohm, 1972); such nuclei are apparently absent in avian orders that lack vocal learning (Karten and Hodos, 1967; Kuenzel and Masson, 1988; Kroodsma and Konishi, 1991).

Some features relevant to the present discussion are immediately evident from the organization of the song control system:

1. There exists both a direct and an indirect projection from HVC to nucleus RA. The indirect projection is part of a pathway involved in the learning and active maintenance of song, and has been postulated to modulate synaptic and cellular plasticity in its target motor nucleus RA (Johnson et al., 1997; Kittelberger and Mooney, 1999; Brainard and Doupe, 2000) (see also Kittelberger and Mooney, Chapter 10, present volume). Such a design may facilitate the implementation of mechanisms for error correction through the tuning of the motor control pathway.
2. Based on its anatomic organization, the song control system consists of a set of nuclei and projections dedicated to vocal motor control, as its output is specifically directed to vocal and respiratory control areas.
3. The song system presents marked sexual dimorphism, with several song nuclei and their projections being very prominent in males, but much smaller or absent in females (Nottebohm and Arnold, 1976).

Interestingly, auditory responses selective for conspecific song (particularly the BOS) can be recorded in the nuclei of the song control system (Margoliash, 1983; Williams and Nottebohm, 1985; Margoliash, 1986; Doupe and Konishi, 1991; Vicario and Yohay, 1993; Volman, 1996). These responses are modulated by exposure to song (Solis and Doupe, 1997, 1999) and may reflect a role of song nuclei in learning the BOS. Studies on mechanisms generating these responses have helped clarify how auditory input modulates activity in the song control system (Mooney, 2000; Rosen and Mooney, 2000). Since these song-evoked responses are mostly seen under anesthesia or during sleep, but are weaker or absent during wakefulness (Dave et al., 1998; Schmidt and Konishi, 1998; Dave and Margoliash, 2000; Nick and Konishi, 2001; Cardin and Schmidt, 2003; Rauske et al., 2003), their role in perceptual processing is unclear. Lesions targeted at song nuclei reportedly affect song-dependent auditory discrimination (Brenowitz, 1991; Del Negro et al., 1998; Scharff et al., 1998; Gentner et al., 2000), but such lesions also affect nearby auditory processing areas or fibers of passage that are part of the auditory projection system (MacDougall-Shackleton et al., 1998). Thus, the evidence for the participation of the song control system in the perceptual processing of birdsong is still inconclusive.

Another important argument against a prominent role of the song system in perceptual discrimination is that female songbirds are typically capable of performing fine song discrimination and recognition both in terms of discriminating conspecific and heterospecific songs, as well as discriminating individual conspecific songs (Ratcliffe and Otter, 1996; Searcy and Yasukawa, 1996; Gentner and Hulse, 2000; Riebel *et al.*, 2002). In that sense, there is no compelling reason to believe that females are intrinsically inferior to males in terms of their auditory processing and discriminatory capabilities. However, their song nuclei and projections are often small or absent (Nottebohm and Arnold, 1976; Arnold *et al.*, 1986), suggesting that areas other than the song control system are primarily involved in song perception in females (see also Williams, 1985).

Much of the search for a neuronal substrate for song perception and discrimination has been dominated by the notion that such processes occur at the level of the song control system. Strictly speaking, however, we still know very little about the exact areas involved in song perception and discrimination, and the formation and storage of song auditory memories. We therefore do not know how the processes outlined in Fig. 11.1a are implemented by neuronal circuits. It has become increasingly clear, however, that both male and female songbirds possess a set of brain areas that are constituent parts of the central auditory pathways and that are involved in various aspects of song auditory processing. The evidence derives from anatomic studies, the expression analysis of activity inducible genes, and electrophysiologic recordings.

Auditory information ascends along a brainstem pathway that is conserved in vertebrates (Butler and Hodos, 1996), reaching the telencephalon through field L (Karten, 1967, 1968; Kelley and Nottebohm, 1979; Brauth *et al.*, 1987). Field L projections to its targets, that is the adjacent caudomedial nidopallium (NCM) and caudomedial mesopallium (CMM), and the shelf and cup areas adjacent to song nuclei HVC and RA, respectively, represent ways for auditory information to reach higher-order telencephalic areas (Bonke *et al.*, 1979; Kelley and Nottebohm, 1979; Fortune and Margoliash, 1995; Vates *et al.*, 1996; Mello *et al.*, 1998). Altogether, these brain areas and projections constitute an avian central auditory system. Similar areas and projections have been described in different avian species (e.g. see Wild *et al.*, 1993; Metzger *et al.*, 1998) regardless of whether they evolved vocal learning and a telencephalic song control system. The nuclei and projections of this central auditory system are pallial and, thus, analogous and possibly homologous to the circuits that constitute the mammalian auditory cortex (Reiner *et al.*, 2004). It is possible that only in vocal learners, which possess telencephalic vocal control nuclei (Brenowitz, 1997; Jarvis *et al.*, 2000), do the vocal control areas have access to song auditory information processed at the pallial/cortical level.

Mapping of activity-dependent gene expression, in particular of the transcription factor *zenk*, has been instrumental in the identification and analysis of areas that respond to song auditory stimulation and thus likely participate in song processing and/or perceptual memorization. Several telencephalic areas distinct from the song control nuclei respond to song presentation with a rapid and robust increase in *zenk* expression (Mello *et al.*, 1992; Mello and Clayton, 1994; Mello and Ribeiro, 1998), whereas the direct motor and anterior forebrain pathways within the song system do not show this response (Mello and Clayton, 1994; Jarvis and Nottebohm, 1997). The *zenk*-expressing areas include field L subfields L1 and L3, NCM, CMM, and the shelf and cup regions, all of which are part of the avian central auditory pathway, as described above. Although the lack of *zenk* induction in specific areas needs to be interpreted with caution, the detection of song-induced *zenk* expression in a given area provides strong and direct evidence for the participation of that area in the sensory processing of song (see further discussion in (see further discussion in Mello, 2002).

Most *zenk* studies have focused on NCM, the area with the most marked *zenk* response to song (Mello *et al.*, 1992). NCM is a field L target (Vates *et al.*, 1996) and is arguably comparable to supragranular layers of the mammalian auditory cortex (Karten and Shimizu, 1989; Mello *et al.*, 1998). Song-induced *zenk* expression in NCM is rapid and transient (Mello and Clayton, 1994; Mello *et al.*, 1995; Kruse *et al.*, 2000), and it is highest for conspecific song, as compared to heterospecific song or tones (Mello *et al.*, 1992). In addition, *zenk* induction decreases markedly upon repeated song presentations (song-specific "habituation"), but it is re-elicited upon presentation of a novel song (Mello *et al.*, 1995). Variations in the spatial distribution of *zenk* expression in NCM correlate with acoustic features of the song stimulus (Ribeiro *et al.*, 1998; Gentner *et al.*, 2001). Altogether, *zenk* expression studies have provided consistent evidence for the participation of NCM in the perceptual processing of birdsong and in song auditory memorization (for reviews, see Ball and Balthazar, 2001; Mello, 2002; Bolhuis and Eda-Fujiwara, 2003).

The evidence from electrophysiologic studies is also consistent with a participation of the areas revealed with *zenk* expression in birdsong auditory processing and perceptual memory formation. For example, robust electrophysiologic responses to song can be recorded in caudal nidopallial areas including NCM, and these responses are of longer latencies and show a higher degree of selectivity towards complex stimuli including conspecific vocalizations than in field L (Bonke *et al.*, 1979; Muller and Leppelsack, 1985; Müller and Scheich, 1985; Ang, 2001). More interestingly, song-evoked responses in NCM decrease, or habituate, in response to repeated song stimulation (Chew *et al.*, 1995, 1996; Stripling *et al.*, 1997, 2001). This habituation is song-specific, as a high response level can be reinstated upon presentation of a novel song stimulus, consistent with the *zenk* studies discussed in the preceding paragraph. The habituated state in NCM can persist for long periods (hours to days), depending on the amount of song stimulation, and its maintenance depends on local gene expression, as indicated by local injections of RNA and protein synthesis inhibitors at specific time windows during and after song stimulation (Chew et al., 1995). Habituation to song in NCM thus consists of an experience-dependent plasticity phenomenon that bears remarkable similarities to hippocampal long-term potentiation, and it has been proposed as a cellular correlate of a song perceptual memory (but also see discussion in Bolhuis and Eda-Fujiwara, 2003).

Electrophysiologic recordings have revealed that song-responsive neurons in CMM also display an experience-dependent plasticity that could be considered a correlate of a perceptual memory trace (Gentner and Margoliash, 2003). More specifically, CMM neurons in starlings show significant selectivity towards songs with which the birds were previously trained in conditioning tasks involving song perceptual discrimination. Thus, the auditory response properties of CMM neurons depend on the previous perceptual history of the bird. Like NCM, CMM presumably represents a higher-order processing station compared with the primary auditory area field L from which it receives a major input, as responses to song in CMM show longer latencies and somewhat higher selectivity towards complex auditory stimuli than those in field L (Heil and Scheich, 1991; Sen et al., 2001). NCM and CMM are highly interconnected (Vates et al., 1996), and can potentially influence each other's response to song and other complex auditory stimuli.

In combination, the anatomic, molecular and physiologic studies indicate that the auditory areas in the caudomedial telencephalon of both sexes play a prominent role in the auditory processing and possibly the memorization of birdsong in the context of perceptual discrimination. These large areas, particularly NCM and CMM, constitute a considerable portion of the telencephalon, indicating that songbirds dedicate a large amount of brain space to the processes above. As argued earlier, because the sensory processing of birdsong is quite likely similar in the contexts of perceptual discrimination and of auditory feedback evaluation for vocal learning, the same or similar pathways/mechanisms may be used in these two contexts. If songbirds evolved an elaborate system to perform song discrimination based on the acoustic properties of song, it would make sense to also utilize that machinery for the process of auditory feedback evaluation of self-vocalizations. NCM and CMM occupy a privileged position within the auditory pathway, as they constitute higher-order areas than the thalamo-recipient field L, and they are in close relationship with vocal control centers. Thus, these auditory areas receive processed sensory input from the ascending auditory pathway, and could later convey the output of their perceptual processing to vocal command centers. For instance, CMM receives a projection from NCM and projects to the caudolateral mesopallium (CLM), which in turn projects to NIf (Vates et al., 1996), providing a possible entry of auditory input from the caudomedial telencephalon into the song control system.

Some evidence that areas like NCM and CMM participate in song auditory feedback processing comes from *zenk* studies. Expression of *zenk* is markedly induced in both structures during singing, but it is abolished in singing birds that have been deafened (Jarvis and Nottebohm, 1997). Thus, *zenk* expression in NCM and CMM during singing is related to the processing of song auditory feedback. In contrast, *zenk* expression in song control nuclei is not abolished by deafening (Jarvis and Nottebohm, 1997), indicating that the activation of the latter areas based on *zenk* expression is more directly related to the motor control of singing than to auditory feedback evaluation (but see also Cardin and Schmidt, 2003; Rauske *et al.*, 2003). The *zenk* expression studies have also provided some evidence that the neuronal circuitry in the caudomedial telencephalon may encode song auditory memories required for both familiar versus unfamiliar song discrimination and auditory feedback evaluation. Song-induced *zenk* expression in NCM is initiated developmentally in association with the onset of the sensitive period for song learning (Jin and Clayton, 1997; Stripling *et al.*, 2001). In addition, the strength of *zenk* induction in adult NCM correlates with the degree to which the bird learned to imitate the stimulus song during vocal learning (Bolhuis *et al.*, 2000, 2001). This effect is independent of the birds' familiarity to the song auditory stimulus (Terpstra *et al.*, 2004). Furthermore, *zenk* expression in NCM of females is modulated depending on the birds' early exposure to song and according to the birds' familiarity to the song stimulus (Maney *et al.*, 2003; Hernandez and MacDougall-Shackleton, 2004). Thus, the available evidence is consistent with a role for NCM in the perceptual processing and discrimination of song in both sexes.

11.5 Implementing feedback evaluation: what do we learn from comparing songbirds and the electric fish

In comparison with the electrosensory system, not enough is known yet about the organization of the circuitry in song processing areas to determine how the comparative processes in Fig. 11.1a are implemented in the brain of songbirds. Based on the discussion above, however, it is apparent that caudomedial telencephalic structures like NCM and CMM likely play a central role in the processing of both external song stimuli and of auditory feedback evaluation during singing. More specifically, these areas appear to be part of an elaborate sensory network involved in the processing of birdsong for both perceptual discrimination and vocal learning. These structures are in a privileged position to access the required auditory information and then send the results of their computations to vocal motor control centers. In addition, in contrast to song control

11.5 Implementing feedback evaluation

nuclei, the output of caudomedial auditory areas is not exclusively dedicated to vocal motor control. Although further studies are needed, the output of these auditory areas is more likely to modulate sensory and/or motor representations and programs involved in broader, non-vocal aspects of songbird behavior (as in Fig. 11.1a, right panel, general motor representations). Thus, as occurs in the cerebellum-like networks of the electric fish, the processing of sensory feedback evaluation in relation to sensory expectations or predictions in songbirds may involve structures that have an eminently sensory/perceptual function.

According to this postulate, the acquired song representations (the song template and the memories of familiar songs) either reside in caudomedial telencephalic structures like NCM and CMM, or these structures have access to such representations. Additionally, neuronal mechanisms for performing perceptual discrimination and auditory feedback evaluation (the comparator in Fig. 11.1a, left panel) would need to be present in these areas. An intriguing possibility is that the phenomenon of habituation observed in NCM plays that role, or at least a contributing one. By decreasing the neuronal responsiveness to a familiar song, habituation sets the auditory system in a state that is mostly sensitive to novelty (i.e. to acoustic features that are present in the incoming auditory input but absent from the internalized memory of the familiar song, and vice versa). This would be somewhat analogous to the negative image mechanism of mormyrids. If such a mechanism were also operative for vocal learning, the strongest responses would be obtained for auditory feedback from vocalizations that differ mostly from the acquired auditory template, whereas vocalizations that match well the template would produce only weak responses. In this regard, the output of an auditory station that habituates would be signaling the extent of a mismatch, as diagrammed in Fig. 11.1a. It is hard to conceive, though, that a mechanism for auditory template formation only makes use of a decrease in responsiveness to song, and further research may reveal correlates of a potentiation-like phenomenon as well. At any rate, habituation could help generate the highly selective responses to song and to specific song features known to occur in areas such as the song control nuclei. In that sense, an acquired song auditory template could be seen as a distributed network with elements recruited from various auditory stations, each with varying degrees of responsiveness to different acoustic features of song.

A very prominent feature of the electrosensory system, at least in mormyrids, is the corollary discharge signal that allows a motor center to create an expectation within the sensory processing system. In songbirds, a modulation of the auditory system by vocal control centers would provide a means for the latter to set the song processing circuits for the comparison between song auditory feedback and the acquired song template. Although no such a mechanism has been directly demonstrated, some limited anatomic evidence suggests the existence of a projection from a song motor nucleus to an auditory structure, namely from the medial extension of HVC, the so-called "paraHVC", to NCM (Foster and Bottjer, 1998). In addition, the finding that *zenk* expression in NCM of singing birds is inversely proportional to the number of song bouts produced (Jarvis and Nottebohm, 1997) suggests that the auditory system can be modulated during vocalizations.

The strong corollary discharge signals in the initial electrosensory processing structure are used for two aspects of electrosensory processing in mormyrids. Firstly, the EOCD gates the electrosensory input to pass only the fish's own EOD and, presumably, to provide a time-reference for decoding the electrosensory afferent spikes. Secondly, the EOCD provides a time-reference in the molecular layer of the ELL for the expectation generator. In songbirds, there is no evidence for a parallel auditory pathway for active sensory processing during singing, so the same pathway is most likely used for active and passive listening. The absence of an AND-gate in the primary auditory pathway allows all auditory information to pass into the system, regardless of the origin. Since the songbird must make the distinction between its own song and external sources late in auditory processing, we have proposed that an AND-gate is present as the sensory information is entering the song motor system as shown in Fig. 11.1a, right panel. The AND-gate then has information about recent motor activity and can be prepared to pass auditory information that shortly follows the generation of song. This mechanism also simplifies the character of the information that is passed to the motor system. Since the auditory information has already been compared with the song template in the auditory pathway, all that needs to be passed to the motor system is the deviation from the expected auditory pattern of the previously learned song.

The corollary discharge signal in the mormyrid fish is also used by the ELL as a time-reference to cancel predictable electrosensory information that immediately follows the EOD. In the songbird auditory system, the absence of a corollary discharge signal implies that the temporal structure of the auditory stimulus itself must be used to form the song template. In addition, neurons in the auditory processing system exhibit habituation in the absence of active singing. Thus, the source of timing information for the generation and recall of the song template might differ from the motor-command driven mechanism in the electrosensory system. One possibility would be that the timing information arises from the song itself, where earlier temporal patterns of the song predict later patterns. Such a hypothesis might lead to experimental tests that could identify which spectro-temporal patterns in the song are the cues upon which the song template is built.

Since bird song is often a very stereotyped sequence of syllables, patterns of syllables early in a song would predict the patterns of later syllables. A mechanism could be proposed where each syllable acts as a surrogate EOCD to provide a series of delayed inputs to generate a template of predictable auditory stimuli. Such a learning system could be based on synaptic plasticity, as in the electric fish, and would be suitable for providing an error signal to the motor system. In addition, this mechanism of template generation would also be precisely what female birds would need to discriminate how consistently male birds are able to repeat their song. Tests for such a hypothesis of auditory learning will involve knowing the details of the song auditory system to characterize responses in candidate substrates for this adaptive template mechanism of auditory learning.

In contrast to the electrosensory processing system in the electric fish, very little is known about the cellular and circuit organization of sensory-processing areas like NCM and CMM. Recent evidence in zebra finches indicates that GABAergic neurons are prevalent and show a marked *zenk* induction response to song in both structures, and that active GABAergic synapses are prevalent in NCM slices (Pinaud et al., 2004). These observations are consistent with previous findings in chickens (Muller and Scheich, 1988), and indicate that inhibitory neurons are likely to play a prominent role in song auditory processing and in the phenomenon of song habituation. It will now be necessary to determine how these neurons are organized into circuits, and to test for their potential role in modulating the response properties and plasticity of song-responsive neurons in NCM and CMM. In testing our postulates, it will also be important to determine the exact outputs of the caudomedial auditory structures, and whether they modulate the physiologic properties of neurons that constitute the song control circuits or other motor representations.

11.6 Conclusion

The comparison between sensory processing systems presented here suggests that similar processing algorithms may be used in systems of widely different sensory modalities and habitats. Data from molecular, anatomic, physiologic and behavioral studies have indicated that brain areas that constitute the central auditory pathways are involved in the perceptual processing and memorization of birdsong. A considerable portion of the songbird telencephalon, distinct from the song control system, may participate in basic perceptual aspects of vocal communication and vocal learning. However, a clearer definition of the sites and mechanisms involved in song auditory memories will require a more refined understanding of how song-responding areas are functionally organized and how they interact with the pathways involved in the motor representation of birdsong. We have attempted to use the well-studied electrosensory processing system to suggest likely mechanisms that account for the observed habituation and learning in the songbird auditory system. It is our expectation that such a comparison will lead to novel hypotheses about songbird auditory learning and suggest new avenues of research.

Acknowledgments

We thank Peter Lovell for insightful comments on this manuscript.

REFERENCES

Ang CW-Y (2001) *Emerging Auditory Selectivity in the Caudomedial Neostriatum of the Zebra Finch Songbird*, The Rockefeller University, New York.

Arnold AP, Bottjer SW, Brenowitz EA, Nordeen EJ, Nordeen KW (1986) Sexual dimorphisms in the neural vocal control system in song birds: ontogeny and phylogeny. *Brain Behav Evol* 28(1–3): 22–31.

Ball GF, Balthazar J (2001) Ethological concepts revisited: immediate early gene induction in response to sexual stimuli in birds. *Brain Behav Evol* 57(5): 252–270.

Bastian J (1995) Pyramidal-cell plasticity in weakly electric fish: a mechanism for attenuating responses to reafferent electrosensory inputs. *J Comp Physiol A* 176: 63–73.

Bell CC (1981) An efference copy in electric fish. *Science* 214: 450–453.

Bell CC (1989) Sensory coding and corollary discharge effects in mormyrid electric fish. *J Exp Biol* 146: 229–253.

Bell CC (1990) Mormyromast electroreceptor organs and their afferent fibers in mormyrid fish. II. Intra-axonal recordings show initial stages of central processing. *J Neurophysiol* 63: 303–318.

Bell CC, Bodznick D, Montgomery J, Bastian J (1997) The generation and subtraction of sensory expectations within cerebellum-like structures. *Brain Behav Evol* 50: 17–31.

Bell CC, Caputi A, Grant K, Serrier J (1993) Storage of a sensory pattern by anti-Hebbian synaptic plasticity in an electric fish. *Proc Natl Acad Sci USA* 90(10): 4650–4654.

Bell CC, Emde Gvd (1995) Electric organ corollary discharge pathways in mormyrid fish II. The medial juxtalobar nucleus. *J Comp Physiol A* 177: 463–479.

Bell CC, Grant K (1992) Sensory processing and corollary discharge effects in the mormyromast regions of the mormyrid electrosensory lobe: II. Cell types and corollary discharge plasticity. *J Neurophysiol* 68: 859–875.

Bell CC, Han V, Sugawara Y, Grant K (1997) Synaptic plasticity in a cerebellum-like structure depends on temporal order. *Nature* 387: 278–281.

Bell CC, Szabo T (1986) Electroreception in mormyrid fish: central anatomy. In: Bullock TH, Heiligenberg W (eds), *Electroreception*, pp. 375–421. Wiley, New York.

Bohner J (1990) Early acquisition of song in the zebra finch. *Taeniopygia guttata. Anim Behav* 39: 369–374.
Bolhuis JJ, Eda-Fujiwara H (2003) Bird brains and songs: neural mechanisms of birdsong perception and memory. *Anim Biol* 53: 129–145.
Bolhuis JJ, Hetebrij E, Den Boer-Visser AM, De Groot JH, Zijlstra GG (2001) Localized immediate early gene expression related to the strength of song learning in socially reared zebra finches. *Eur J Neurosci* 13(11): 2165–2170.
Bolhuis JJ, Zijlstra GG, den Boer-Visser AM, Van Der Zee EA (2000) Localized neuronal activation in the zebra finch brain is related to the strength of song learning. *Proc Natl Acad Sci USA* 97(5): 2282–2285.
Bonke BA, Bonke D, Scheich H (1979) Connectivity of the auditory forebrain nuclei in the guinea fowl (*Numida meleagris*). *Cell Tis Res* 200(1): 101–121.
Bonke BA, Scheich H, Langner G (1979) Responsiveness of units in the auditory neostriatum of the guinea fowl (*Numida meleagris*) to species-specific calls and synthetic stimuli. I. Tonotopy and functional zones of field L. *J Comp Physiol* 132: 243–255.
Bottjer SW, Halsema KA, Brown SA, Miesner EA (1989) Axonal connections of a forebrain nucleus involved with vocal learning in zebra finches. *J Comp Neurol* 279(2): 312–326.
Brainard MS, Doupe AJ (2000) Interruption of a basal ganglia-forebrain circuit prevents plasticity of learned vocalizations. *Nature* 404(6779): 762–766.
Brauth SE, McHale CM, Brasher CA, Dooling RJ (1987) Auditory pathways in the budgerigar. I. Thalamo-telencephalic projections. *Brain Behav Evol* 30(3–4): 174–199.
Brauth SE, Heaton JT, Shea SD, Durand SE, Hall WS (1997) Functional anatomy of forebrain vocal control pathways in the budgerigar (*Melopsittacus undulatus*). *Ann N Y Acad Sci* 807: 368–385.
Brenowitz EA (1991) Altered perception of species-specific song by female birds after lesions of a forebrain nucleus. *Science* 251(4991): 303–305.
Brenowitz EA (1997) Comparative approaches to the avian song system. *J Neurobiol* 33(5): 517–531.
Brenowitz EA, Margoliash D, Nordeen KW (1997) The neurobiology of birdsong. *J Neurobiol* 33(5).
Bullock TH, Heiligenberg W (1986) *Electroreception*, Wiley, New York.
Butler AB, Hodos W (1996) *Comparative Vertebrate Neuroanatomy: Evolution and Adaptation*, Wiley-Liss, New York.
Cardin JA, Schmidt MF (2003) Song system auditory responses are stable and highly tuned during sedation, rapidly modulated and unselective during wakefulness, and suppressed by arousal. *J Neurophysiol* 90(5): 2884–2899.
Catchpole CK, Slater PJB (1995) *Bird Song: Biological Themes and Variations*, Cambridge University Press, Cambridge, UK.
Chew SJ, Mello C, Nottebohm F, Jarvis E, Vicario DS (1995) Decrements in auditory responses to a repeated conspecific song are long-lasting and require two periods of protein synthesis in the songbird forebrain. *Proc Natl Acad Sci USA* 92(8): 3406–3410.
Chew SJ, Vicario DS, Nottebohm F (1996) A large-capacity memory system that recognizes the calls and songs of individual birds. *Proc Natl Acad Sci USA* 93(5): 1950–1955.

Chew SJ, Vicario DS, Nottebohm F (1996) Quantal duration of auditory memories. *Science* 274(5294): 1909–1914.

Dave AS, Margoliash D (2000) Song replay during sleep and computational rules for sensorimotor vocal learning. *Science* 290(5492): 812–816.

Dave AS, Yu AC, Margoliash D (1998) Behavioral state modulation of auditory activity in a vocal motor system. *Science* 282(5397): 2250–2254.

Del Negro C, Gahr M, Leboucher G, Kreutzer M (1998) The selectivity of sexual responses to song displays: effects of partial chemical lesion of the HVC in female canaries. *Behav Brain Res* 96(1–2): 151–159.

Doupe AJ, Konishi M (1991) Song-selective auditory circuits in the vocal control system of the zebra finch. *Proc Nat Acad Sci USA* 88(24): 11339–11343.

Durand SE, Heaton JT, Amateau SK, Brauth SE (1997) Vocal control pathways through the anterior forebrain of a parrot (*Melopsittacus undulatus*). *J Comp Neurol* 377: 179–206.

Eales LA (1985) Song learning in zebra finches: some effects of song model availability on what is learnt and when. *Anim Behav* 33: 1293–1300.

Farries MA, Perkel DJ (2002) A telencephalic nucleus essential for song learning contains neurons with physiological characteristics of both striatum and globus pallidus. *J Neurosci* 22(9): 3776–3787.

Fortune ES, Margoliash D (1995) Parallel pathways and convergence onto HVc and adjacent neostriatum of adult zebra finches (*Taeniopygia guttata*). *J Comp Neurol* 360(3): 413–441.

Foster EF, Bottjer SW (1998) Axonal connections of the high vocal center and surrounding cortical regions in juvenile and adult male zebra finches. *J Comp Neurol* 397(1): 118–138.

Gentner TQ, Hulse SH (2000) Female european starling preference and choice for variation in conspecific male song. *Anim Behav* 59: 443–458.

Gentner TQ, Hulse SH, Bentley GE, Ball GF (2000) Individual vocal recognition and the effect of partial lesions to HVc on discrimination, learning, and categorization of conspecific song in adult songbirds. *J Neurobiol* 42(1): 117–133.

Gentner TQ, Hulse SH, Duffy D, Ball GF (2001) Response biases in auditory forebrain regions of female songbirds following exposure to sexually relevant variation in male song. *J Neurobiol* 46(1): 48–58.

Gentner TQ, Margoliash D (2003) Neuronal populations and single cells representing learned auditory objects. *Nature* 424(6949): 669–674.

Hahnloser RH, Kozhevnikov AA, Fee MS (2002) An ultra-sparse code underlies the generation of neural sequences in a songbird. *Nature* 419(6902): 65–70.

Hall C, Bell CC, Zelick R (1995) Behavioral evidence of a latency code for stimulus intensity in mormyrid electric fish. *J Comp Physiol A* 177: 29–39.

Han V, Bell CC, Grant K, Sugawara Y (1999) Mormyrid electrosensory lobe *in vitro*: I. Morphology of cells and circuits. *J Comp Neurol* 404: 359–374.

Heil P, Scheich H (1991) Functional organization of the avian auditory cortex analogue. II. Topographic distribution of latency. *Brain Res* 539(1): 121–125.

Hernandez AM, MacDougall-Shackleton SA (2004) Effects of early song experience on song preferences and song control and auditory brain regions in female house finches (*Carpodacus mexicanus*). *J Neurobiol* 59(2): 247–258.

Jarvis ED, Mello CV (2000) Molecular mapping of brain areas involved in parrot vocal communication. *J Comp Neurol* 419(1): 1–31.

Jarvis ED, Nottebohm F (1997) Motor-driven gene expression. *Proc Natl Acad Sci USA* 94(8): 4097–4102.

Jarvis ED, Ribeiro S, da Silva ML, Ventura D, Vielliard J, Mello CV (2000) Behaviourally driven gene expression reveals song nuclei in hummingbird brain. *Nature* 406(6796): 628–632.

Jin H, Clayton DF (1997) Localized changes in immediate-early gene regulation during sensory and motor learning in zebra finches. *Neuron* 19(5): 1049–1059.

Johnson F, Hohmann SE, DiStefano PS, Bottjer SW (1997) Neurotrophins suppress apoptosis induced by deafferentation of an avian motor-cortical region. *J Neurosci* 17(6): 2101–2111.

Karten HJ (1967) The organization of the ascending auditory pathway in the pigeon (*Columba livia*). I. Diencephalic projections of the inferior colliculus (nucleus mesencephali lateralis, pars dorsalis). *Brain Res* 6(3): 409–427.

Karten HJ (1968) The ascending auditory pathway in the pigeon (*Columba livia*). II. Telencephalic projections of the nucleus ovoidalis thalami. *Brain Res* 11(1): 134–153.

Karten HJ, Hodos W (1967) *A Stereotaxic Atlas of the Brain of the Pigeon* (*Columba livia*), Johns Hopkins Press, Baltimore, MD.

Karten HJ, Shimizu T (1989) The origins of neocortex: connections and lamination as distinct events in evolution. *J Cogn Neurosci* 1: 291–301.

Kelley DB, Nottebohm F (1979) Projections of a telencephalic auditory nucleus-field L-in the canary. *J Comp Neurol* 183(3): 455–469.

Kittelberger JM, Mooney R (1999) Lesions of an avian forebrain nucleus that disrupt song development alter synaptic connectivity and transmission in the vocal premotor pathway. *J Neurosci* 19(21): 9385–9398.

Konishi M (1965) The role of auditory feedback in the control of vocalization in the white-crowned sparrow. *Zeitung für Tierpsychologie* 22: 770–783.

Konishi M, Akutagawa E (1985) Neuronal growth, atrophy and death in a sexually dimorphic song nucleus in the zebra finch brain. *Nature* 315(6015): 145–147.

Kroodsma DE, Konishi M (1991) A suboscine bird (eastern phoebe, *Sayornis phoebe*) develops normal song without auditory feedback. *Anim Behav* 42: 477–487.

Kroodsma DE, Miller EH (1996) *Ecology and Evolution of Acoustic Communication in Birds*, (London: Cornell University Press).

Kruse AA, Stripling R, Clayton DF (2000) Minimal experience required for immediate-early gene induction in zebra finch neostriatum. *Neurobiol Learn Mem* 74(3): 179–184.

Kuenzel WJ, Masson M (1988) *A Stereotaxic Atlas of the Brain of the Chick* (*Gallus domesticus*), Johns Hopkins University Press, Baltimore, MD.

Leonardo A, Konishi M (1999) Decrystallization of adult birdsong by perturbation of auditory feedback. *Nature* 399(6735): 466–470.

Luo M, Ding L, Perkel DJ (2001) An avian basal ganglia pathway essential for vocal learning forms a closed topographic loop. *J Neurosci* 21(17): 6836–6845.

MacDougall-Shackleton SA, Hulse SH, Ball GF (1998) Neural bases of song preferences in female zebra finches (*Taeniopygia guttata*). *NeuroReport* 9(13): 3047–3052.

Maney DL, MacDougall-Shackleton EA, MacDougall-Shackleton SA, Ball GF, Hahn TP (2003) Immediate early gene response to hearing song correlates with receptive behavior and depends on dialect in a female songbird. *J Comp Physiol A Neuroethol Sens Neural Behav Physiol* 189(9): 667–674.

Margoliash D (1983) Acoustic parameters underlying the responses of song-specific neurons in the white-crowned sparrow. *J Neurosci* 3(5): 1039–1057.

Margoliash D (1986) Preference for autogenous song by auditory neurons in a song system nucleus of the white-crowned sparrow. *J Neurosci* 6(6): 1643–1661.

Marler P (1970) A comparative approach to vocal learning: song development in white-crowned sparrows. *J Comp Physiol Psychol* 71: 1–25.

Marler P (1997) Three models of song learning: evidence from behavior. *J Neurobiol* 33(5): 501–516.

Marler P, Peters S (1977) Selective vocal learning in a sparrow. *Science* 198: 519–521.

Marler P, Peters S (1981) Sparrows learn adult song and more from memory. *Science* 213: 780–782.

Marler P, Peters S (1982) Structural changes in song ontogeny in the swamp sparrow. *Melospiza georgiana*. *The Auk* 99: 446–458.

Marler P, Peters S (1987) A sensitive period for song acquisition in the song sparrow, *Melospiza melodia*: a case of age-limited learning. *Ethology* 76: 89–100.

Marler P, Peters S (1988) The role of song phonology and syntax in vocal learning preferences in the song sparrow, *Melospiza melodia*. *Ethology* 77: 125–149.

Meek J, Grant K, Bell CC (1999) Structural organization of the mormyrid electrosensory lateral line lobe. *J Exp Biol* 202: 1291–1300.

Mello C, Nottebohm F, Clayton D (1995) Repeated exposure to one song leads to a rapid and persistent decline in an immediate early gene's response to that song in zebra finch telencephalon. *J Neurosci* 15(10): 6919–6925.

Mello CV (2002) Mapping vocal communication pathways in birds with inducible gene expression. *J Comp Physiol A Neuroethol Sens Neural Behav Physiol* 188(11–12): 943–959.

Mello CV, Clayton DF (1994) Song-induced ZENK gene expression in auditory pathways of songbird brain and its relation to the song control system. *J Neurosci* 14(11 Pt 1): 6652–6666.

Mello CV, Ribeiro S (1998) ZENK protein regulation by song in the brain of songbirds. *J Comp Neurol* 393(4): 426–438.

Mello CV, Vates GE, Okuhata S, Nottebohm F (1998) Descending auditory pathways in the adult male zebra finch (*Taeniopygia guttata*). *J Comp Neurol* 395(2): 137–160.

Mello CV, Vicario DS, Clayton DF (1992) Song presentation induces gene expression in the songbird forebrain. *Proc Natl Acad Sci USA* 89(15): 6818–6822.

Metzger M, Jiang S, Braun K (1998) Organization of the dorsocaudal neostriatal complex: a retrograde and anterograde tracing study in the domestic chick with special emphasis on pathways relevant to imprinting. *J Comp Neurol* 395(3): 380–404.

Montgomery JC, Bodznick D (1994) An adaptive filter that cancels self-induced noise in the electrosensory and lateral line mechanosensory systems of fish. *Neurosci Lett* 174: 145–148.

Mooney R (2000) Different subthreshold mechanisms underlie song selectivity in identified HVc neurons of the zebra finch [published erratum appears in

J Neurosci 2000 Aug 1; 20(15): following table of contents]. *J Neurosci* 20(14): 5420–5436.

Muller CM, Leppelsack HJ (1985) Feature extraction and tonotopic organization in the avian auditory forebrain. *Exp Brain Res* 59(3): 587–599.

Muller CM, Scheich H (1988) Contribution of GABAergic inhibition to the response characteristics of auditory units in the avian forebrain. *J Neurophysiol* 59(6): 1673–1689.

Müller SC, Scheich H (1985) Functional organization of the avian auditory field L: a comparative 2DG study. *J Comp Physiol* 156: 1–12.

Nick TA, Konishi M (2001) Dynamic control of auditory activity during sleep: correlation between song response and EEG. *Proc Natl Acad Sci USA* 98(24): 14012–14016.

Nordeen KW, Nordeen EJ (1992) Auditory feedback is necessary for the maintenance of stereotyped song in adult zebra finches. *Behav Neural Biol* 57(1): 58–66.

Nottebohm F (1972) The origins of vocal learning. *Am Nat* 106: 116–140.

Nottebohm F, Arnold AP (1976) Sexual dimorphism in vocal control areas of the songbird brain. *Science* 194(4261): 211–213.

Nottebohm F, Kelley DB, Paton JA (1982) Connections of vocal control nuclei in the canary telencephalon. *J Comp Neurol* 207(4): 344–357.

Nottebohm F, Stokes TM, Leonard CM (1976) Central control of song in the canary, Serinus canarius. *J Comp Neurol* 165(4): 457–486.

Oertel D, Young ED (2004) What's a cerebellar circuit doing in the auditory system? *TINS* 27: 104–110.

Okuhata S, Saito N (1987) Synaptic connections of thalamo-cerebral vocal nuclei of the canary. *Brain Res Bull* 18(1): 35–44.

Paton JA, Manogue KR, Nottebohm F (1981) Bilateral organization of the vocal control pathway in the budgerigar, *Melopsittacus undulatus*. *J Neurosci* 1(11): 1279–1288.

Pinaud R, Velho TA, Jeong JK, Tremere LA, Leao RM, von Gersdorff H, Mello CV (2004) GABAergic neurons participate in the brain's response to birdsong auditory stimulation. *Eur J Neurosci* 20(5): 1318–1330.

Ratcliffe L, Otter K (1996) Sex differences in song recognition. In: Kroodsma DE, Miller EH (eds), *Ecology and Evolution of Acoustic Communication in Birds*, pp. 340–355. Cornell University Press, Ithaca, NY.

Rauske PL, Shea SD, Margoliash D (2003) State and neuronal class-dependent reconfiguration in the avian song system. *J Neurophysiol* 89(3): 1688–1701.

Reiner A, Perkel DJ, Bruce LL, Butler AB, Csillag A, Kuenzel W, Medina L, Paxinos G, Shimizu T, Striedter G, Wild M, Ball GF, Durand S, Guturkun O, Lee DW, Mello CV, Powers A, White SA, Hough G, Kubikova L, Smulders TV, Wada K, Dugas-Ford J, Husband S, Yamamoto K, Yu J, Siang C, Jarvis ED (2004) Revised nomenclature for avian telencephalon and some related brainstem nuclei. *J Comp Neurol* 473(3): 377–414.

Ribeiro S, Cecchi GA, Magnasco MO, Mello CV (1998) Toward a song code: evidence for a syllabic representation in the canary brain. *Neuron* 21(2): 359–371.

Riebel K, Smallegange IM, Terpstra NJ, Bolhuis JJ (2002) Sexual equality in zebra finch song preference: evidence for a dissociation between song recognition and production learning. *Proc R Soc Lond B Biol Sci* 269(1492): 729–733.

Roberts PD, Bell CC (2000) Computational consequences of temporally asymmetric learning rules. II. Sensory image cancellation. *J Comp Neurosci* 9: 67–83.

Rosen MJ, Mooney R (2000) Intrinsic and extrinsic contributions to auditory selectivity in a song nucleus critical for vocal plasticity. *J Neurosci* 20(14): 5437–5448.

Scharff C, Nottebohm F, Cynx J (1998) Conspecific and heterospecific song discrimination in male zebra finches with lesions in the anterior forebrain pathway. *J Neurobiol* 36(1): 81–90.

Schmidt MF, Konishi M (1998) Gating of auditory responses in the vocal control system of awake songbirds. *Nat Neurosci* 1(6): 513–518.

Searcy WA, Yasukawa K (1996) Song and female choice. In: Kroodsma DE, Miller EH (eds), *Ecology and Evolution of Acoustic Communication in Birds*, pp. 455–473. Cornell University Press, Ithaca, NY.

Sen K, Theunissen FE, Doupe AJ (2001) Feature analysis of natural sounds in the songbird auditory forebrain. *J Neurophysiol* 86(3): 1445–1458.

Solis MM, Doupe AJ (1997) Anterior forebrain neurons develop selectivity by an intermediate stage of birdsong learning. *J Neurosci* 17(16): 6447–6462.

Solis MM, Doupe AJ (1999) Contributions of tutor and bird's own song experience to neural selectivity in the songbird anterior forebrain. *J Neurosci* 19(11): 4559–4584.

Striedter GF (1994) The vocal control pathways in budgerigars differ from those in songbirds. *J Comp Neurol* 343(1): 35–56.

Stripling R, Kruse AA, Clayton DF (2001) Development of song responses in the zebra finch caudomedial neostriatum: role of genomic and electrophysiological activities. *J Neurobiol* 48(3): 163–180.

Stripling R, Volman SF, Clayton DF (1997) Response modulation in the zebra finch neostriatum: relationship to nuclear gene regulation. *J Neurosci* 17(10): 3883–3893.

Tchernichovski O, Mitra PP (2002) Towards quantification of vocal imitation in the zebra finch. *J Comp Physiol A Neuroethol Sens Neural Behav Physiol* 188(11–12): 867–878.

Tchernichovski O, Mitra PP, Lints T, Nottebohm F (2001) Dynamics of the vocal imitation process: how a zebra finch learns its song. *Science* 291(5513): 2564–2569.

Terpstra NJ, Bolhuis JJ, den Boer-Visser AM (2004) An analysis of the neural representation of birdsong memory. *J Neurosci* 24(21): 4971–4977.

Thorpe WH (1958) The learning of song patterns by birds with special reference to the song of the chaffinch. *Fringilla coelebs Ibis* 100: 535–570.

Vates GE, Broome BM, Mello CV, Nottebohm F (1996) Auditory pathways of caudal telencephalon and their relation to the song system of adult male zebra finches. *J Comp Neurol* 366(4): 613–642.

Vicario DS (1991) Organization of the zebra finch song control system: II. Functional organization of outputs from nucleus Robustus archistriatalis. *J Comp Neurol* 309(4): 486–494.

Vicario DS, Nottebohm F (1988) Organization of the zebra finch song control system: I. Representation of syringeal muscles in the hypoglossal nucleus. *J Comp Neurol* 271(3): 346–354.

Vicario DS, Yohay KH (1993) Song-selective auditory input to a forebrain vocal control nucleus in the zebra finch. *J Neurobiol* 24(4): 488–505.

Volman SF (1996) Quantitative assessment of song-selectivity in the zebra finch high vocal center. *J Comp Physiol A* 178(6): 849–862.

Wild JM (1993) Descending projections of the songbird nucleus robustus archistriatalis. *J Comp Neurol* 338(2): 225–241.

Wild JM (1997) Neural pathways for the control of birdsong production. *J Neurobiol* 33(5): 653–670.
Wild JM, Karten HJ, Frost BJ (1993) Connections of the auditory forebrain in the pigeon (Columba livia). *J Comp Neurol* 337(1): 32–62.
Williams H (1985) Sexual dimorphism of auditory activity in the zebra finch song system. *Behav Neural Biol* 44(3): 470–484.
Williams H, Nottebohm F (1985) Auditory responses in avian vocal motor neurons: a motor theory for song perception in birds. *Science* 229(4710): 279–282.
Woolley SM, Rubel EW (2002) Vocal memory and learning in adult Bengalese Finches with regenerated hair cells. *J Neurosci* 22(17): 7774–7787.
Yu AC, Margoliash D (1996) Temporal hierarchical control of singing in birds. *Science* 273(5283): 1871–1875.
Zeigler HP, Marler P (2004) Behavioral Neurology of birdsong. In: Annals of the New York Academy of Sciences (New York: The New York Academy of Sciences).
Zipser B, Bennet MVL (1976) Interaction of electrosensory and electromotor signals in lateral line lobe of a mormyrid fish. *J Neurophysiol* 39: 713–721.

12
Cortical Plasticity and Auditory Communication

Jun Yan and Jos J. Eggermont

12.1 Introduction

Auditory communication often relates to species-specific vocalization in a small group of individuals who share almost the same auditory experience. By such experience, acoustic signals may gain behavioral significance and animals may be stimulated to produce these signals so that they can be used for communication (Kraus *et al.*, 1998; Wang, 2000; White, 2001). Therefore, it seems that an important characteristic of many communication sounds is their dependency on experience.

The neural processing of species-specific vocalizations presents a big challenge for neuroscientists because vocalizations have a complex acoustical, spectrotemporal structure and serve an important social function. Vocalizations, such as the "meow" produced by cats (Fig. 12.1), consist of multiple segments including constant frequency components, frequency and amplitude modulations, and noise bursts, all with particular durations, timing, or overlap of components. Sound perception and cognition are based on an entire sound object, that is the integration of a time-ordered sequence of acoustic segments or elements (Grossberg, 1999). The neurophysiological mechanisms underlying the neural processing and/or coding of vocalizations are still far from understood. However, it appears that the central auditory system represents vocalizations by integrating various sets of elements of the signals that are perceived and decoded by the auditory periphery (Rauschecker *et al.*, 1995; Geissler and Ehret, 2002; Medvedev *et al.*, 2002). From this point of view, neural processing of individual components is the essential key for the neural processing of vocalizations. Alteration of one or more components would change not only the acoustical context of the vocalization but also its behavioral significance. On the other hand, it suggests that the auditory system would form an enhanced representation of an entire vocalization if the neural processing of one or more

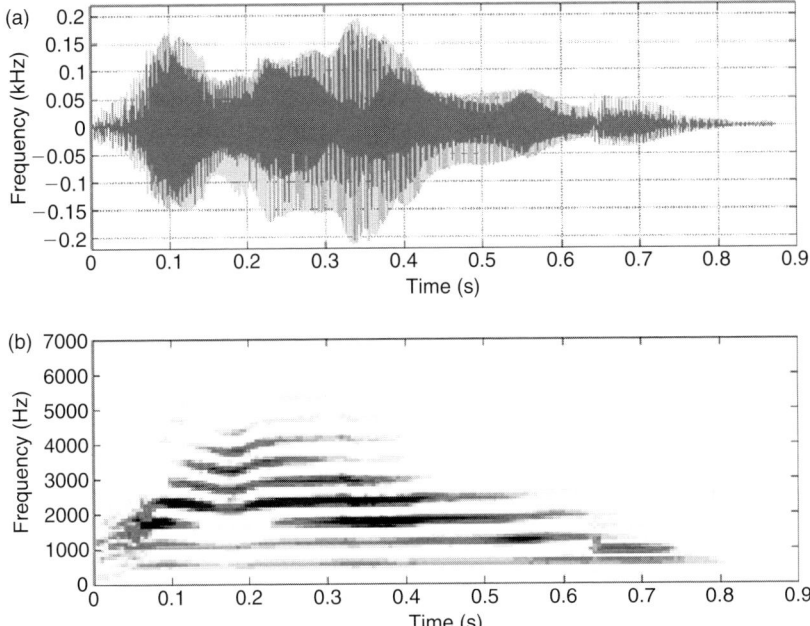

Figure 12.1. Kitten vocalization that illustrates IBEs and IBPs. The waveform (a) and spectrogram (b) of a kitten meow is presented. The duration of this meow is 0.87 s. It has a slow amplitude modulation. Distinct downward and upward frequency modulations occur simultaneously in all harmonics between about 100 and 200 ms after onset. (Reprinted from *Hearing Research* 150: 27–42, Gehr DD, Komiya H, Eggermont JJ (2000) Neuronal responses in cat primary auditory cortex to natural and altered species-specific calls, with permission from Elsevier).

components would be enhanced. For example, language-learning disorders can result from deficits in the processing of fundamental components of the acoustic signals, such as deficits in discriminating the temporal order of rapidly delivered acoustic signals, in elevated thresholds for frequency discrimination and for amplitude and frequency modulation (Bailey and Snowling, 2002). Therefore, the neural coding of communication sounds in the central auditory system can be understood from knowledge of the neural processing or coding of individual elements or components of a communication sound.

Complex acoustic signals such as human speech and animal communication sounds share major components such as constant frequency, frequency-modulated parts and noise bursts which, if used for discriminating sounds or their messages, are called information-bearing elements (IBE; Suga, 1988). Animal vocalizations are generally a stereotyped combination of IBEs. Human speech sounds have more complexity. The parameters of each IBE (information-bearing parameters, IBP) and the way in which the IBEs are combined show large inter-species variability and characterize species-specific vocalizations. The IBPs characterize the differences in sounds by describing their physical properties such as frequency, amplitude, the rate and depth of frequency and amplitude modulations, duration, and time intervals between sound elements. As the IBEs and IBPs constitute the building blocks of complex acoustic signals, the investigation of their neural coding gives an insight into the neural processing of complex acoustic signals including communication sounds (Suga, 1988; Suga et al., 1998; Wang, 2000; Eggermont, 2001).

The auditory cortex is a crucial level of processing of communication sounds. In monkeys, the ablation of the auditory cortex completely abolished the ability to discriminate species-specific "coo" vocalizations while the animals were still able to discriminate sound frequencies (Heffner and Heffner, 1984; Harrington et al., 2001). In rats, reversible inactivation of primary auditory cortex by muscimol (a GABA agonist) even led to the inability to detect tones of different frequency (Talwar et al., 2001). The fundamental components of acoustical signals, for example frequency and amplitude, are systematically represented in the auditory cortex (Reale and Imig, 1980; Heil et al., 1994; Sutter and Schreiner, 1995). Within this frequency and amplitude frame, cortical neurons show vigorous tuning to other IBPs such as rate and depth of frequency and amplitude modulation or sound duration (Suga, 1978; Ehret, 1997; Eggermont, 2001). More importantly, the auditory cortex also comprises many kinds of combination-sensitive neurons that show facilitative responses to combined IBEs and may show specific tuning to time intervals or overlaps between sounds (Margoliash and Fortune, 1992; DeCharms et al., 1998; Suga et al., 1998). One may consider the combination-sensitivity of cortical neurons as a prototypical requirement for the processing of vocalizations because it reveals the integration of information from different aspects of sounds.

Cortical representations of IBEs/IBPs are adaptable to alterations in auditory information (Ehret, 1997; Weinberger, 1998; Eggermont and Komiya, 2000; Suga et al., 2002). This plasticity occurs in response to changes in sound structure, in response to processing sounds in the auditory periphery, and as the result of learning (Weinberger, 1998). In this chapter, we will review the importance of auditory experience for cortical plasticity during early development and adulthood. We will further investigate the important role of the cortex in the plasticity of the subcortical auditory system. Finally, the cortical representation of species-specific vocalizations within the perspective of experience-dependent plasticity for auditory communication is discussed.

12.2 Experience-dependent plasticity of the auditory cortex during early development

12.2.1 Critical role of sensory experience in early cortical development

The primary auditory cortex, like the primary visual and somatosensory cortex, is topographically organized; it systematically represents distance along the sensory epithelium of the cochlea (King, 1995; Ehret, 1997). The cochleotopic organization in the cortex is mainly a consequence of the topographical organization of the ascending pathways. This statement is most likely correct when it pertains to the early stage of development (i.e. perinatally) before the arrival of sensory-driven activity at the cortical level (Yuste and Sur, 1999). Distinct functional areas and some tuning properties of the cortex start to be detectable when thalamocortical afferents arrive. During this period, cortical development depends on intrinsic factors and the prototype of sensory cortex develops based on the topography of thalamocortical projections. Further functional shaping of the prototypical sensory cortex, however, is initiated with the arrival of sensory-driven activities, that is according to auditory experience (Rubenstein, 2000).

Increasing evidence shows that sensory experience plays a more important role in development of cortical representations or functional organization than the initial hard-wired projections. It is striking that sensory experience is not only able to determine the topographic organization of sensory cortex but also to change its anatomic connections. The pioneering studies by Wiesel and Hubel (1963) have demonstrated that the formation of the ocular dominance columns in the visual cortex in kittens is strongly influenced by monocular deprivation. Interestingly, deprivation of both eyes did not lead to a plastic change in ocular dominance, suggesting that plasticity occurred in response to an imbalance in activation. The findings unveil that visual experience, guiding the competition for synaptic connections between the thalamocortical afferents from both eyes, plays a critical role in organization of ocular dominance in the visual cortex (Hubel and Wiesel, 1998). Further, findings from so-called cross-modal plasticity studies clearly demonstrated the predominant role of sensory experience on cortical development. For example, redirection of visual inputs to the auditory thalamus of the ferret in the first postnatal day made the neurons in the auditory cortex respond to visual stimulation comparable to that of neurons in the visual cortex (Gao and Pallas, 1999; Sharma *et al.*, 2000). Such redirection dramatically changed horizontal connections in the auditory cortex, which no longer were restricted to isofrequency bands but extended in a more isotropic pattern as seen in the visual cortex. Therefore, sensory experience does not simply fill up the topographical frame that is established by the anatomic connections. Instead, cortical circuits are actively reformed and cortical function is determined by sensory experiences according to the acoustic signals encountered in external world.

12.2.2 Normal development of cortical IBEs/IBPs and tonotopic maps

In the primary auditory cortex of the adult cat, neurons generally show sharp frequency tuning and a well-defined characteristic frequency (the frequency to which a neuron shows the lowest response threshold), and normal minimum neural thresholds are reached by day 30 (Eggermont, 1996). Figure 12.2 shows the minimum threshold (MT) as a function of the characteristic frequency for young (less than 30 days old), juvenile (30–100 days) and adult cats (more than 100 days). Clearly, young and juvenile cats have, on average, higher MTs than older ones, especially at frequencies below 4 kHz, and the represented frequency range of adult cats is more extended to lower frequencies. This implies that sensitivity of

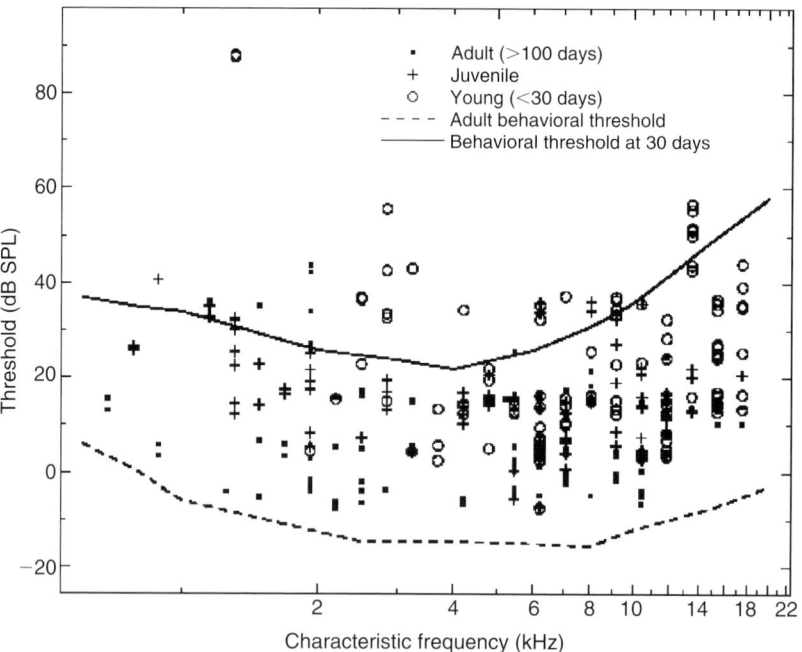

Figure 12.2. Scatterplot of individual unit thresholds as a function of characteristic frequency. Behavioral threshold curves obtained in adult cats and 30 day old kittens (after Ehret and Romand, 1981) are shown for comparison (modified from Eggermont, 1996).

auditory cortical neurons to the IBPs of sounds improves from young to adult animals. Tone-evoked neuronal responses in the cortex of the cat can be detected as early as postnatal day 9 (P9), however only in a restricted central frequency range. With increasing age, responsiveness rapidly expands to lower and higher frequencies as has been shown in behavioral tests (Ehret and Romand, 1981). In C57 mice, the cortical representation of frequencies expands from low to high frequencies with age (Willott *et al.*, 1993), in rats an initially small area of cortical responsiveness quickly increases and finally is compressed again (Zhang *et al.*, 2001). Raised in a normal acoustic environment, cat cortical neurons show maturity of threshold and most IBPs at around postnatal day 40 (Eggermont, 1996).

There is no doubt that the representational plasticity of the auditory cortex during early development is largely associated with the maturation of the entire auditory system. This is true for the development of the response thresholds of cortical neurons in cat, which parallel developments in the auditory nerve (Kettner *et al.*, 1985), cochlear nucleus (Brugge *et al.*, 1978) and inferior colliculus (Moore and Irvine, 1979). Whether and how auditory experience interferes with intrinsic developmental patterns, has been highlighted in experiments focusing on cortical development in distorted acoustic environments.

12.2.3 Cortical development in a distorted acoustic environment

For individuals growing up in a natural environment, the cortical representation of frequency information is similar since they have acquired similar auditory experience. By manipulation of the acoustic environment during early life, one can examine the role of auditory experience in auditory cortical development. Studies in the visual system (Blakemore and Cooper, 1971) point to what can be expected in the auditory system. The experience-dependent modification of the tonotopic representation in the auditory cortex was recently demonstrated by the exposure of neonatal animals to distorted acoustic environments. Stanton and Harrison (1996) reared newborn kittens in continuous 8 kHz frequency-modulated (± 1 kHz) sound from birth to 3 months of age. Mapping the primary auditory cortex in mature animals revealed an expanded representation of the 6–12 kHz area. Zhang *et al.* (2001, 2002) exposed rat pups for 10–16 h/day from postnatal day 9 to 28 to 4 or 19 kHz tone pips. The cortical frequency representation was measured after day 22 when the tonotopic map normally would be adult like, and most cortical neurons were now tuned to either 4 or 19 kHz. Exposing animals during early development to white noise, reduced the frequency selectivity of cortical neurons and disrupted the tonotopic representation of the auditory cortex (Zhang *et al.*, 2002). Thus, auditory experience during a critical period in ontogeny is extremely important for the development of frequency tuning in the auditory cortex. These studies suggest that sounds frequently perceived during early life selectively enhance the neural representation of their acoustical parameters in the auditory cortex. In other words, the development of the auditory cortex is biased by frequently perceived sounds so that it is able to better process IBPs of sounds of obvious biological importance.

12.2.4 Impact of inner ear ablation on cortical development

Studies have shown that the tonotopic representation in the primary auditory cortex is reorganized in newborn animals with high-frequency hearing loss caused by ototoxic drugs (Harrison *et al.*, 1991). Neurons normally tuned to high frequencies became tuned to the highest unaffected frequencies. This shows that cortical neurons do not degenerate when their corresponding main input information degenerates; instead they start to be driven by input from the vicinity of what has been lost in the auditory periphery. The consequences for cortical development and plasticity after complete deprivation of one ear are intriguing. Due to several chiasmas (cross projections) in the auditory system, the dominant input to the auditory cortex is from the contralateral ear. About 65% of cortical neurons in cats also receive signals from the ipsilateral ear, although with higher response thresholds (Reale *et al.*, 1987). According to the impact of auditory experience on early auditory cortical development and synaptic competition, as originally demonstrated by monocular dominance plasticity in the visual cortex, one expects that the development of the auditory cortex will be dominated by the intact ipsilateral ear if the contralateral input is completely removed. This is indeed the case. In cats with neonatal deprivation of one ear the tonotopic representation of the contralateral auditory cortex appears to be normal, however the response thresholds in ipsilateral cortex are now reduced to the usually lower level of the contralateral input (Reale *et al.*, 1987). Similar findings have been obtained in the ferret inferior colliculus after neonatal unilateral cochlear removal (Moore *et al.*, 1993). These data suggest that the processing in the auditory midbrain and cortex, when being deprived from the normal inputs of one ear, is completely driven by the acoustic signals from the intact ear. Morphological studies showing a 27% increase in dendrite length in layers III/IV of the auditory cortex contralateral to the deprived ear (McMullen *et al.*, 1988) support these findings.

Acoustic signals are also required for normal cortical maturation. In children, deafened in early life, the application of a cochlear implant, which typically restores speech perception to near normal levels, allows the testing of the subsequent maturation of the auditory cortex (Ponton *et al.*, 1996; Ponton and

Eggermont, 2001). Following the restoration of hearing after a deaf period of several years, the latencies of cortical auditory-evoked potential components were delayed compared to those of age-matched controls. However, when latency was related to the hearing age, that is chronological age minus duration of deafness, the values overlapped with those in normal hearing children. This suggests that human auditory cortical maturation stops at the onset of deafness but can be reactivated by adequate stimulation.

12.3 Experience-dependent plasticity of the auditory cortex in adult animals

After the most dynamic period of development, the representation of IBEs/IBPs by cortical neurons reaches its adult form (Eggermont, 1996; Zhang *et al.*, 2001) which, however, does not become static. Although the scale of plasticity becomes smaller during adulthood, the mature cortex is subject to continuing plastic change (Weinberger, 1998; Eggermont and Komiya, 2000; Suga *et al.*, 2002). This is an intriguing feature of the sensory cortex, which allows learning throughout life.

12.3.1 Enhancement of cortical representation induced by acoustic signals alone

As discussed above, over-exposure of neonatal animals to particular sounds leads to the enhancement of cortical representation of these sounds (Zhang *et al.*, 2001). This phenomenon is, however, also seen in adult animals. One example is the rapid shift of frequency tuning of auditory midbrain neurons in a big brown bat toward the frequency of a particular sound that was repetitively perceived (Chowdhury and Suga, 2000). Although the range of frequency shift was small, the data suggest that even in adult animals, acoustic stimulation alone is able to enhance the neural representation of a frequently presented signal.

12.3.2 Enhanced cortical representation as a result of learning

As early as 1956, it was demonstrated in a number of species that neuronal activity in the central auditory system including auditory cortex is modifiable in response to tone-related behavioral conditioning (Galambos *et al.*, 1956; Disterhoft and Olds, 1972). However, the most significant progress in learning-induced auditory plasticity was made in the past two decades, specifically regarding the frequency specificity of plasticity in the auditory cortex and subcortical nuclei after associative learning (Weinberger, 1998). Classic conditioning has two important properties. One is that behavioral changes of the animal are specific to the conditioned stimulus (CS) and the other is the association of the unconditioned stimulus (US) with the CS. Hence, it is necessary to examine whether the learning-induced neural plasticity is specific to CS sound and whether the association of US with CS is required for inducing the CS-specific changes in response properties of auditory neurons. The initial study by Diamond and Weinberger (1984) demonstrated that plastic changes in auditory responses induced by classical conditioning, such as tone-shock, were specific to the frequency of the conditioned sound. When the CS and US were paired, cortical neurons showed significant increases in response to the CS frequency and decreases to other frequencies. The effect could be diminished after extinction training, that is by presenting the CS sound alone. Thus, enhancing the biological significance of a tone by classical conditioning induces frequency-specific plasticity in the auditory cortex (Bakin and Weinberger, 1990). If a CS tone- and US foot-shock are not paired or are delivered randomly, the CS tone evoked only a general decrease in the auditory responses of cortical neurons (Edeline and Weinberger, 1993). As a result of the shift in the best frequencies (BFs) of cortical neurons, the cortical representation of the CS frequency is enhanced (Kilgard and Merzenich, 1998). The CS-specific change can be quickly induced in as few as five training trials (Edeline *et al.*, 1993) and can last for more than 2 months after conditioning (Weinberger *et al.*, 1993). These findings strongly suggest that the auditory cortex is able to quickly reorganize itself through learning, and maintain this reorganization, in order to retain and to better process the auditory information associated with particular biological events.

Adult monkeys are able to enhance their ability for sound discrimination and to acquire new skills or behaviors as a result of repetitive auditory training over an extensive period (Prosen *et al.*, 1990). This type of training might be related to human's ability to acquire new communication sounds, that is learn to understand and to speak new words. Adult monkeys show progressive improvement in the discrimination of the frequency differences of two tones during a training period of several weeks. Most interestingly, at the time that the monkeys can successfully distinguish the frequency differences, the cortical representation of these frequencies and the frequency selectivity of cortical neurons is significantly enhanced (Recanzone *et al.*, 1993). Clearly, the adult auditory cortex can be shaped to enhance the processing of frequency differences of biological significance, which obviously underlies the accompanying improvement of frequency discrimination during the training.

12.3.3 Cortical plasticity after localized hearing loss

Increasing the presence or significance of particular auditory inputs leads to the expansion of the respective cortical representations as discussed above. What happens to the auditory cortex if particular inputs are eliminated on a scale that is much smaller than the elimination of one ear (see above). A restricted lesion of the cochlea can be induced with ototoxic drugs (Harrison *et al.*, 1991; Takeno *et al.*, 1994) or high-intensity sounds (Eggermont and Komiya, 2000). Such restricted lesions in both cochleae do lead to profound reorganization of the auditory cortex. As shown by Eggermont and Komiya (2000), exposure of cats to a loud 6 kHz tone (126 dB SPL for 1 h) causes mild to moderate binaural high-frequency hearing loss, and a reorganization of the cortical tonotopic map. Neurons originally tuned to frequencies above about 4–6 kHz shift their tuning to the highest undamaged frequencies (Fig. 12.3). Similarly, a minor-to-mild pure-tone induced hearing loss for a small frequency range increases MTs and the bandwidth of frequency tuning curves for neurons with characteristic frequencies

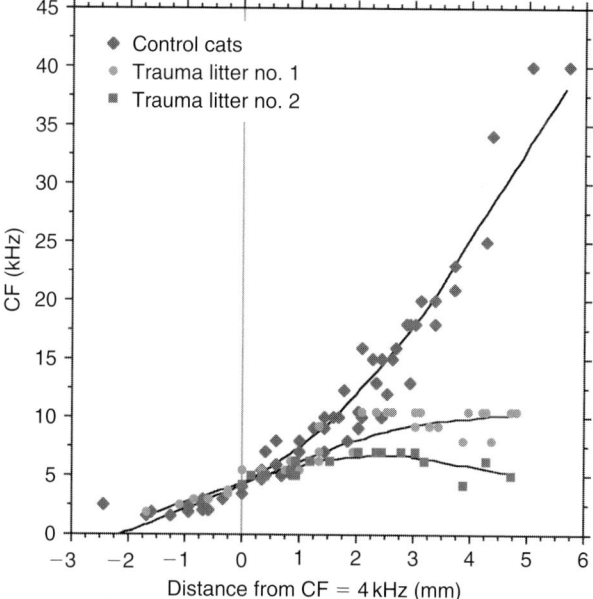

Figure 12.3. Characteristic frequency (CF) of recording sites in primary auditory cortex as a function of rostro-caudal distance from the 4 kHz site for control and trauma animals. Litter no. 1 comprised three exposed kittens and two controls, litter no. 2 comprised two exposed kittens and two controls. Both litters received the same exposure. Locally weighted average curves are drawn in for the control and two trauma cat litters separately (modified from Eggermont and Komiya, 2000).

in the range of the hearing loss (Seki and Eggermont, 2002). Due to the restricted lesion in a particular frequency band, the frequency tuning curve of cortical neurons across the damaged frequency band may show double peaks at the normal edge frequencies of the hearing loss range with increased thresholds between the peaks in the damaged frequency range. Similar findings of broad, double-tuned, frequency tuning curves were obtained in inferior colliculus immediately after mechanical lesions to the spiral ganglion in the region of the neurons' characteristic frequencies (Snyder *et al.*, 2000; Snyder and Sinex, 2002).

These data strongly suggest that the deprivation of particular inputs causes a degeneration of these afferents to cortical neurons and an unmasking of the nearest normal inputs that then dominate the response property of the affected neurons. Hence, the degenerative type of cortical plasticity may follow a "nearby" rule. One may ask what changes in the adult auditory cortex will happen after a restricted ablation in only one cochlea. If the "nearby" rule is true, the deprived frequency band should be replaced by nearest normal frequency bands and not by the same frequency band from the normal ear. Importantly, this turns out to be the case. Neurons in the deprived frequency region tune to the nearest normal frequency in the damaged contralateral cochlea so that the cortical representation of the nearest non-deprived frequency in that cochlea is largely increased. The tonotopic map of the ipsilateral cortex measured by stimulation of the non-damaged ear is not affected (Robertson and Irvine, 1989; Rajan *et al.*, 1993).

12.4 Cortex-oriented plasticity in the central auditory system: corticofugal modulation

The functional role of corticofugal projections in auditory information processing and plasticity had largely been ignored, although corticofugal projections were already known (e.g. Huffman and Henson, 1990). The auditory cortex plays a critical role in auditory cognitive processing such as recognizing communication sounds and even in simple auditory perception tasks such as frequency discrimination (Jancke *et al.*, 1999; Talwar *et al.*, 2001). Therefore, the auditory cortex must be actively involved in these tasks instead of passively perceiving information from the periphery. How is this effectuated? A large number of topographic descending projections from the auditory cortex to subcortical nuclei have been found in a variety of species (Andersen *et al.*, 1980; Huffman and Henson, 1990; Herbert *et al.*, 1991; Saldana *et al.*, 1996). What is the function of these projections? Frequency-specific neural plasticity evoked by associative learning has been found not only in the auditory cortex but also in subcortical nuclei such as auditory midbrain and thalamus (Gao and Suga,

1998; Edeline *et al.*, 1993). The cholinergic projection of the basal forebrain to the auditory cortex is known to be an important pathway mediating learning induced and frequency specific cortical plasticity. The question is whether cortical plasticity can feed back to lower levels of the auditory system to induce representational plasticity there. Recent progress in the study of the corticofugal function has started to give answers to these questions. As described below, one breakthrough from recent studies indicates corticofugal modulation to be functionally related or specific to the IBE/IBP tunings of neurons in the central auditory system (Suga *et al.*, 1997, 1998). The studies suggest an active role of the auditory cortex through the corticofugal projections in priming lower-level information processing to enhance cognitive processing and plasticity.

12.4.1 Frequency-specific modulation of frequency tuning and tonotopic maps

Tonotopic maps in the auditory cortex and subcortical nuclei are constructed according to the systematical arrangement of characteristic frequencies of auditory nerve fibers innervating haircells along the cochlear tonotopy. Therefore, the fundamental property of most auditory neurons is their distinct frequency tuning. The physiological importance of corticofugal modulation has been unveiled by the correlation of frequency tuning of activated cortical neurons and those recorded from subcortical neurons. Thus, it has been shown in house mice (Yan and Ehret, 2001, 2002) that focal electrical stimulation of cortical neurons reduces auditory responses in the central nucleus of the inferior colliculus (ICC) when the stimulated cortical and the recorded ICC neurons are tuned to different frequencies. In contrast, it enhances the auditory responses of ICC neurons that are tuned to the same frequency as the stimulated cortical neurons. At the same time, cortical activation shifts ICC frequency tuning upward when the BFs of ICC neurons are lower than cortical BFs and downward when they are higher than cortical BFs. When the difference in the BFs between cortical and ICC neurons are within 2 kHz, the ICC BFs are rarely changed. The shift in ICC BFs is significantly correlated to the BF difference between cortical and ICC neurons, that is, the larger the difference in BFs the larger the shift in ICC BFs is (Fig. 12.4a). As a result of the tuning shifts of ICC neurons, the tonotopic map in the ICC is reorganized and centered on the stimulated cortical BF. The neural representation of the frequency corresponding to the BF of the stimulated cortical neurons is enhanced and that of neighboring frequencies reduced. Apparently, the corticofugal adjustment and improvement is highly specific to the BF of the stimulated cortical neurons.

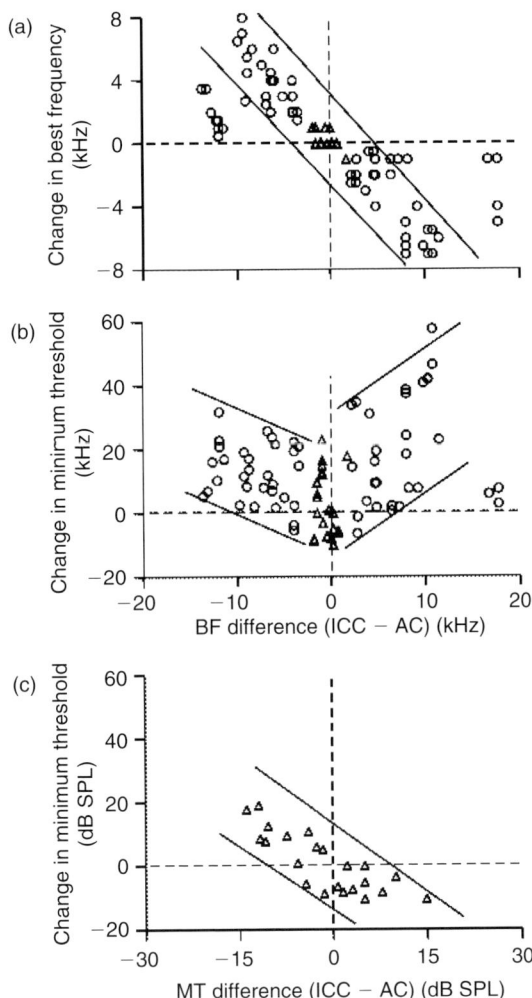

Figure 12.4. Frequency-specific and amplitude-specific changes in the BF and MTs of collicular neurons evoked by focal cortical activation. (a) The amount of BF change is plotted as a function of the difference in the BFs between the cortical and collicular neurons. The BF change is significantly correlated with the BF difference when the BF difference ranges from +12 to −10 kHz (between two solid lines). The amount of MT change is plotted as function of the differences in both the (b) BFs and (c) MTs between cortical and collicular neurons. The range of change is about 30 dB. It is most likely that the variation of cortically evoked changes for particular BF differences and matched neurons is due to the MT-dependent modulation as shown in (c). Circles indicate physiologically unmatched neurons. Triangles indicate physiologically matched neurons. Dashed lines indicate zero crossings. AC: auditory cortex (modified from Yan and Ehret, 2002).

Interestingly, the maximum BF range in the ICC that is affected by the corticofugal modulation is related to the stimulated cortical BF; the higher the cortical BF the larger is the maximum affected range. This range corresponds to the critical bandwidths of the same species of mice, which changes as a function of the center frequencies of these filters (Ehret, 1976; Yan and Ehret, 2001) suggesting that the frequency bandwidths of corticofugal adjustment are related to the frequency resolution of the auditory system. Another property of corticofugal effects is that they are rapidly induced and long lasting. When the cortical neurons are stimulated with 0.5 µA monophasic current paired with a tone burst at a rate of 4/s for 7 min, the changes in the auditory responses and frequency tuning can be detected immediately after the stimulation and last more than 8 h (Yan and Ehret, 2001).

Corticofugal effects due to cortical activation have been found even in the cochlea as frequency-specific changes in the amplitude of the microphonic potential (Xiao and Suga, 2002). Such effects seem to be larger the closer the affected auditory centers are to the stimulation site, that is they are larger in the thalamus and other cortical areas than in the midbrain, however with similar pattern of changes induced (Zhang *et al.*, 1997; Yan and Suga, 1998; Chowdhury and Suga, 2000; Zhang and Suga, 2000). Altogether, these findings show that cortical neurons are able to modulate the auditory responses and to reorganize the tonotopic maps in the central auditory system through a highly focused positive feedback to physiologically "matched" subcortical neurons with a widespread inhibition of physiologically "unmatched" neurons (Suga *et al.*, 1997).

12.4.2 Modulation of IBEs/IBPs in different domains

Recent studies indicate that corticofugal modulation is not restricted to effects on the frequency organization, but also acts in the amplitude and time domains in an IBE/IBP-specific manner. In mice, cortical stimulation adjusts the MTs of ICC neurons in both frequency-specific and amplitude-specific manner (Fig. 12.4b and c; Yan and Ehret, 2002). MTs of ICC neurons increase after cortical microstimulation when the BFs of cortical and ICC neurons are different. The larger the BF difference is, the larger is the MT increase. For ICC neurons of BFs similar ("matched") to those of the stimulated cortical location, MTs either increase or decrease according to the MT difference between cortical and ICC neurons (Fig. 12.4c). Cortical activation also reduces the dynamic ranges of rate-intensity function of ICC neurons in a manner specific to the BF and MT relationship between cortical and ICC neurons.

Duration-tuned neurons have been found in the central auditory system in various species (Pinheiro *et al.*, 1991; He *et al.*, 1997; Brand *et al.*, 2000). In big brown bats, focal stimulation of cortical duration-tuned neurons sharpens the duration tuning and increases the response magnitude of ICC neurons whose best-duration matches that of the stimulated cortical neurons. Differences in best-duration tuning of stimulated cortical and tested ICC neurons lead to broadening of ICC duration tuning and shifts of best durations toward those of the cortical neurons. Similar to MTs, the shifts in best duration are systematically related to the difference in best durations between cortical and ICC neurons (Ma and Suga, 2001).

Together, these studies show that corticofugal modulation affects the whole configuration of sounds (frequencies, amplitudes of frequency components via MTs, and durations). In other words, corticofugal modulation most likely enhances the central processing of IBEs/IBPs of a given sound, that is the whole sound as an acoustic object. This concept is very important as an approach to the biological significance of auditory plasticity for the neural processing of complex acoustic signals including communication sounds.

12.4.3 Corticofugal modulation of response properties of combination-sensitive neurons

One of the important features of the neural processing of communication sounds is integration; neurons frequently show combination sensitivity when there is more than one IBE/IBP in a sound. In mustached bats, for example, the central auditory system contains neurons that are sensitive to the combination of two frequency-modulated sounds (FM–FM neurons) separated by the time interval between the echolocation pulse and its echo (Yan and Suga, 1996a). The FM–FM neurons show strong tuning to the time delay of the returning high harmonic components (FM_{2-4}) from the low harmonic component (FM_1) in the pulse. Combination-sensitive neurons in the auditory midbrain show plasticity to modulation by corticofugal projections. The plastic changes of the bat's FM–FM neurons are systematic and specific only to the time delay, that is the best delay, of neurons (Yan and Suga, 1996b). This type of plasticity may be part of the neural plasticity for communication sound processing.

12.4.4 Species-specific differences in corticofugal modulation

Corticofugal modulation is a common mechanism in the central sensory processing, contributing to plasticity in subcortical centers (Yan and Suga, 1996b; Murphy *et al.*, 1999; Ma and Suga, 2001; Jen *et al.*, 2002; Suga *et al.*, 2002; Yan and Ehret, 2002). For physiologically "matched" subcortical and cortical neurons, auditory corticofugal modulation shows identical effects across species, namely an increase of response and a sharpening of tuning. However, corticofugal modulation displays different forms across species for physiologically "unmatched" neurons. In specialized neurons of the mustached bat inferior colliculus such as FM–FM neurons (for distance measurement) and Doppler-shifted constant frequency neurons (for velocity measurement), cortical activation shifts their tunings away from that of the stimulated cortical location (Yan and Suga, 1996b; Suga *et al.*, 1997; Zhang *et al.*, 1997) when the tuning of the cortical and collicular neurons is different. In big brown bats, cortical activation shifts BFs of ICC neurons toward the stimulated cortical BF when ICC BFs are higher than cortical BFs (Yan and Suga, 1998). In mice, however, cortical activation shifts ICC tunings toward the stimulated cortical BFs when ICC BFs are either higher or lower than cortical BFs. Therefore, the pattern of corticofugal modulation is species specific.

The pattern and direction of the shift in frequency tuning of unmatched neurons evoked by focal cortical activation is identical to that evoked by auditory learning (Edeline *et al.*, 1993; Gao and Suga, 1998; Yan and Suga, 1998; Yan and Ehret, 2002). The comparable shift patterns of frequency tuning for corticofugal modulation and associative learning suggest that corticofugal modulation may play an important role in learning-induced auditory plasticity (Gao and Suga, 1998). Combined with the specific role of the basal forebrain in associative learning (Bakin and Weinberger, 1996; Bjordahl *et al.*, 1998; Kilgard and Merzenich, 1998), we hypothesize that cholinergic inputs from the basal forebrain facilitate responses of particular cortical neurons that are tuned to the frequency of the target sound during auditory learning. Such facilitation further induces the frequency-specific shift in the tuning of "unmatched" cortical and subcortical neurons by triggering the intrinsic mechanism of corticofugal modulation.

12.5 Perspective of cortical plasticity for cortical coding of vocalizations

Human speech and animal communication sounds are complex compounds of IBEs and IBPs displaying rich and rapid alteration in frequency, amplitude, and time. Therefore, a large population of neurons tuned to different types of IBEs and IBPs of animal calls or human speech may be activated synchronously and sequentially in the time window of a vocalization. It is believed that both the temporal and spatial pattern of neuronal population activity are important for the neural representation or coding of the vocalization (Rauschecker and Tian, 2000; Wang, 2000; Eggermont, 2001). The response pattern of individual neurons is only understandable if their role in the overall pattern is clear.

310   *12 Cortical plasticity and auditory communication*

It is likely that plasticity of cortical IBE/IBP tuning, resulting from auditory experience, will in turn change the neural processing of vocalizations. Most of our knowledge about cortical plasticity has been obtained by studying changes in IBE/IBP tuning. To understand the suggested impact of perceiving vocalizations on cortical plasticity, and what that means, in reverse, for the neural processing of vocalizations, one first has to analyze and identify the IBEs/IBPs of vocalizations and, subsequently, the IBE/IBP tuning of cortical neurons.

Figure 12.1 shows the waveform (a) and spectrogram (b) of a natural meow produced by kittens. The meow consists of at least eight harmonics. The frequency of the first or fundamental harmonic is about 500 Hz, equal to the repetition pitch of the meow. The increment in frequency of each harmonic is about 500 Hz. The frequency of the eighth harmonic is about 4.2 kHz. The strongest harmonics are the third and the fourth (Fig. 12.1b). Most components of these harmonics are of constant frequency. However, there is a frequency modulation, initially downward then upward, between 100 and 220 ms after the onset of the meow. The entire call is strongly amplitude modulated (Fig. 12.1a) and the frequency-modulated components are relatively weak in intensity compared with the constant frequency parts.

Gehr *et al.* (2000) studied the response properties of cortical neurons to stimulation with normal and "morphed" meows presented at 65 dB SPL. The morphing consisted of either changing the frequency content of the meow, while leaving the envelope the same, or by compressing or stretching the envelope while keeping the frequency content the same. The three envelopes are shown in Fig. 12.5a together with the response of a cortical neuron tuned to 2.5 kHz. The neuron shows a robust response to the onset of the meow and again about 100 ms later. This is a general response pattern to the meow shown by most neurons. Further analysis indicates that the responses to the meow are related to the frequency tuning of the neurons. Figure 12.6 shows the same nine combinations of waveform and frequency content as in Fig. 12.5, now after filtering by the neuronal filter (Fig. 12.6c), mimicking the frequency-tuning curve of the studied neurons. This shows that the response to the amplitude modulation of the envelope of the forward meow depends on the meow's frequency content. Thus, the frequency tuning of the neuronal filters provides a basis for neuronal responses to sharp intensity increases (onsets of amplitude modulations) in vocalizations. In other words, when frequencies of a given stimulus fit the frequency tuning of a given neuron, the neuron shows more clear responses to other IBEs/IBPs such as amplitude modulation in this case. Cortical neurons locking their responses to amplitude-modulated noise or tone burst, usually have best modulation frequencies between 6 and 20 Hz (tuning to rhythm). This suggests that the response pattern and response strength of cortical neurons to

12.5 Perspective of cortical plasticity for cortical coding

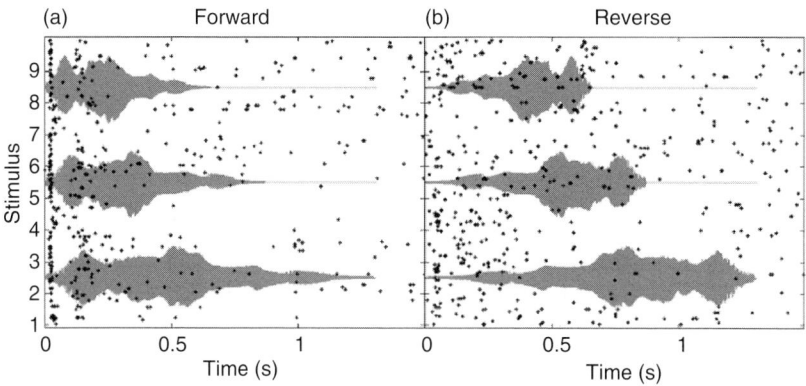

Figure 12.5. Dot displays of the auditory cortical responses to a particular meow of a kitten played forward (a) or time-reversed (b). The cortical multi-unit recording site with a characteristic frequency of 2.5 kHz. Time (x-axis) is specified in seconds. The different morphed stimuli are numbered from 1 to 9 (y-axis). Stimuli 1 to 3 are the time-expanded meows with lowered (stimulus 1), unaltered (stimulus 2), and increased (stimulus 3) frequency components. Stimuli 4 to 6 are unaltered lengthwise and stimuli 7 to 9 are compressed in time both sets with the same three frequency variations as in 1–3. The waveform envelopes of the stimuli 2, 5, and 8 are shown in gray shade. The dots (spikes) responding to, for example stimulus 5 are those between the dash marking 5 and the dash marking 6, etc. (Reprinted from *Hearing Research* 150: 27–42, Gehr DD, Komiya H, Eggermont JJ (2000) Neuronal responses in cat primary auditory cortex to natural and altered species-specific calls, with permission from Elsevier.)

vocalizations with amplitude-modulated envelopes are determined by both the frequency and temporal tuning of the neurons.

In monkey cortex, the spectrotemporal response patterns to a conspecific twitter call are a reflection of the spectrotemporal acoustic pattern of the twitter sound (Wang *et al.*, 1995). Cortical neurons show excitatory responses if the frequency spectrum extends into the frequency-tuning curve of a given neuron. For example, neurons tuned to frequencies lower than about 7 kHz do not respond to the first high-frequency (>10 kHz) phrase of the twitter, whereas the reverse is true for responses to other phrases with frequencies mostly below 10 kHz. Note that the cortical responses to species-specific vocalizations are not simply the result of addition or subtraction of the responses to distinct acoustic elements or parameters, that is IBE/IBP of a vocalization, but show a more complex integrative form representing the vocalization and species specificity. This is obvious from the much weaker onset response to a time reversed compared to the natural vocalization in awake monkeys (Wang *et al.*, 1995), and to some extent also in the anesthetized cat (Figs. 12.5b, 12.6b; Gehr *et al.*, 2000).

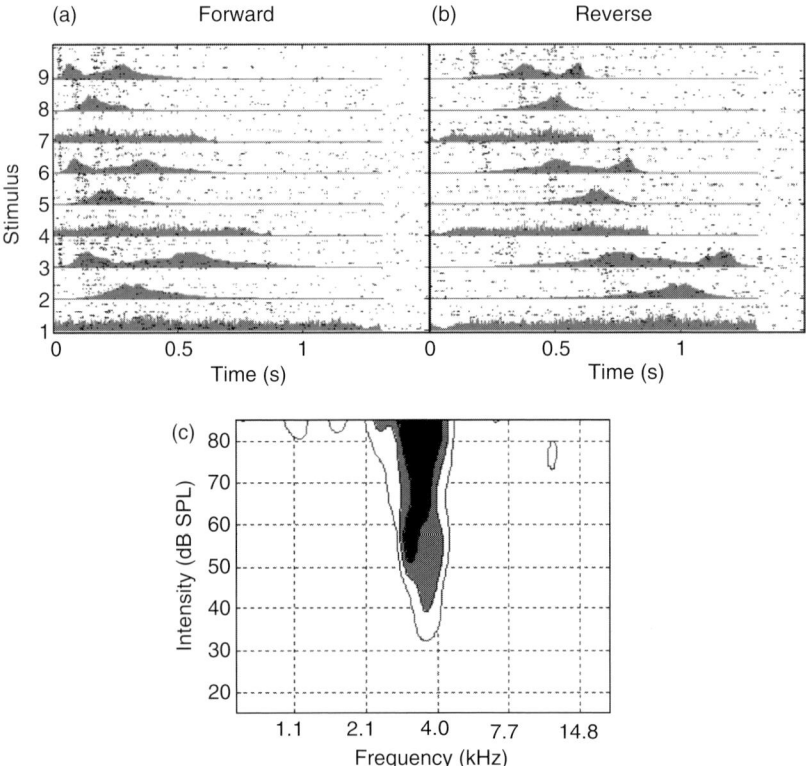

Figure 12.6. Dot displays of the cortical responses to the forward meow (a) and the reverse meow (b) as in Fig. 12.5. The multi-unit recording site has a characteristic frequency of 3.7 kHz. Time (x-axis) is specified in seconds. The different stimuli are numbered from 1 to 9 as in Fig. 12.5. The stimuli are filtered through the tuning curve (c) and the positive parts of the waveform are shown superimposed on the dot rasters. The envelope waveforms are scaled individually to the maximum. (Reprinted from *Hearing Research* 150: 27–42, Gehr DD, Komiya H, Eggermont JJ (2000). Neuronal responses in cat primary auditory cortex to natural and altered species-specific calls, with permission from Elsevier.)

Furthermore, cortical neurons in the cat show poor responses to a monkey twitter call even if the frequency content of the stimuli falls into the excitatory frequency tuning curve of the studied neurons (Wang and Kadia, 2001).

Considering the properties of cortical plasticity discussed above, we can predict that the shaping of the auditory cortex into its adult processing form is normally dominated by the IBEs/IBPs of species-specific vocalizations to which humans and animals are most exposed most during early life. Training or learning during adulthood should improve the cortical processing or representation of vocalizations. It is also predictable that training with one call may facilitate the cortical processing of another call if the two calls share some IBEs/IBPs. Indeed, listening training in humans can improve the discrimination and neural representation of acoustic signals that are not used in the training (Tremblay *et al.*, 1997). Ontogenetic plasticity and learning in adults shapes the auditory cortex in multiple domains and for complex combination sensitivities of its neurons, so that cortical representation of relevant IBPs/IBEs is enhanced. In turn, cortical plasticity, reflected in the enhanced cortical representation of IBPs/IBEs of vocalizations, will improve the neural processing of the heard and learnt vocalization and of those which share their IBEs/IBPs.

12.6 Conclusion

Our knowledge of the auditory cortical processing of complex acoustic signals such as species-specific communication sounds is still poor. Most communication sounds are complex acoustic compounds of IBEs and IBPs. Considering that acoustical communication in birds and mammals depends much on auditory experience, species-specific vocalizations are the most important sound source as basis for such experience and for the development of the functional organization of the auditory cortex during early life. We hypothesize that learning can constantly reshape the functional organization of the auditory cortex according to the IBE/IBP tunings of the learnt vocalizations. This cortical plasticity further improves the cortical processing and representation of the learnt vocalizations and others that share major IBEs/IBPs with the learnt ones. This suggests that cortical coding or representation of IBEs/IBPs of complex, often species specific, vocalizations is one important key toward the understanding of the functional role of cortical plasticity.

REFERENCES

Andersen RA, Synder RL, Merzenich MM (1980) The topographic organization of corticocollicular projections from physiologically identified loci in the AI, AII, and anterior auditory cortical field of the cat. *J Comp Neurol* 191: 479–494.

Bailey PJ, Snowling MJ (2002) Auditory processing and the development of language and literacy. *Brit Med Bull* 63: 135–146.

Bakin JS, Weinberger NM (1990) Classical conditioning induces CS-specific receptive field plasticity in the auditory cortex of the guinea pig. *Brain Res* 536: 271–286.

Bakin JS, Weinberger NM (1996) Induction of a physiological memory in the cerebral cortex by stimulation of the nucleus basalis. *Proc Natl Acad Sci USA* 93: 11219–11224.

Bjordahl TS, Dimyan MA, Weinberger NM (1998) Induction of long-term receptive field plasticity in the auditory cortex of the waking guinea pig by stimulation of the nucleus basalis. *Behav Neurosci* 112: 1–13.

Blakemore C, Cooper GF (1971) Modification of the visual cortex by experience. *Brain Res* 31: 366–367.

Brand A, Urban R, Grothe B (2000) Duration tuning in the mouse auditory midbrain. *J Neurophysiol* 84: 1790–1799.

Brugge JF, Javel E, Kitzes LM (1978) Signs of functional maturation of peripheral auditory system in discharge patterns of neurons in anteroventral cochlear nucleus of kitten. *J Neurophysiol* 41: 1557–1559.

Chowdhury SA, Suga N (2000) Reorganization of the frequency map of the auditory cortex evoked by cortical electrical stimulation in the big brown bat. *J Neurophysiol* 83: 1856–1863.

DeCharms RC, Blake DT, Merzenich MM (1998) Optimizing sound features for cortical neurons. *Science* 280: 1439–1443.

Diamond DM, Weinberger NM (1984) Classical conditioning rapidly induces specific changes in frequency receptive fields of single neurons in secondary and ventral ectosylvian auditory cortical fields. *Brain Res* 372: 357–360.

Disterhoft JF, Olds J (1972) Differential development of conditioned unit changes in thalamus and cortex of rat. *J Neurophysiol* 35: 665–679.

Edeline J-M, Weinberger NM (1993) Receptive field plasticity in the auditory cortex during frequency discrimination training: selective returning independent of task difficulty. *Behav Neurosci* 107: 82–103.

Edeline J-M, Pham P, Weinberger NM (1993) Rapid development of learning-induced receptive field plasticity in the auditory cortex. *Behav Neurosci* 107: 539–557.

Eggermont JJ (1996) Differential maturation rates for responses parameters in cat primary auditory cortex. *Audit Neurosci* 2: 309–327.

Eggermont JJ (2001) Between sound and perception: reviewing the search for a neural code. *Hear Res* 157: 1–42.

Eggermont JJ, Komiya H (2000) Moderate noise trauma in juvenile cats results in profound cortical topographic map changes in adulthood. *Hear Res* 142: 89–101.

Ehret G (1976) Critical bands and filter characteristics in the ear of the housemouse (*Mus musculus*). *Biol Cybern* 24: 35–42.

Ehret G (1997) The auditory cortex. *J Comp Physiol A* 181: 547–557.

Ehret G, Romand R (1981) Postnatal development of absolute auditory thresholds in kittens. *J Comp Physiol Psychol* 95: 304–311.

Galambos R, Sheatz G, Vernier V (1956) Electrophysiologic correlates of a conditioned response in cats. *Science* 123: 376–377.

Gao W, Pallas SL (1999) Cross-modal reorganization of horizontal connectivity in auditory cortex without altering thalamocortical projections. *J Neurosci* 19: 7940–7950.

Gao E, Suga N (1998) Experience-dependent corticofugal adjustment of midbrain frequency map in bat auditory system. *Proc Natl Acad Sci USA* 95: 12663–12670.

Gehr DD, Komiya H, Eggermont JJ (2000) Neuronal responses in cat primary auditory cortex to natural and altered species-specific calls. *Hear Res* 150: 27–42.

Geissler DB, Ehret G (2002) Time-critical integration of formants for perception of communication calls in mice. *Proc Natl Acad Sci USA* 99: 9021–9025.

Grossberg S (1999) The link between brain learning, attention, and consciousness. *Conscious Cogn* 8: 1–44.

Harrington IA, Heffner RS, Heffner HE (2001) An investigation of sensory deficits underlying the aphasia-like behavior of macaques with auditory cortex lesions. *NeuroReport* 12: 1217–1221.

Harrison RV, Nagasawa A, Smith DW, Stanton S, Mount RJ (1991) Reorganization of auditory cortex after neonatal high frequency cochlear hearing loss. *Hear Res* 54: 11–19.

He J, Hashikawa T, Ojima H, Kinouchi Y (1997) Temporal integration and duration tuning in the dorsal zone of cat auditory cortex. *J Neurosci* 17: 2615–2625.

Heffner HE, Heffner RS (1984) Temporal lobe lesions and perception of species-specific vocalizations by macaques. *Science* 226: 75–76.

Heil P, Rajan R, Irvine DR (1994) Topographic representation of tone intensity along the isofrequency axis of cat primary auditory cortex. *Hear Res* 76: 188–202.

Herbert H, Aschoff A, Ostwald J (1991) Topography of projections from the auditory cortex to the inferior colliculus in the rat. *J Comp Neurol* 304: 103–122.

Hubel DH, Wiesel TN (1998) Early exploration of the visual cortex. *Neuron* 20: 401–412.

Huffman RF, Henson Jr OW (1990) The descending auditory pathway and acoustico-motor system: connections with the inferior colliculus. *Brain Res Rev* 15: 295–323.

Jancke L, Mirzazade S, Shah NJ (1999) Attention modulates activity in the primary and the secondary auditory cortex: a functional magnetic resonance imaging study in human subjects. *Neurosci Lett* 266: 125–128.

Kettner RE, Feng JZ, Brugge JF (1985) Postnatal development of the phase-locked response to low frequency tones of auditory nerve fibers in the cat. *J Neurosci* 5: 275–283.

Kilgard MP, Merzenich MM (1998) Cortical map reorganization enabled by nucleus basalis activity. *Science* 279: 1714–1718.

King AJ (1995) Asking the auditory cortex the right question. *Curr Biol* 5: 1110–1113.

Kraus N, McGee TJ, Koch DB (1998) Speech sound representation, perception, and plasticity: a neurophysiologic perceptive. *Audiol Neurootol* 3: 168–182.

Jen PH, Zhou X, Zhang J, Chen QC, Sun X (2002) Brief and short-term corticofugal modulation of acoustic signal processing in the bat midbrain. *Hear Res* 168: 196–207.

Ma X, Suga N (2001) Corticofugal modulation of duration-tuned neurons in the midbrain auditory nucleus in bats. *Proc Natl Acad Sci USA* 98: 14060–14065.

Margoliash D, Fortune ES (1992) Temporal and harmonic combination-sensitive neurons in the zebra finch's HVc. *J Neurosci* 12: 4309–4326.

McMullen NT, Goldberger B, Suter CM, Glaser EM (1988) Neonatal deafening alters nonpyramidal dendrite orientation in auditory cortex: a computer microscope study in the rabbit. *J Comp Neurol* 267: 92–106.

Medvedev AV, Chiao F, Kanwal JS (2002) Modeling complex tone perception: grouping harmonics with combination-sensitive neurons. *Biol Cybern* 86: 497–505.

Moore DR, Irvine DR (1979) The development of some peripheral and central auditory responses in the neonatal cat. *Brain Res* 163: 49–59.

Moore DR, King AJ, McAlpine D, Martin RL, Hutchings ME (1993) Functional consequences of neonatal unilateral cochlear removal. *Prog Brain Res* 97: 127–133.

Murphy PC, Duckett SG, Sillito AM (1999) Feedback connections to the lateral geniculate nucleus and cortical response properties. *Science* 286: 1552–1554.

Pinheiro AD, Wu M, Jen PH (1991) Encoding repetition rate and duration in the inferior colliculus of the big brown bat, *Eptesicus fuscus. J Comp Physiol A* 169: 69–85.

Ponton CW, Eggermont JJ (2001) Of kittens and kids. Altered cortical maturation following profound deafness and cochlear implant use. *Audiol Neurootol* 6: 363–380.

Ponton CW, Don M, Eggermont JJ, Waring MD, Kwong B, Masuda A (1996) Plasticity of the auditory system in children after long periods of complete deafness. *NeuroReport* 8: 61–65.

Prosen CA, Moody DB, Sommers MS, Stebbins WC (1990) Frequency discrimination in the monkey. *J Acoust Soc Am* 88: 2152–2158.

Rajan R, Irvine DR, Wise LZ, Heil P (1993) Effect of unilateral partial cochlear lesions in adult cats on the representation of lesioned and unlesioned cochleas in primary auditory cortex. *J Comp Neurol* 338: 17–49.

Rauschecker JP, Tian B (2000) Mechanisms and streams for processing of "what" and "where" in auditory cortex. *Proc Natl Acad Sci USA* 97: 11800–11806.

Rauschecker JP, Tian B, Hauser M (1995) Processing of complex sounds in the macaque nonprimary auditory cortex. *Science* 268: 111–114.

Reale RA, Imig TJ (1980) Tonotopic organization in auditory cortex of the cat. *J Comp Neurol* 192: 265–291.

Reale RA, Brugge JF, Chan JC (1987) Maps of auditory cortex in cats reared after unilateral cochlear ablation in the neonatal period. *Brain Res* 431: 281–290.

Recanzone GH, Schreiner CE, Merzenich MM (1993) Plasticity in the frequency representation of primary auditory cortex following discrimination training in adult owl monkeys. *J Neurosci* 13: 87–103.

Robertson D, Irvine DR (1989) Plasticity of frequency organization in auditory cortex of guinea pigs with partial unilateral deafness. *J Comp Neurol* 282: 456–471.

Rubenstein JL (2000) Intrinsic and extrinsic control of cortical development. *Novartis Fnd Symp* 228: 67–75.

Saldana E, Feliciano M, Mugnaini E (1996) Distribution of descending projections from primary auditory neocortex to inferior colliculus mimics the topography of intracollicular projections. *J Comp Neurol* 371: 15–40.

Seki S, Eggermont JJ (2002) Changes in cat primary auditory cortex after minor-to-moderate pure-tone induced hearing loss. *Hear Res* 173: 172–186.

Sharma J, Angelucci A, Sur M (2000) Induction of visual orientation modules in auditory cortex. *Nature* 404: 841–847.

Snyder RL, Sinex DG (2002) Immediate changes in tuning of inferior colliculus neurons following acute lesions of cat spiral ganglion. *J Neurophysiol* 87: 434–452.

Snyder RL, Sinex DG, McGee JD, Walsh EW (2000) Acute spiral ganglion lesions change the tuning and tonotopic organization of cat inferior colliculus neurons. *Hear Res* 147: 200–220.

Stanton SG, Harrison RV (1996) Abnormal cochleotopic organization in the auditory cortex of the cats reared in a frequency augmented environment. *Audit Neurosci* 2: 97–107.

Suga N (1978) Specialization of the auditory system for reception and processing of species-specific sounds. *Fed Proc* 37: 2342–2354.

Suga N (1988) What does single-unit in the auditory cortex tell us about information processing in the auditory system. In: Raku P, Singer W (eds), *Neurobiology of Neocortex*, pp. 331–350.Wiley, New York.

Suga N, Yan J, Zhang Y (1997) Cortical maps for hearing and egocentric selection for self-organization. *Trends Cogn Sci* 1: 1–14.
Suga N, Yan J, Zhang Y (1998) The processing of species-specific complex sounds by the ascending and descending auditory system. In: Poon PWF, Brugge JF (eds), *Central Auditory Processing and Neural Modeling*, pp. 55–70. Plenum Press, New York.
Suga N, Xiao Z, Ma X, Ji W (2002) Plasticity and corticofugal modulation for hearing in adult animals. *Neuron* 36: 9–18.
Sutter ML, Schreiner CE (1995) Topography of intensity tuning in cat primary auditory cortex: single-neuron versus multiple-neuron recordings. *J Neurophysiol* 73: 190–204.
Takeno S, Harrison RV, Ibrahim D, Wake M, Mount RJ (1994) Cochlear function after selective inner hair cell degeneration induced by carboplatin. *Hear Res* 75: 93–102.
Talwar SK, Musial PG, Gerstein GL (2001) Role of mammalian auditory cortex in the perception of elementary sound properties. *J Neurophysiol* 85: 2350–2358.
Tremblay K, Kraus N, Carrell TD, McGee T (1997) Central auditory system plasticity: generalization to novel stimuli following listening training. *J Acoust Soc Am* 102: 3762–3773.
Wang X (2000) On cortical coding of vocal communication sounds in primates. *Proc Natl Acad Sci USA* 97: 11843–11849.
Wang X, Kadia SC (2001) Differential representation of species-specific primate vocalizations in the auditory cortices of marmoset and cat. *J Neurophysiol* 86: 2616–2620.
Wang X, Merzenich MM, Beitel R, Schreiner CE (1995) Representation of a species-specific vocalization in the primary auditory cortex of the common marmoset: temporal and spectral characteristics. *J Neurophysiol* 74: 2685–2706.
Weinberger NM (1998) Physiological memory in primary auditory cortex: characteristics and mechanisms. *Neurobiol Learn Mem* 70: 226–251.
Weinberger NM, Javid R, Lepan B (1993) Long-term retention of learning-induced receptive-field plasticity in the auditory cortex. *Proc Natl Acad Sci USA* 90: 2394–2398.
White SA (2001) Learning to communicate. *Curr Opin Neurobiol* 11: 510–520.
Wiesel TN, Hubel DH (1963) Single-cell responses in striate cortex of kittens deprived of vision in one eye. *J Neurophysiol* 26: 1003–1017.
Willott JF, Aitkin LM, McFadden SL (1993) Plasticity of auditory cortex associated with sensorineural hearing loss in adult C57BL/6j mice. *J Comp Neurol* 329: 402–411.
Xiao Z, Suga N (2002) Modulation of cochlear hair cells by the auditory cortex in the mustached bat. *Nat Neurosci* 5: 57–63.
Yan J, Ehret G (2001) Corticofugal reorganization of midbrain tonotopic map in mice. *NeuroReport* 12: 3313–3316.
Yan J, Ehret G (2002) Corticofugal modulation of midbrain sound processing in the house mouse. *Eur J Neurosci* 16: 119–128.
Yan J, Suga N (1996a) The midbrain creates and the thalamus sharpens echo-delay tuning for the cortical representation of target-distance information in the mustached bat. *Hear Res* 93: 102–110.
Yan J, Suga N (1996b) Corticofugal modulation of time-domain processing of biosonar information in bats. *Science* 273: 1100–1103.
Yan W, Suga N (1998) Corticofugal modulation of the midbrain frequency map in the bat auditory system. *Nat Neurosci* 1: 54–58.

Yuste R, Sur M (1999) Development and plasticity of the cerebral cortex: from molecules and maps. *J Neurobiol* 41: 1–6.

Zhang Y, Suga N (2000) Modulation of responses and frequency tuning of thalamic and collicular neurons by cortical activation in mustached bats. *J Neurophysiol* 84: 325–333.

Zhang Y, Suga N, Yan J (1997) Corticofugal modulation of frequency-processing in bats auditory system. *Nature* 387: 900–903.

Zhang LI, Bao S, Merzenich MM (2001) Persistent and specific influences of early acoustic environments on primary auditory cortex. *Nat Neurosci* 4: 1123–1130.

Zhang LI, Bao S, Merzenich MM (2002) Disruption of primary auditory cortex by synchronous auditory inputs during a critical period. *Proc Natl Acad Sci USA* 99: 2309–2314.

13
Mesoscopic Neurodynamics in Auditory Cortex During Auditory Concept Learning

Frank W. Ohl, Henning Scheich and Walter J. Freeman

13.1 Introduction

13.1.1 Traditional theoretical underpinnings of the study of category learning

We define concept learning or category learning as the process by which categorization of stimuli, or, more generally, of situations experienced by a subject is acquired. The term categorization describes the phenomenon that under suited conditions a (human or non-human) subject might behaviorally respond to a multitude of stimuli or situations with a considerably smaller number of response types although the stimuli themselves could, in principle, be discriminated by the subject. The nature of this phenomenon is subject to intellectual debates at least since Aristotle. A brief review of some of the more traditional problems which have repeatedly emerged in the context of characterizing the nature of categorization will be used to motivate the viewpoint that has been taken in planning the experiments focused on in this chapter.

The Aristotelian view (often referred to as the "Classical Theory") considers having or establishing a set of necessary and/or sufficient criteria to be met by stimuli or situations as being the essence of determining their membership to a category. This view, enriched by an appropriate formal framework, is also maintained by some "artificially intelligent" approaches to the categorization problem. A main objection that was brought up against this view is that for many "natural" categories such sets of criteria cannot be found (e.g. Rosch, 1973); neither can category membership always be defined by features which can be unequivocally attributed to all members, nor can features always be evaluated to indicate membership to a category. Rather, varying degrees of "typicalness" of features and/or category members seem to be the rule. Consequently, a number of accounts have been proposed which can be summarized under the heading of "prototype theories" (e.g. Posner and Keele, 1968). Prototype theories hold that some form of idealized representation of category members exists and actual membership of a given stimulus is determined by some scaling of "similarity" of this member to the prototype. Prototype theories in some instances have faced the problem of providing a convincing theory for establishing the required similarity relations (e.g. Ashby and Perrin, 1988) or could not account for non-prototype members having more pronounced effects on categorization performance than the prototype itself (e.g. Brooks, 1978; Medin and Schaffer, 1978). Both studies on human (Rosch, 1975) and animal categorization (Lea and Ryan, 1990) have argued that some of these problems can be attenuated by lifting the requirement for local comparisons with a singular prototype and instead requiring global comparisons with multiple (in the extreme form with all) category members (Estes, 1986; Hintzman, 1986; Nosofsky, 1986) for which reason such theories are summarized under the heading "exemplar theories". While historically these (and other) approaches have been initiated and put forward by their authors with very different rationales in mind they can be more objectively compared to each other using a suitably formalized reformulation (Ashby and Maddox, 1993) which can be derived from general recognition theory (Ashby and Townsend, 1986). It is particularly disturbing that many experimentally achievable observations can be accounted for by any of the theoretical viewpoints provided, at least if appropriate modifications to the extreme positions of such theories are allowed (Pearce, 1994). This is notwithstanding the fact that category learning, rather than being a coherently definable mechanism, seems to include several separable types of cognitive processes probably mediated by separable physiologic processes in separable anatomical structures of the brain (for review see Kéri, 2003).

13.1.2 Eco-ethologic aspects of categorization

While in human studies, the viewpoints sketched above (and others) are maintained without questioning the existence of concepts, as mental constructs, as a possible basis for our categorizations, this option is not so clear for non-human species, simply because they cannot report on having a concept that might guide their categorization behavior. It has been argued that defining a concept would require reference to language so that non-human species could not have concepts at all (Chater and Heyes, 1994). A less extreme viewpoint suggests that some instances of animal categorization might reflect mere discriminations albeit with very complex stimuli (Wasserman and Astley, 1994). It has been elaborated that the demonstration of concepts in non-human species might be possible by suitably designed transfer experiments with appropriately constructed control stimulus sets (Lea, 1984). Transfer of learned discriminations to novel stimuli is a relevant phenomenon because it demonstrates that during an instantiation of category learning more has happened than mere associations of responses to trained particular stimuli (Lea and Ryan, 1990).

It is reasonable to assume that not all possible categorizations can be equally established in different species. For example, while pigeons could not be trained to categorize (and in some cases even discriminate) photographs showing two different individual pigeons, or showing two different individual chickens, bantam cockerels could be trained easily using the same stimulus material (Ryan, 1982a, b). While in particular cases it might be difficult to trace species-specific differences in categorization behavior back to differences in the social behavior of different species (e.g. it can be excluded that the discrimination of individual conspecifics were simply irrelevant to pigeons (Ryan, 1982b)), it seems clear that category learning might serve species-specific needs. In the vervet monkey it has been demonstrated, for example, that maturing individuals are able to develop fine categorization of the production of and response to three different types of alarm calls (each one signaling a different predator category) while young individuals might be wrong, for example, by uttering an "eagle alarm call" in response to a harmless bird (Seyfarth et al., 1980a, b). More generally, it is reasonable to assume that categorization of stimuli on privately learned exemplar experiences is advantageous to categorization based on a species-specific prototype (Nelson and Marler, 1990).

13.1.3 Categorization and generalization

It is appropriate to consider the terms "categorization" and "generalization" at this point in more detail as they are inconsistently used in the literature and have a bearing on the experiments to be described in this chapter.

When a (human or non-human) subject is trained to discriminate a stimulus A from a stimulus B the discrimination performance typically develops gradually over some time as is manifest in the various forms of functions usually referred to as "learning curves". A typical form is schematized in Fig. 13.1a which displays the temporal evolution of the hit rate and false alarm rate in a GO/(NO–GO) discrimination experiment. Other depictions of the discrimination learning behavior are possible depending on the kind of experiment performed (be these symmetric choice experiments, signal detection approaches, etc.) and the choice of behavioral observables (e.g. the various transformations of hit rate and false alarm rates which are suitable under given conditions). Generally, however, the changing conditional rate of occurrence of some behavior must be assessed at some point in the analysis. The typically asymptotic response rate depends on various parameters (vigilance, internal response biases, strength of the learned

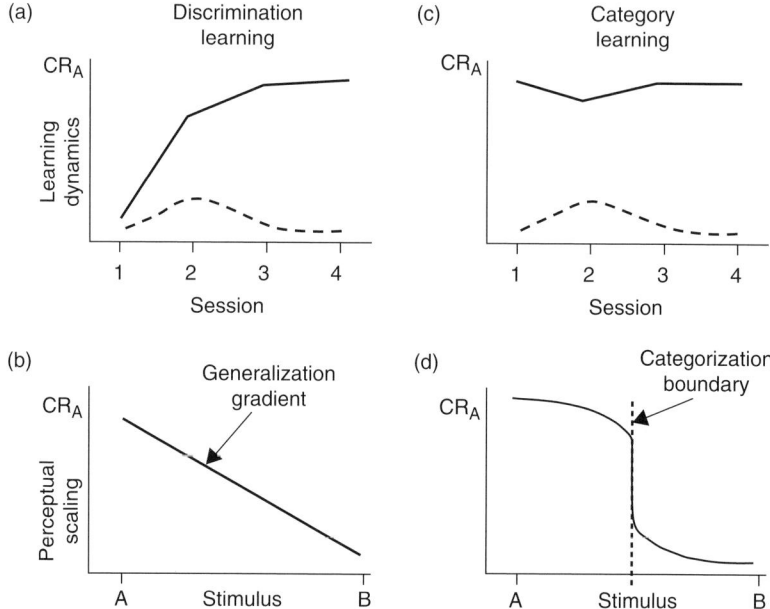

Figure 13.1. Suitable observables differentiating discrimination learning and category learning. Learning curves ((a), (c)) and psychometric functions ((b), (d)) after discrimination learning ((a), (b)) and after category learning ((c), (d)) are shown. The A-specific conditioned response rate CR_A is typically measured by the rate of occurrence or by the strength of the response that has been trained to be elicited by stimulus A. Solid and dashed curves in (a) and (c) depict the hit rate and false alarm rate, respectively. See text for details.

association, etc.) but also shows some degree of stimulus specificity. This can be demonstrated by generalization gradients: in a GO/(NO–GO) experiment in which a subject is trained to show the GO reaction in response to stimulus A and the NO–GO reaction in response to stimulus B, a generalization gradient can be assessed by measuring the conditioned response (quantified by its frequency of occurrence across trials, by the strength of its expression, or by some other suited observable) as a function of a physical distance between parameters characterizing stimuli A and B. The latter defines a path through the parameter space connecting stimuli A and B. When physical stimulus parameters are varied along this path a more or less gradual fall-off in the A-specific response amplitude is typically observed and referred to as a generalization gradient (Fig. 13.1b).

Conversely, when a subject has formed categories it can recognize even novel stimuli as representatives of the learned categories. A depiction of behavioral variables analogous to a learning curve would therefore indicate a high discrimination performance even in the first training session (Fig. 13.1c). The experimental demonstration of this behavior is critical for assessing category learning and is sometimes referred to as the criterion of the "transfer of learned behaviors to novel stimuli". Therefore, category learning is distinguished from simple or discriminative conditioning also by the psychometric functions it produces. Instead of gradual generalization gradients we find sigmoid psychometric functions with a more or less sharp boundary at some location in the stimulus parameter space, called the categorization boundary (Fig. 13.1d). Categories develop as cognitive constructs; they epitomize subjective "hypotheses" that are expressible as parcellations of the set of actually perceived or imagined stimuli, conditions or actions, into equivalence classes of meaning in particular contexts. Transfer of learned behaviors to novel stimuli therefore follows the subjective laws of this parcellation, rather than being guided by (physical) similarity relations between stimuli or stimulus features. These are important criteria for cognitive structures that have to be accounted for by physiologic models of learning.

In a nutshell, generalization is a general feature of learned stimulus-cued behaviors reflecting the converse of stimulus specificity, while categorization is a cognitive process based on the parcellation of the represented world into equivalence classes of meaning, valid for an individual in a particular context and in a particular time.

13.2 A new animal model of category learning

In this chapter, the Mongolian gerbil (*Meriones unguiculatus*) is presented as a new model of category learning (Wetzel *et al.*, 1998; Ohl *et al.*, 2001; Ohl *et al.*, 2003a, b; Ohl and Scheich, 2005). With respect to the background summarized in the first section of this chapter, the focus will be on the transfer of learned behaviors to novel stimuli. Category learning "goes beyond the information given" (Kommatsu, 1992); the experiment, then, has to be designed to shed light on the behavioral process and its physiologic basis which thus transcend the simpler forms of associative learning and generalization typically studied in behavioral neuroscience (for review see Miller *et al.*, 2003). The occurrence of the transfer of the learned discrimination behaviors to novel stimuli will be used as a marker event for the study of the physiologic correlates of category learning. The particular focus of the experiment described here is the investigation of the neuronal processes underlying the construction of that "which goes beyond the information given". It will be emphasized that beyond a mere extraction of statistical invariants (which could be done on the basis of classic feature detectors retaining the metric of evoked activity reflecting physical stimulus features), this construction involves recruitment of novel patterns of neuronal activity establishing a new metric of evoked activity which reflects the subjective perceptual scaling of stimuli instead.

With respect to the background discussed in the section on eco-ethologic aspects of categorization, a stimulus set was used that is demonstratively behaviorally neutral to the naïve gerbil. Moreover, stimuli were so selected that they did not differ in the ease with which they could be associated with the particular behaviors used in the training experiments. Care was taken to separate generalization behavior from categorization behavior.

13.2.1 Categorization of modulation direction in frequency-modulated tones

13.2.1.1 Stimuli

As stimuli, we used linearly frequency-modulated tones traversing a frequency range of 1 octave in 250 ms in rising or falling fashion, played at an intensity of 70 dB SPL as measured at a distance of 10 cm in front of the speaker. Due to reflections of the sound wave in the shuttle box and various possible head positions of the animal during the experiment, a considerable variance of effective stimulus intensity across trials is predictable. In frequency-modulated tones a number of parameters can be (co-)varied, like duration, intensity, frequency ranges covered, and modulation rate; that is, the rate of change of the tones' instantaneous frequency. Most of the stimuli (all, except those with zero-modulation rate, i.e. pure tones) can be categorized as either "rising" or "falling" frequency-modulated tones, depending on whether the instantaneous frequency changes from low to high or from high to low, respectively.

For the experiment, stimuli were so designed that they fell outside the generalization gradients established by training naïve animals to the two neighboring (in the parameter space) stimuli. This ensures that observed learning curves would resemble the schemes in Fig. 13.1a and c as indicating that the subject has or has not categorized the stimuli, respectively.

13.2.1.2 Apparatus and training paradigm

Training consisted of a GO/(NO–GO) avoidance paradigm carried out in a 2-compartment shuttle box, with the two compartments separated by a little hurdle, ensuring a low rate of spontaneous hurdle crossings. Animals were trained to cross the hurdle in response to a rising frequency-modulated tone and to stay in the current compartment in response to a falling frequency-modulated tone. Training was organized in so-called "training blocks" during which the discrimination of one particular rising frequency-modulated tone from a tone traversing the identical frequency range in falling direction was trained. A training block consisted of a number of training sessions, with one session per day, and was continued until no further changes in conditioned response rates were achieved in three consecutive sessions. Then another training block was initiated in which the discrimination of another pair of a rising and falling frequency-modulated tone was trained. A training session encompassed the randomized presentation of 30 rising and 30 falling frequency-modulated tones with the animals' false responses (misses and false alarms) being negatively reinforced by a mild electrodermal stimulation through a metal grid forming the cage floor. Control groups were run with the opposite contingencies to test for potential biases in this behavior which have been reported for the dog (McConnell, 1990). We showed that for the stimulus parameters tested no such biases existed for the gerbil.

13.2.1.3 Behavioral results

Animals trained on one or more training blocks never generalized to pure tones of any frequency (e.g. start or stop frequencies of the modulated tone, or frequencies traversed by the modulation or extrapolated from the modulation). This could be demonstrated by direct transfer experiments (Ohl *et al.*, 2001, supplementary material) or by measuring generalization gradients for modulation rate which never encompassed zero modulation rates (Ohl *et al.*, 2001).

Categorization was demonstrated by the transfer of the conditioned response behavior (changing compartment in response to tones of one category and remaining in the current compartment in response to tones of the other) to novel stimuli as measured by the response rates in the first session of a new training block. This sequential design allowed the experimenter to determine the moment in which an individual would change its response behavior from a "discrimination phase" (Fig. 13.1a and b) to a "categorization phase" (Fig. 13.1c and d). All animals tested were able to categorize novel frequency-modulated tones but, most notably, different individuals showed the transition from the discrimination phase to the categorization phase at different points in time in their training histories, although all had been trained with the same sequence of training blocks (Ohl *et al.*, 2001). Also, the transition from the discrimination phase to the categorization phase occurred abruptly rather than gradually; that is, in the first session of a training block either no discrimination performance (discrimination phase) or the full performance was observed (categorization phase). A third property of the transition was that after its occurrence the discrimination performance remained stable for the rest of the training blocks. These three properties of the transition, individuality of the time point of occurrence, abruptness of occurrence and behavioral stability after its occurrence, make it resemble a state transition in dynamic systems. We used this transition as a marker in the individual learning history of a subject to guide our search for physiologic correlates of this behavioral state transition.

13.2.2 Physiologic correlates of category learning

13.2.2.1 General aspects of the search for physiologic correlates of learning

In the search for physiologic correlates of category learning it has to be taken into account that the relationship between learning and its physiologic correlates, in general, is nontrivial, because the former is a psychologic/ethologic concept while the latter is a physiologic concept. Conceptual difficulties in interrelating these two domains are therefore predictably similar to other situations in science where conceptually different levels have to be linked, like in the case of relationship between Newtonian physics and thermodynamics, or in the case of the "mind-body problem". In the former case, a partial solution to the problem of linking the two levels has been achieved via the development of statistical physics, which worked by introducing scientific concepts that were not motivated by either Newtonian physics or thermodynamics alone. The solution must be considered only partial because it is sufficiently developed only for the equilibrium state of matter (cf. Haken, 1983). In this section it will be argued that traditional physiologic accounts for learning phenomena are insufficient to characterize the physiologic basis of category learning.

Traditional physiologic accounts for learning phenomena view the capacity for "re-routing" the flow of excitation through a neuronal system as the key element of a physiologic correlate of a learning process. This capacity is thought to underlie a stimulus processing which is altered by the learning experience. This concept is very much encouraged by the study of Pavlovian conditioning, probably the one learning phenomenon that has been most intensively studied by physiologists. Pavlovian conditioning can be described, and has in fact traditionally been defined, as a process by which an initially behaviorally neutral stimulus can elicit a particular behavior, after having been paired with a stimulus, the "unconditional stimulus" (US), which unconditionally triggers this behavior. Figure 13.2a shows a scheme of the flow of information in Pavlovian conditioning. For the case of Pavlovian conditioning it has proved successful to build models of the physiologic basis of learning by translating the flow of information within the conditioning scheme into a flow of neuronal excitation through a neuronal substrate (Fig. 13.2b). In the example of the conditioned gill withdrawal reflex in *Aplysia*, this concept is manifest in the feedforward convergence of (in the simplest case) two sensory neurons on an interneuron which projects onto a motor neuron. The straightforwardness of the translation of the

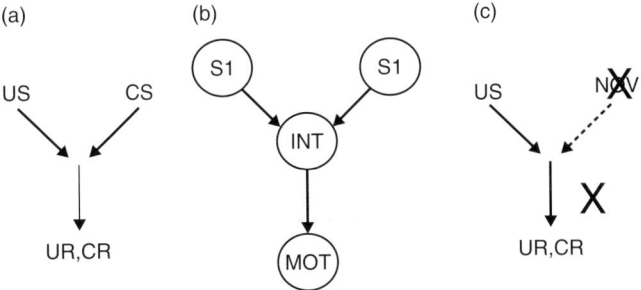

Figure 13.2. (a) General scheme of flow of information before and after Pavlovian conditioning. Before conditioning an unconditioned stimulus (US) will elicit a particular behavior, then referred to as unconditioned response (UR). After conditioning, a previously neutral stimulus can elicit this behavior as a conditioned response (CR) and is then referred to as the conditioned stimulus (CS). (b) A straightforward translation of the flow of information during Pavlovian conditioning into a flow of neuronal excitation within a neuronal substrate. (c) The architecture in (b) cannot explain responses to novel stimuli (NOV), because novel stimuli have not been associated with unconditional triggers and, consequently, cannot be conditioned. See text for details.

flow scheme of information into a flow scheme of neuronal excitation has led to consider the latter as elemental to physiologic accounts of learning (see however Schouten and De Long, 1999). This is expressed, for example, in the metaphors of the "cellular alphabet" (Hawkins and Kandel, 1984) or the "molecular alphabet" of learning.

In contrast to Pavlovian conditioning category learning encompasses aspects which go beyond mere stimulus–response associations. In particular, the transfer of learned responses to novel stimuli (Lea, 1984), outside of generalization gradients established by associative training, implies the insufficiency of a simple scheme as in Fig. 13.2b, because novel stimuli not encountered before cannot be associated with unconditional triggers (Fig. 13.2c).

13.2.2.2 Neuronal representation of utilized stimuli

A suitable level for studying electrophysiologic correlates of perceptual organization is the mesoscopic level of neurodynamics (Freeman, 2000), which defines the spatial scale of phenomena observed and provides focus on electrical phenomena emergent from the mass action of ensembles of some 10^4–10^5 neurons. Since in the gerbil the discrimination of the direction of frequency-modulated tones requires a functional auditory cortex (Ohl et al., 1999; Kraus et al., 2002) the training procedures were combined with the parallel measurement of the neurodynamics in auditory cortex. For cortical structures, this level of description is accessible by measurement of the electrocorticogram. We therefore combined the described category learning paradigm with the measurement of the electrocorticogram using arrays (3 × 6) of microelectrodes chronically implanted on the dura over the primary auditory cortex. The spatial configuration and interelectrode distance (600 μm) of the recording array were designed to cover the tonotopic representation of the frequency modulated stimuli used and avoid spatial aliasing of electrocorticogram activity (Ohl *et al.*, 2000a).

The spatial organization of the thalamic input into the auditory cortex can be studied by averaging electrocorticograms across multiple stimulus presentations, yielding the well-known auditory evoked potential (Barth and Di, 1990, 1991). Our studies of evoked potentials in primary auditory cortex, field AI, induced by pure tones (Ohl *et al.*, 2000a) or frequency-modulated tones (Ohl *et al.*, 2000b) revealed topographically organized early components (P1 and N1) located at positions within the tonotopic gradient of the field that corresponded to the frequency interval traversed by the frequency modulation, while their late components (P2 and N2) were not. On a finer spatial scale, the location of the early components of rising and falling frequency-modulated tones was found to be shifted towards tonotopic representations of the respective end frequencies of the modulations; that is, towards higher frequencies for rising modulations and towards lower frequencies for falling modulations. These "tonotopic shifts" (Ohl *et al.*, 2000b) could be explained by the finding that single neurons are usually activated more strongly when the frequency modulation is towards the neuron's best frequency than when it is away from it (Phillips *et al.*, 1985). In the former case, the activation of frequency channels in the neighborhood of the best frequency of a single neuron are recruited more synchronously than in the latter case, due to the increasing response latency with increasing spectral distance from the neuron's best frequency. If this asymmetry is transferred onto a tonotopically organized array of neurons the described tonotopic shift is the result. Comparable tonotopic shifts have previously been reported in the cortex analog of the chick (Heil *et al.*, 1992).

13.2.2.3 Single trial analysis of electrocorticograms

Since physiologic correlates of category learning could not be expected to occur time-locked to stimulus presentation we analyzed electrocorticograms recorded during the training with a single trial type of analysis. Instantaneous spatial patterns in the ongoing cortical activity were described by state vectors (Ohl *et al.*, 2001). State vectors were formed from estimates of signal power in 120 ms time windows obtained for each channel. A state vector moved through the state space along a trajectory according to the temporal evolution of the spatial pattern over a 6 s period starting 2 s before stimulus delivery. For each trial, the Euclidean distance (parameterized by time) to a reference trajectory was calculated and termed "dissimilarity function". In each case, the reference trajectory was calculated as the centroid over trajectories measured during presentation of stimuli belonging to the respective other category in the same training session. Thus, each trajectory associated with a rising frequency-modulated tone was compared to the centroid over all trajectories associated falling frequency-modulated tones in the same session, and vice versa. Comparison of single trajectories with centroids of trajectories, rather than other single trajectories, ensured that, on a statistical basis, transient increases in the pattern dissimilarity (peaks in the dissimilarity function) were due to pattern changes in the observed trajectory rather than in the centroid. In naïve animals, dissimilarity functions showed a "baseline behavior" with a sharp peak (2–7 standard deviations of baseline amplitude) after stimulus onset. This peak occurred predictably because of the topographically dissimilar patterns (tonotopic shifts) of early evoked responses that rising and falling frequency-modulated tones produce (Ohl et al., 2000b). With learning, additional peaks emerged from the ongoing activity. Those labeled spatial activity patterns in single trials with transiently increased dissimilarity to the reference trajectory indicating a potential relevance for representing category-specific information processing. These patterns were therefore termed "marked states".

To test whether marked states do in fact represent processing of category-specific information, we analyzed the similarity and dissimilarity relations among them in the entire course of the training. While animals were in their discrimination phases (prior to the formation of categories), we observed that dissimilarities between marked states within categories were of the same order of magnitude than between categories. After an individual animal had entered its categorization phase dissimilarities within a category were significantly smaller than between categories (Ohl et al., 2001). This indicated the existence of a metric which reflected the parcellation of stimuli into equivalence classes of meaning. This type of metric reflects subjective aspects of stimulus meaning, namely its belonging to categories formed by previous experience and, therefore, it is different from the known tonotopic one, which reflects similarity relations of physical stimulus parameters, namely spectral composition.

The spatial organization of the emerging marked states was analyzed in more detail and compared to that of the early evoked activity (also yielding peaks in the dissimilarity function) by a multivariate discriminant analysis, identifying the regions in the recording area which maximally contributed to the dissimilarity between the observed pattern and the reference pattern (Ohl et al., 2003b), or identifying the regions which contribute most information about the pattern (Ohl et al., 2003a).

13.3 Conclusions

It was possible to develop an animal model of auditory category learning which demonstrated the formation of categories as a process with three main behavioral characteristics: First, categorization developed abruptly rather than gradually. Second, it developed at a point in time that was specific for each individual subject in its learning history, and was typically different from subject to subject. Third, when categorization had occurred in a subject it remained stable for the rest of the subject's training experiences (unless a change in the reinforcement schedule forced a change in the meaning attributed to stimuli). A process which conforms to these characteristics is sometimes termed "Aha"-event to indicate a sudden change in the cognitive state of a subject.

The neurophysiologic analysis revealed that the process of associating meaning to acoustic stimuli as indicated by an increasing discrimination performance in the behavioral data was paralleled by the emergence of transient activity states in the ongoing cortical activity, that could be identified on the basis of their dissimilarity to patterns found in trials associated with stimuli not belonging to the category. These activity states represent a "constructive aspect" of neural activity during category learning which must clearly be distinguished from neural representations mapping stimulus features. It is noteworthy, that spatial patterns of electrocortical activity in single trials were already observed by Lilly in his toposcopic studies (Lilly and Cherry, 1954). To him it was already apparent that the long lasting dynamics was not just random "noise", but was better described by "figures" moving in time and space across the cortical surface. At that time the majority of research programs had already turned to the analysis of averaged data in which such spatiotemporal structures are no longer detectable. The few research programs pursuing analysis of activity patterns in single trials (e.g. DeMott, 1970; Livanov, 1977) had faced a major problem for the interpretation of such patterns: the lack of invariance with the applied stimuli. A large body of data accumulated over the last decades (summarized in Freeman, 2000) showed that such patterns might remain stable when repetitively evoked by sensory stimulation for a certain period in time, but might typically vary with behavioral context, particularly in learning situations when stimuli were associated with particular meanings. This lack of invariance with the mere physical parameters challenged their interpretation as "sensory representations". The observed metastability of the patterns was hypothesized to reflect context aspects of the stimulation as well as the perceptual history of the individual, and it was inferred that such patterns reflect subjectively relevant cognitive structures (for summary see Freeman, 2000, and references therein). The results described above critically confirm this interpretation: The category learning paradigm, first, allows determination of the point in time when a particular cognitive structure (the formation of the categories "rising frequency-modulated tone" and "falling frequency-modulated tone") emerges, and second, predicts that the main source of variance in the stimuli (the spectral interval traversed by the frequency modulation) is no longer a relevant feature after a subject's transition to categorization. Consequently, it was found that the dissimilarity between marked states associated with stimuli belonging to the same category was significantly reduced after the transition to categorization, although the physical dissimilarity of the corresponding stimuli was still high, as also reflected in the topographic organization of the stimulus-locked peaks in the dissimilarity function and the fact the dissimilarities remained high in individuals that had not yet formed categories.

In this sense, the utilized paradigm and analysis strategy provided an objective (for the experimenter) window of a subjective cognitive structure (that of the animal).

Acknowledgments

This work was supported by grants from the German Ministry of Education and Research, BMBF, by grants from the Land Sachsen-Anhalt, and the Bio-Future grant No. 0311891 of the BMBF to F.W.O. We thank Brian Burke and Daniela Labra Cardero (Berkeley), as well as Kathrin Ohl and Thomas Wagner (Magdeburg) for their skilled technical assistance.

REFERENCES

Ashby FG, Townsend JT (1986) Varieties of perceptual independence. *Psychol Rev* 93: 154–179.

Ashby FG, Perrin NA (1988) Toward a unified theory of similarity and recognition. *Psychol Rev* 95: 124–150.

Ashby FG, Maddox WT (1993) Relations between prototype, exemplar, and decision bound models of categorization. *J Math Psychol* 37: 372–400.

Barth DS, Di S (1990) Three-dimensional analysis of auditory-evoked potentials in rat neocortex. *J Neurophysiol* 64: 1527–1636.

Barth DS, Di S (1991) The functional anatomy of middle latency auditory evoked potentials. *Brain Res* 565: 109–115.

Brooks L (1978) Nonanalytic concept formation and memory for instances. In: Rosch E, Lloyd BB (eds), *Cognition and Categorization*, pp. 169–211. Erlbaum, Hillsdale, NJ.

Chater N, Heyes C (1994) Animal concepts: content and discontent. *Mind Lang* 9: 209–247.
DeMott DW (1970) *Toposcopic Studies of Learning.* Thomas Books, Springfield, IL.
Estes WK (1986) Array models for category learning. *Cogn Psychol* 18: 500–549.
Freeman WJ (2000) *Neurodynamics. An Exploration in Mesoscopic Brain Dynamics.* Springer-Verlag, London.
Haken H (1983) *Synergetics. An Introduction. Non-equilibrium phase transitions and self-organization in physics, chemistry and biology.* Springer-Verlag, Berlin.
Hawkins RD, Kandel ER (1984) Is there a cell-biological alphabet for simple forms of learning? *Psychol Rev* 91: 375–391.
Heil P, Langner G, Scheich H (1992) Processing of frequency-modulated stimuli in the chick auditory cortex analogue: evidence for topographic representations and possible mechanisms of rate and directional sensitivity. *J Comp Physiol A* 171: 583–600.
Hintzman DL (1986) "Schema abstraction" in a multiple trace memory model. *Psychol Rev* 93: 411–428.
Kéri S (2003) The cognitive neuroscience of category learning. *Brain Res Rev* 43: 85–109.
Kommatsu LK (1992) Recent views of conceptual structure. *Psychol Bull* 112: 500–526.
Kraus M, Schicknick H, Wetzel W, Ohl F, Staak S, Tischmeyer W (2002) Memory consolidation for the discrimination of frequency-modulated tones in Mongolian gerbils is sensitive to protein-synthesis inhibitors applied to auditory cortex. *Learn Mem* 9: 293–303.
Lea SEG (1984) In what sense do pigeons learn concepts. In: Terrace HS, Bever TG, Roitblat HL (eds), *Animal Cognition*, pp. 263–276. Erlbaum, Hillsdale, NJ.
Lea SEG, Ryan CME (1990) Unnatural concepts and the theory of concept discrimination in birds. In: Commons ML, Herrnstein RJ, Kosslyn SM, Mumford DB (eds), *Quantitative Analyses of Behaviour. Vol. VIII. Behavioral Approaches to Pattern Recognition and Concept Formation*, pp. 165–185. Erlbaum, Hillsdale, NJ.
Lilly JC, Cherry RB (1954) Surface movements of click responses from acoustic cerebral cortex of cat: leading and trailing edes of a response figure. *J Neurophysiol* 17: 531–537.
Livanov MN (1977) *Spatial Organization of Cerebral Processes.* Wiley, New York.
McConnell PB (1990) Acoustic structure and receiver response in domestic dogs. *Canis familiaris Anim Behav* 39: 897–904.
Medin DL, Schaffer MM (1978) Context theory of classification learning. *Psychol Rev* 85: 207–238.
Miller EK, Nieder A, Freedman DJ, Wallis JD (2003) Neural correlates of categories and concepts. *Curr Opin Neurobiol* 13: 198–203.
Nelson DA, Marler P (1990) The perception of birdsong and an ecological concept of signal space. In: Stebbins WC, Berkeley MA (eds), *Comparative Perception: Complex Signals*, Vol. 2, pp. 443–477. John Wiley & Sons, NY.
Nosofsky RM (1986) Attention and learning processes in the identification and categorization of integral stimuli. *J Exp Psychol: Learn Mem Cogn* 13: 87–108.
Ohl FW, Wetzel W, Wagner T, Rech A, Scheich H (1999) Bilateral ablation of auditory cortex in Mongolian gerbil affects discrimination of frequency modulated tones but not of pure tones. *Learn Mem* 6: 347–362.
Ohl FW, Scheich H, Freeman WJ (2000a) Topographic analysis of epidural pure-tone-evoked potentials in gerbil auditory cortex. *J Neurophysiol* 83: 3123–3132.

Ohl FW, Scheich H, Freeman, WJ (2000b) Spatial representation of frequency-modulated tones in gerbil auditory cortex revealed by epidural electrocorticography. *J Physiol (Paris)* 94: 549–554.

Ohl FW, Scheich H, Freeman WJ (2001) Change in pattern of ongoing cortical activity with auditory category learning. *Nature* 412: 733–736.

Ohl FW, Deliano M, Scheich H, Freeman WJ (2003a) Early and late patterns of stimulus-related activity in auditory cortex of trained animals. *Biol Cybern* 88: 374–379.

Ohl FW, Deliano M, Scheich H, Freeman WJ (2003b) Analysis of evoked and emergent patterns of stimulus-related auditory cortical activity. *Rev Neurosci* 14: 35–42.

Ohl FW, Scheich H (2005) Learning–induced Plasticity in animal and human auditory cortex. *Curr Opin Neurobiol*, in Press.

Pearce JM (1994) Discrimination and categorization. In: Mackintosh NJ (ed.), *Animal Learning and Cognition*, pp. 109–134. Academic Press, San Diego.

Phillips DP, Mendelson JR, Cynader MS, Douglas RM (1985) Response of single neurons in the cat auditory cortex to time-varying stimuli: frequency-modulated tones of narrow excursion. *Exp Brain Res* 58: 443–454.

Posner MI, Keele SW (1968) On the genesis of abstract ideas. *J Exp Psychol* 77: 353–363.

Rosch E (1973) Natural categories. *Cogn Psychol* 4: 328–350.

Rosch E (1975) Cognitive reference points. *Cogn Psychol* 7: 192–238.

Ryan CME (1982a) Concept formation and individual recognition in the domestic chicken (*Gallus gallus*). *Behav Anal Lett* 2: 213–220.

Ryan CME (1982b) *Mechanisms of individual recognition in birds*. Master Thesis, University of Exeter, UK.

Schouten MKD, De Long L (1999) Reduction, elimination, and levels: the case of the LTP-learning link. *Phil Psychol* 12: 237–262.

Seyfarth RM, Cheney DL, Marler P (1980a) Monkey responses to three different alarm calls: evidence of predator classification and semantic communication. *Science* 210: 801–803.

Seyfarth RM, Cheney DL, Marler P (1980b) Vervet monkey alarm calls: semantic communication in a free-ranging primate. *Anim Behav* 28: 1070–1094.

Wasserman EA, Astley SL (1994) A behavioural analysis of concepts: its application to pigeons and children. *Psychol Learn Motivation* 31: 73–132.

Wetzel W, Wagner T, Ohl FW, Scheich H (1998) Categorical discrimination of direction in frequency-modulated tones by Mongolian gerbils. *Behav Brain Res* 91: 29–39.

Neural Adaptations and Plasticity: Summary and Discussion

Jagmeet S. Kanwal and Günter Ehret

Neurophysiologic studies on auditory communication are inherently complex and require behavioral and acoustic analyses in addition to the usual neurophysiologic questions, skills and techniques. After an initial move in the mid-1970s towards understanding neuronal coding and representation of complex acoustic stimuli, including communication sounds, in the auditory system (summary in Bullock, 1977), the field was at a standstill for nearly two decades. The chapters in this section of the book prove that the new technologies, improved experimental approaches and behavioral and acoustic analyses of communication sounds are once again moving this field ahead. With the new and exciting ideas that have been put forth in recent years, there is promise of enormous progress, especially in relation to neural coding, mechanisms of adaptation and plasticity of the brain. Thus, our earlier perception of the auditory and vocal systems as static systems that do not and cannot change much after they are fully developed is no longer valid.

Auditory communication and hormones

It is becoming increasingly clear that hormones play a major role in the seasonal abilities of fish, frogs and birds who use sounds for audiovocal communication (e.g. Brantley *et al.*, 1993; Doupe, 1994; Panna et al., 1997; Goodson and Bass, 2000; Remage-Healey and Bass, 2004). The significance of the role of hormones in audiovocal communication in mammals and particularly humans who use sounds in a season-independent manner, however, is still unclear.

Direct evidence for a role of hormones on audition and on the vocal apparatus come from studies on fishes and amphibians. In the midshipman fish, *Porichthys notatus*, a steroid-dependent auditory plasticity leads to adaptive coupling of the sender and receiver of communication sounds. In the absence of enough estrogen-like hormone in their systems, females turn a deaf ear to loud vocalizations of the males (Sisneros et al., 2004). Increasing steroid hormone levels in non-reproductive females of this species, temporarily alter their inner-ear auditory mechanism, but do not change the reproductive status of the females. Rather, the hormone estradiol acts as a trigger for generally activating the auditory system and, thus, increasing the auditory sensitivity of the females.

Bass (Chapter 5, this volume) describes recent work on the midshipman fish that use loud humming sounds for mating. Males loaded with the hormone testosterone produce the most energetic humming call. In the toadfish, *Opsanus beta*, playback of boatwhistles evokes rapid increases in calling rate and duration of advertisement calls that are accompanied by increases in plasma levels of 11-ketotestosterone (Remage-Healey and Bass, 2005). Earlier, Kelley (1980) showed that in the male African clawed frog "*Xenopus laevis*" steroids regulate the final common pathways, including the vocal motor nucleus, for vocal behavior. A recent study on the laryngeal neuromuscular synapses, which are the final neural elements that control sexually differentiated call production, has shown that males have stronger laryngeal synapses than females and the synapse strength is estrogen dependent (Wu et al., 2003). Estrogen-induced increases in synaptic strength require at least 3 weeks of exposure, suggesting that the hormone acts via a classical genomic mechanism involving estrogen receptors.

Hormonal changes are also intimately related to the production of song in birds via replaceable neurons in the telencephalon (Nottebohm, 2004). In particular, testosterone results in a dramatic (90%) increase in size of the HVC (formerly called higher vocal center or hyperstriatum ventrale pars caudale), and a 53% increase in the nucleus robustus arcopallialis (RA) that control the production of song (Nottebohm, 1980). Besides decreasing the fundamental frequency of the voice of males compared to females during puberty (e.g. Kent, 1976), the role of sex hormones on audiovocal behavior in mammals has not been thoroughly investigated yet. Hormones, however, strongly influence brain and behavior (e.g. Becker et al., 2002) and can be expected to significantly influence vocal production and sound perception. For example, in mice, ultrasound recognition and the process of parental pup-retrieving is enhanced by the presence of estradiol (Koch and Ehret, 1989). Further, estrogen receptors have been demonstrated in the human inner ear implying a role of hormones on peripheral hearing mechanisms (Stenberg et al., 1999, 2001). Women with Turner's syndrome, a genetic aberration that results in loss of estrogen production in the ovary, show an early onset in progressive high-frequency hearing loss. Other experiments have suggested that steroid hormones may play a role in causing some of the reported changes in hearing sensitivity of human females at differing stages of their menstrual cycle (Elkind-Hirsch *et al.*, 1992). Hormones may also facilitate the onset of tinnitus, a mentally debilitating disorder where tones and more complex sounds are heard because of a disturbance in the activity of inner hair cells and of sound-independent activity of cortical neurons (e.g. 335Jastreboff, 1990). Recent research on oxytocin and vasopressin in mustached bats shows that oxytocin-containing cell bodies and fiber terminals are abundant in the cochlear nucleus and are present throughout the auditory system up to the level of the frontal cortex (Kanwal and Rao, 2002; Rao and Kanwal, 2004). These hormones may function as neuromodulators and, thus, influence audiovocal communication in the central nervous system. It is evident that the main progress on the role of hormones in shaping neural activities and coding of species-specific calls for audiovocal communication is still ahead of us.

Auditory representations

How does the brain accomplish a robust and, at the same time, plastic representation of complex sounds? Research on this question requires analyses of the spatiotemporal distribution of neural activity in the brain on the level of single neurons, groups of neurons and mass actions, such as oscillations within large populations of neurons. Furthermore, a principle of the central auditory system, the processing of information in parallel pathways (e.g. Ehret and Romand, 1997), has to be extended to levels beyond the auditory cortex in order to interface auditory perception with the generation of complex adaptive behavior such as the production of certain vocalizations. Clearly, common properties of neurons in higher auditory centers that represent aspects of communication sounds, for example, combination-sensitivity, have to be separated from species-specific adaptations. With the advent of speech in humans, we have to test whether processing and perception mechanisms for speech are special or can be derived from a general understanding of the neural machinery underlying audiovocal communication in vertebrates, especially mammals. Numerous studies have addressed these points (e.g. Narins and Capranica, 1980; Mudry and Capranica, 1987; Schwartz and Simmons, 1990; Rauschecker *et al.*, 1995; Ohlemiller *et al.*, 1996; Esser *et al.*, 1997; Kanwal, 1999; Kanwal *et al.*, 1999; Tian *et al.*, 2001; Ehret and Riecke, 2002; Geissler and Ehret, 2002, 2004; Suga *et al.*, 2002; Wysocki and Ladich, 2003).

Three chapters in this book examine call processing at different levels of the brain. Klug *et al.* (Chapter 6, this volume) report recent findings about neural representations of species-specific vocalizations in the auditory brain stem and midbrain of Mexican free tailed bats, *Tadarida brasilensis*. They discuss how complex sounds are initially coded and represented by the auditory periphery and lower brainstem and how this representation may change in the central nucleus of the inferior colliculus of the auditory midbrain. Neurons just below the level of the inferior colliculus, that is, in the dorsal nucleus of the lateral lemniscus, respond to complex signals in a rather simplistic and stereotypic way, whereas responses in the inferior colliculus are much more selective, with a much larger degree of variation and a lower level of predictability of responses between neurons. The significant increase in the diversity of neuronal coding, especially in response to complex communication sounds, in the auditory midbrain is most probably due to two distinct properties of the inferior colliculus. First, more than sixteen ascending input pathways that converge within the inferior colliculus may create local clusters of neurons of different combination-sensitivity and, second, the inhibitory inputs to and intrinsic inhibition within the inferior colliculus may play an important role in shaping neuronal responses (e.g. Ehret, 1997b; Egorova *et al.*, 2001; Hage and Ehret, 2003). The topic of how communication sounds are encoded in the auditory midbrain has not received much attention so far (Scheich *et al.*, 1977; Ehret and Moffat, 1985; Muller-Preuss,1986; Portfors and Wenstrup, 2002; Šuta *et al.*, 2003; Portfors, 2004); it is, however, important to see where and how a rather uniform peripheral code for complex sounds is transformed into a code related to combinations of specific sound properties for perception and recognition.

The next two chapters examine call processing in the cortex using single unit electrophysiologic (Kanwal, Chapter 7, this volume) and more global imaging techniques (Horikawa *et al.*, Chapter 8, this volume). The analysis of call responses of neurons in the auditory cortex of mustached bats suggests that several cortical minicolumns across different cortical areas constitute the functional unit for mapping a call within the cortex. As proposed in Kanwal (Chapter 7, this volume), the representational scheme for calls may be best described as a patchy, distributed representation of key acoustic features in different cortical areas containing functionally specialized neurons. Simultaneously triggered "hot spots" of activity in these areas may be temporally continuous and functionally united within the call percept they signify (see Horikawa *et al.*, Chapter 8, this volume). The data presented in this chapter support the proposed scheme (Ehret, 1997a) for neural coding of complex sounds. Whether the spatiotemporal spread of cortical activity in response to a sound stimulus represents an essential component of neural coding or is an epiphenomenon resulting from the complex excitatory and inhibitory interactions remains controversial. Important findings about encoding "phonetic-like" syntax of composite calls and of trains of simple syllables in the mustached bat's cortex (Ohlemiller *et al.*, 1996; Esser et al., 1997;

Kanwal, Chapter 7, this volume), nevertheless indicate that a detailed knowledge of species-specific communication sounds for use in neural stimulation tests (e.g. Ohlemiller *et al.*, 1994) is necessary to ask the auditory cortex the "right" questions. This is a decisive step in our approach of using animal models to explain the representation of speech in the human auditory cortex.

Horikawa *et al.* (Chapter 8, this volume) demonstrate how optic recording methods detect complex waves of excitation and suppression in the auditory cortex even to simple stimuli such as pure tones. The spatiotemporal activity generated by presenting complex sounds is even more complex. Further, they show that auditory information spreads from core fields to the belt fields via three distinct (dorsocaudal, ventrorostral and caudal) pathways. The response properties in belt fields of the guinea pig (a larger frequency overlap, longer latency and longer duration of their activation) are similar to the data reported in primates and exhibit different temporal response characteristics from those present in the core fields. In principal, the different temporal response characteristics and overall complexity of responses to complex sounds matches the electrophysiologic data on call responses in the mustached bat's cortex, although the fields themselves are not necessarily homologous. At least, it is clear that responses of auditory cortical fields develop in a highly dynamic way over time and cortical space. This knowledge is important for practical and theoretical approaches to solve the "binding problem" of the brain (e.g. Singer, 1995), that is, to resolve the issue of how the brain binds together temporally coherent neural activations from several places so that, finally, an auditory percept can emerge.

Finally, Rauschecker and Tian (Chapter 9, this volume) go beyond neural coding in the auditory cortex to report on the "what" and "where" pathways in the brain of primates. They present evidence suggesting that communication sounds are processed along an anteroventral axis in the superior temporal gyrus of nonhuman and possibly human primates.

Neurophysiologic single-unit studies show that the response selectivity for species-specific vocalizations is relatively high in the anterolateral (AL) area of the auditory belt cortex in the rhesus monkey. Anatomical pathways from area AL project to the orbitofrontal cortex, which has previously been implicated in visual object working memory. Finally, functional magnetic resonance imaging (fMRI) and positron emission tomography (PET) studies in humans also show a similar pattern of activation in that areas in the anterior superior temporal gyrus are activated by species-specific sounds, that is, human voices, speech and speech-related sounds.

These and other studies (e.g. Kaas and Hacket, 2000; Poremba *et al.*, 2003) are providing a growing body of evidence for an anteriorly directed processing stream dedicated to the identification and recognition of behaviorally relevant auditory patterns. Activity of neurons in this pathway signals "what" a complex sound represents. Areas in the lateral belt, in particular area AL, constitute a relatively early station in this process, and may participate in the extraction and integration of auditory features relevant for communication. Thus, studies in several different species (bats, mice, guinea pigs, monkeys and humans) suggest the presence of a roughly common scheme for the flow of auditory information within and beyond the auditory cortex. The functional specificity and across-species similarity of auditory information present withing single neurons in any one stream, however, is difficult to quantify at present. It is also unclear how many levels of processing and/or cortical loops need to be simultaneously activated to generate a meaningful and robust percept of communication sound.

Learning, memory and plasticity

The unique human ability to speak is the basis for learning a language during the development of each individual. How is language learning accomplished? Which parts of the audiovocal system are involved? Which neural mechanisms are important? Ever since the discovery that many birds learn their songs and that seasonal developments in the bird HVC are associated with changes in the song, songbirds have been used extensively as model organisms for experiments on auditory learning and memory, especially on sensitive periods of learning and on comparative aspects between birds and humans (e.g. Nottebohm, 1980; Marler and Sherman, 1983; Kuhl, 1994, 2003; Kuhl and Meltzoff, 1996; Doupe and Kuhl, 1999).

Kittelberger and Mooney (Chapter 10, this volume) report that synaptic connections in RA reach a peak density at a developmental time corresponding with the height of song plasticity. Initially, HVC–RA synaptic connectivity is numerically exuberant, but individual synapses are functionally weak. Over the course of song learning, certain synapses active in neural circuits of song learning and reproduction undergo consolidation, that is, their strength increases as the total number of synapses declines. The authors show that the maintenance of high numbers of relatively weak synapses in the juvenile RA depends on the presence of inputs from the lateral magnocellular nucleus of the anterior nidopallium (LMAN). Furthermore, injections of the brain-derived neurotrophic factor (BDNF) into RA destabilize the song phenotype, transiently increasing the variability of syllable sequencing and causing permanent changes in song structure. Kittelberger and Mooney postulate that BDNF plays a role in enhancing song plasticity and variability by increasing the number of synaptic boutons and connections between HVC and RA. They suggest that endogenous, LMAN-dependent signals enable vocal plasticity during the sensitive period for sensorimotor learning by maintaining high numbers of weak excitatory synaptic connections within RA and between HVC and RA neurons. Developmental changes in LMAN-dependent signals trigger synaptic consolidation in RA, resulting in diminished vocal plasticity and song crystallization. Neurotrophin-dependent synaptic overproduction seems to be an important regulator of sensitive periods for vocal plasticity in songbirds.

The perceptual processing of song and evaluation of auditory feedback plays a key role in the formation, storage and recall of auditory memories and vocal learning. Auditory memories are easily measured and quantified by spectrographic analyses of the sounds produced and therefore are ideal for examining the neural mechanisms involved in learning and memory in general. Even in humans, sound in the form of music is frequently used to facilitate recall in patients with memory loss due to trauma or neurodegenerative disorders (Schulkind et al., 1999; Foster and Valentine, 2001; Cuddy and Duffin, 2005). Therefore, an understanding of auditory processing in the context of audiovocal communication provides a unique opportunity of solving the problem of how memories are processed and stored for efficient recall.

Mello and Roberts (Chapter 11, this volume) propose neural mechanisms of audiovocal learning and auditory memory in songbirds based on what is known about the processing and perception of the electric organ discharge in weakly electric fish (mormyrids). The key idea is to apply the reafference principle (v. Holst and Mittelstaedt, 1950; Gallistel, 1980) to the songbird's brain according to the finding in electric fish that the electric organ corollary discharge, which provides a copy of the efferent outgoing signal (efference copy), is compared with the afferent incoming signal. During song learning in birds, the bird's own neural motor patterns for song production may function as an efference copy or neural template or perceptual "expectation" to be matched with the afferent sensory patterns generated by a song from a tutor individual. As long as the afferent patterns deviate from the efference copy, the learning bird has to change its own song in order to sufficiently minimize the differences. Further experiments have to show whether the reafference principle as a general mechanism of motor-sensory adjustment can explain all forms of audiovocal learning in birds.

The final two chapters focus on the role of corticofugal (efferent) influences and on neurodynamics that contribute to shaping neural plasticity in the neocortex of mammals for the optimal perception of behaviorally relevant sounds. Yan and Eggermont (Chapter 12, this volume) discuss how learning the significance of certain sounds can constantly alter the functional organization of the auditory cortex with regard to the representation of the information-bearing elements such as frequency components and information-bearing parameters such as inter-syllable intervals in these sounds. Further, changes of neuronal response properties in the auditory cortex due to learning can feed back to lower auditory centers and even outside the auditory system via the multiple efferent (corticofugal) pathways from the auditory cortex (e.g. Spangler and Warr, 1991; Winer, 1992). Thus, an effective communication loop through afferent and efferent auditory systems of the brain can help to adjust auditory processing, perceptions and vocalizations according to what the higher auditory centers of the neocortex have learned to be important. Learning to hear what is important, especially in auditory communication, memorizing it, and using the memory again in an adaptive way for future communication in different situations is the main profit birds and mammals can draw from their highly plastic audiovocal systems. This plasticity culminates in the speech system of humans. It goes far beyond the plasticity in audiovocal systems of fish and frogs, which seems to be controlled and driven mainly by hormonal changes.

Calls in animals and speech sounds in humans may be learned to belong to different categories to which different meanings and behaviors are associated. Once this categorization has developed, novel sounds can then immediately be classified in one of the categories. Unlike generalization, categorization allows a subject to recognize novel stimuli as representatives of learned categories. Whether based on innate or learned mechanisms, auditory categories represent perceptual concepts which help to structure the acoustic world and provide a basis for a fast and secure audiovocal or other response behavior to sounds and, thus, is a helpful adaptation for the survival of the individual. Hence, the understanding of category formation in auditory perception is central to our understanding of audiovocal communication.

Ohl *et al.* (Chapter 13, this volume) demonstrate auditory category learning in the Mongolian gerbil. They show that the formation of categories appears to be similar to an "Aha" event indicating a sudden change in the cognitive state of a subject, which is paralleled by the emergence of transient spatial patterning of evoked and ongoing cortical activity. Previous research has established that spatial activity patterns are not invariant with fixed physical stimulus parameters, but are affected by the particular context of stimulus presentation and the perceptual history of the animal (Freeman, 2000). Categorization paradigms can be used to positively demonstrate that this lack of invariance does indeed reflect subjective attributes of perception rather than uncontrolled variable context effects. This is because during category formation the perceived equivalence of category members arises as only a subjectively valid invariance. It was demonstrated earlier that spatial activity patterns reflected the sorting of a stimulus into a category rather than its raw physical attributes (Ohl *et al.*, 2001). This organizational principle of brain dynamics for category formation may represent a first building block for constructing a neural substrate underlying cognition.

REFERENCES

Becker JB, Breedlove SM, Crews D, McCarthy MM (eds) (2002) *Behavioral Endocrinology.* MIT Press, Cambridge Mass.

Brantley RK, Marchaterre MA, Bass AH (1993) Androgen effects on vocal muscle structure in a teleost fish with inter- and intra-sexual dimorphism. *J Morphol* 216: 305–318.

Bullock TH (ed.) (1977), *Recognition of Complex Acoustic Signals.* Abacon Verlagsgesellschaft, Berlin.

Cuddy LL, Duffin J (2005) Music, memory, and Alzheimer's disease: is music recognition spared in dementia, and how can it be assessed? *Med Hypotheses* 64: 229–235.

Doupe AJ (1994) Songbirds and adult neurogenesis: a new role for hormones. *Proc Natl Acad Sci USA* 91: 7836–7838.

Doupe AJ, Kuhl PK (1999) Birdsong and human speech: common themes and mechanisms. *Annu Rev Neurosci* 22: 567–631.

Egorova M, Ehret G, Vartanian I, Esser KH (2001) Frequency response areas of neurons in the mouse inferior colliculus. I. Threshold and tuning characteristics. *Exp Brain Res* 140: 45–161.

Ehret G (1997a) The auditory cortex. *J Comp Physiol A* 181: 547–557.

Ehret G (1997b) The auditory midbrain, a "shunting yard" of acoustical information processing. In: Ehret G, Romand R (eds), *The Central Auditory System*, pp. 259–316. Oxford University Press, New York.

Ehret G, Moffat AJM (1985) Inferior colliculus of the house mouse. III. Response probabilities and thresholds of single units to synthesized mouse calls compared to tone and noise bursts. *J Comp Physiol A* 156: 637–644.

Ehret G, Riecke S (2002) Mice and humans perceive multiharmonic communication sounds in the same way. *Proc Natl Acad Sci USA* 99: 475–482.

Ehret G, Romand R (eds) (1997) *The Central Auditory System.* Oxford University Press, New York.

Elkind-Hirsch KE, Stoner WR, Stach BA, Jerger JF (1992) Estrogen influences auditory brainstem responses during the normal menstrual cycle. *Hearing Res* 60: 143–148.

Esser KH, Condon CJ, Suga N, Kanwal JS (1997) Syntax processing by auditory cortical neurons in the FM–FM area of the mustached bat *Pteronotus parnellii. Proc Natl Acad Sci USA* 94:14019–14024.

Foster NA, Valentine ER (2001) The effect of auditory stimulation on autobiographical recall in dementia. *Exp Aging Res* 27: 215–228.

Freeman WJ (2000) *Neurodynamics. An exploration in Mesoscopic Brain Dynamics.* Springer-Verlag, London.

Gallistel CR (1980) *The Organization of Action: A New Synthesis.* Lawrence Erlbaum, Hillsdale, NJ.

Geissler DB, Ehret G (2002) Time-critical integration of formants for perception of communication calls in mice. *Proc Natl Acad Sci USA* 99: 9021–9025.

Geissler DB, Ehret G (2004) Auditory perception vs. recognition: representation of complex communication sounds in the mouse auditory cortical fields. *Eur J Neurosci* 19: 1027–1040.

Goodson JL, Bass AH (2000) Forebrain peptides modulate sexually polymorphic vocal circuitry. *Nature* 403: 769–772.

Hage SR, Ehret G (2003) Mapping responses to frequency sweeps and tones in the inferior colliculus of house mice. *Eur J Neurosci* 18: 2301–2312.

v Holst E, Mittelstaedt H (1950) Das Reafferenzprinzip. Wechselwirkung zwischen Zentralnervensystem und Peripherie. *Naturwissenschaften* 37: 464–476.

Jastreboff PJ (1990) Phantom auditory perception (tinnitus): mechanisms of generation and perception. *Neurosci Res* 8: 221–254.

Kaas JH, Hacket TA (2000) Subdivisions of auditory cortex and processing streams in primates. *Proc Natl Acad Sci USA* 97: 11793–11799.

Kanwal JS (1999) Processing species-specific calls by combination-sensitive neurons in an echolocating bat. In: Hauser MD, Konishi M (eds), *The Design of Animal Communication*, pp. 135–157. MIT Press, Cambridge, Mass.

Kanwal JS, Rao PD (2002) Oxytocin within auditory nuclei: a neuromodulatory function in sensory processing? *NeuroReport* 13: 2193–2197.

Kanwal JS, Fitzpatrick DC, Suga N (1999) Facilitatory and inhibitory frequency tuning of combination-sensitive neurons in the primary auditory cortex of mustached bats. *J Neurophysiol* 82: 2327–2345.

Kelley DB (1980) Auditory and vocal nuclei in the frog brain concentrate sex hormones. *Science* 207: 553–555.

Kent RD (1976) Anatomical and neuromuscular maturation of the speech mechanism: evidence from acoustic studies. *J Speech Hearing Res* 19: 421–447.

Koch M, Ehret G (1989) Estradiol and parental experience, but not prolactin are necessary for ultrasound recognition and pup-retrieving in the mouse. *Physiol Behav* 45: 771–776.

Kuhl PK (1994) Learning and representation in speech and language. *Curr Opin Neurobiol* 4: 812–822.

Kuhl PK (2003) Human speech and birdsong: communication and the social brain. *Proc Natl Acad Sci USA* 100: 9645–9646.

Kuhl PK, Meltzoff AN (1996) Infant vocalizations in response to speech: vocal imitation and developmental change. *J Acoust Soc Am* 100: 2425–2438.

Marler P, Sherman V (1983) Song structure without auditory feedback: emendations of the auditory template hypothesis. *J Neurosci* 3: 517–531.

Mudry KM, Capranica RR (1987) Correlation between auditory evoked responses in the thalamus and species-specific call characteristics. I. Rana catesbeiana (Anura: Ranidae). *J Comp Physiol A* 160: 477–489.

Muller-Preuss P (1986) On the mechanisms of call coding through auditory neurons in the squirrel monkey. *Eur Arch Psychiat Neurol Sci* 236: 50–55.

Narins PM, Capranica RR (1980) Neural adaptations for processing the two-note call of the Puerto Rican treefrog, *Eleutherodactylus coqui*. *Brain Behav Evol* 17: 48–66.

Nottebohm F (1980) Testosterone triggers growth of brain vocal control nuclei in adult female canaries. *Brain Res* 189: 429–436.

Nottebohm F (2004) The road we travelled: discovery, choreography, and significance of brain replaceable neurons. *Ann NY Acad Sci* 1016: 628–658.

Ohl FW, Scheich H, Freeman WJ (2001) Change in pattern of ongoing cortical activity with auditory category learning. *Nature* 412: 733–736.

Ohlemiller KK, Kanwal JS, Butman JA, Suga N (1994) Stimulus design for auditory neuroethology: synthesis and manipulation of complex communication sounds. *Audit Neurosci* 1: 19–37.

Ohlemiller KK, Kanwal JS, Suga N (1996) Facilitative responses to species-specific calls in cortical FM–FM neurons of the mustached bat. *NeuroReport* 7: 1749–1755.

Portfors CV, Wenstrup JJ (2002) Excitatory and facilitatory frequency response areas in the inferior colliculus of the mustached bat. *Hear Res.* 168: 131–138.

Portfors CV (2004) Combination sensitivity and processing of communication calls in the inferior colliculus of the moustahced bat, *Pteronotus parnellii*. *Ann Brazil Acad Sci* 76: 253–257.

Penna M, Feng AS, Narins PM (1997) Temporal selectivity of evoked vocal responses of *Batrachyla antartandica* (Amphibia: Leptodactylidae). *Anim Behav* 54: 833–848

Poremba A, Saunders RC, Crane AM, Cook M, Sokoloff L, Mishkin M (2003) Functional mapping of primate auditory system. *Science* 299: 568–572.

Rao PD, Kanwal JS (2004) Oxytocin and vasopressin immunoreactivity within the forebrain and limbic-related areas in the mustached bat, *Pteronotus parnellii*. *Brain Behav Evol* 63: 151–158.

Rauschecker JP, Tian B, Hauser M (1995) Processing of complex sounds in the macaque nonprimary auditory cortex. *Science* 268: 111–114.

Remage-Healey L, Bass AH (2004) Rapid, hierarchical modulation of vocal patterning by steroid hormones. *J Neurosci* 24: 5892–5900.

Remage-Healey L, Bass AH (2005) Rapid elevations in both steroid hormones and vocal signaling during playback challenge: a field experiment in Gulf toadfish. *Horm Behav* 47: 297–305.

Scheich H, Langner G, Koch R (1977) Coding of narrow-band and wide-band vocalizations in the auditory midbrain nucleus (MLD) of the Guinea fowl (*Numida meleagris*). *J Comp Physiol A* 117: 245–265.

Schulkind MD, Hennis LK, Rubin DC (1999) Music, emotion, and autobiographical memory: they're playing your song. *Mem Cognit* 27: 948–955.

Schwartz JJ, Simmons AM (1990) Encoding of a spectrally-complex communication sound in the bullfrog's auditory nerve. *J Comp Physiol A* 166: 489–499.

Singer W (1995) Time as coding space in neocortical processing: a hypothesis. In: Gazzaniga MS (ed.) *The Cognitive Neurosciences*, pp. 91–104. MIT Press, Cambridge, Mass.

Sisneros JA, Forlano PM, Deitcher DL, Bass AH (2004) Steroid-dependent auditory plasticity leads to adaptive coupling of sender and receiver. *Science* 305: 404–407.

Spangler KM, Warr WB (1991) The descending auditory system. In: Altschuler RA, Bobbin RP, Clopton BM, Hoffman DW (eds), *Neurobiology of Hearing. The Central Auditory System*, pp. 27–45. Raven Press, New York.

Stenberg AE, Wang H, Sahlin L, Hultcrantz M (1999) Mapping of estrogen receptors alpha and beta in the inner ear of mouse and rat. *Hearing Res* 136: 29–34.

Stenberg AE, Wang H, Fish J, Schrott-Fischer A, Sahlin L, Hultcrantz M (2001) Estrogen receptors in the normal adult and developing human inner ear and in Turner's syndrome. *Hearing Res* 157: 87–92.

Suga N, Xiao Z, Ma X, Ji W (2002) Plasticity and corticofugal modulation for hearing in adult animals. *Neuron* 36: 9–18.

Šuta D, Kvašňák E, Popelář J, Syka J (2003) Representation of species-specific vocalizations in the inferior colliculus of the guinea pig. *J Neurophysiol* 90: 3794–3808.

Tian B, Reser D, Durham A, Kustov A, Rauschecker JP (2001) Functional specialization in rhesus monkey auditory cortex. *Science* 292: 290–293.

Winer JA (1992) The functional architecture of the medial geniculate body and the primary auditory cortex. In: Webster DB, Popper AN, Fay RR (eds), *The Mammalian Auditory Pathway: Neuroanatomy*, pp. 222–409. Springer-Verlag, New York.

Wu KH, Tobias ML, Kelley DB (2003) Estrogen receptor expression in laryngeal muscle in relation to estrogen-dependent increases in synapse strength. *Neuroendocrinology* 78: 72–80.

Wysocki LE, Ladich F (2003) The representation of conspecific sounds in the auditory brainstem of teleost fishes. *J Exp Biol* 206: 2229–2240.

Appendix

Basics of acoustic signal processing

The study of sound communication in animals requires a knowledge of acoustics in addition to animal biology and physiology. This knowledge concerns the use of proper methods for the recording, digitization, analysis, synthesis and playback of sounds. Acoustic communication signals in vertebrates are typically complex, both in the spectral and temporal domain. They frequently consist of combinations of two types of sound elements in the spectral domain, namely tones and noise. The fundamental frequency or the first harmonic often is the predominant frequency, that is, has the most energy, and in the great majority of communication sounds, multiple higher harmonics are also present. Adding time as a variable (spectrotemporal domain), frequencies and noise can be constant over time or modulated in frequency and/or intensity with the result of a time-dependent frequency spectrum and a time-varying intensity distribution across the frequency components. Amplitude modulations with a low repetition rate lead to rhythmic sounds. At high repetition rates of amplitude modulations, side bands to frequency components are created, which lead to the perception of a so-called periodicity pitch (approximately the pitch of the modulation frequency). Finally, single sounds, separated by temporal gaps or intervals of certain durations, may be put together as a series of sounds that are perceived as an acoustic gestalt. This is sometimes referred to as auditory grouping. When sound series of two alternating frequencies are presented together it can lead to streaming or auditory stream segregation. Here, depending upon the time interval and the difference in the frequencies the acoustic gestalt may be heard as a galloping rhythm or as two separate streams of sounds.

Glossary of terms   Below is a glossary of terms occurring in studies of sound communication at the behavioral and neurophysiologic levels.

Aliasing   Undersampling of a frequency band while digitizing a sound. This leads to the generation of low-frequency peaks of energy that are absent in the original signal.

Amplitude envelop   Changes in peak amplitude of sound waves over time. Often used to describe the time contour of the integrated spectral energy of a sound.

Audiogram   A graph showing the absolute hearing threshold for pure tones as a function of frequency. In other words, the absolute auditory threshold curve.

Aural harmonic   A harmonic generated in the auditory system.

Best frequency (BF)   The frequency at which the response of a given auditory neuron is strongest, i.e. frequency at which the maximum maximum number of action potentials can be elicited in the response to a single tone. See also "characteristic frequency".

Call   Nonverbal or animal vocalization that is emitted to elicit a response from a receiving animal.

Characteristic frequency (CF)   The frequency at which the threshold of a given auditory neuron is the lowest, that is, the frequency at which it is most sensitive. For many neurons in the auditory brainstem, the CF is identical to the BF

Click   A wide spectrum of frequencies present within a very short (<1 ms) time segment.

Combination-sensitivity   A nonlinear facilitation of the response of a neuron to the combination of two or more stimuli; the response to the stimuli presented together is greater than the sum of the responses to each stimulus presented separately.

Constant frequency   A periodic wave that does not change in the duration of its period over time.

dB SPL   Decibel (dB) sound pressure level (see also "sound pressure level").

Fall time   The time it takes for a sound-wave amplitude to fall from peak to baseline.

Frequency modulation   Upward, downward or sinusoidal changes in frequency of a tone over time.

Frequency tuning curve   The envelope function of an auditory neuron's receptive field in a frequency (x-axis) amplitude (y-axis) plot, that is, the curve following the minimum tone amplitude (threshold) for a neuron's response as a function of tone frequency.

Fundamental frequency   Usually the lowest frequency in the Fourier spectrum of a complex tone. In vocalizations, the vibration frequency of the sound-generating substrate, for example, of the vocal chords.

Harmonics Fourier components of a complex tone with frequencies that are multiples of the fundamental frequency.
Infrasound Sound frequencies below the frequency range of human hearing ($f < 20\,\text{Hz}$).
Loudness The quality of sound that ranges from soft to loud. The loudness of a tone at a given frequency is measured in "phons" relative to the loudness of a 1 kHz tone given in dB SPL.
Minimum threshold The minimum detectable level of a sound in the absence of any other external sounds.
Motif A salient thematic sound pattern that is repeated in variable forms within a vocalization sequence (song).
Noise burst A short duration segment of noise.
Note The smallest segment of sound surrounded by silence intervals in a song.
Octave Frequency interval between two tones that have a frequency ratio of 2:1.
Oscillogram A plot of a sound waveform in amplitude (y-axis) versus time (x-axis) coordinates.
Partials A mathematically inclined term for frequency components, often harmonics, of a complex sound.
Perception Unconscious or conscious representation of sensory input in a receiving animal or human.
Period Time taken to complete one cycle of a sound wave.
Periodic Having a constant or regular period.
Periodicity pitch The pitch that approximately corresponds to the frequency of sinusoidal amplitude modulation of a single tone or complex sound. Harmonics to a given fundamental frequency lead, in the absence of the fundamental, also to the perception of a periodicity pitch equal to the pitch of the fundamental. This is called the pitch of the missing fundamental.
Phase A time point at a particular energy level in a periodic waveform usually measured in relation to the node (zero energy) point in the same or a time point in another waveform.
Phase angle The angle relative to the horizontal made by the tangent on the curvature at a specific point in a wave form.
Phase locking This is the tendency of a neuronal response to occur at a particular phase of the stimulating sound waveform or, for a nerve fiber, the waveform on the basilar membrane within the ear.
Phoneme The shortest unit of a speech sound (e.g. a consonant, a vowel or a consonant–vowel pair) that leads to a contrast in the perceived sound. A phoneme by itself does not usually convey a meaning.
Pink noise Broadband noise in which the intensity per hertz declines with increasing frequency at a rate of 10 dB/decade or 3 dB/octave.

Pitch   A perception that is most closely related to the fundamental frequency of a complex sound.
Polytonal   Made up of multiple frequencies or tones that are not harmonically related.
Power   Power is a unit of energy per unit time. It is difficult to measure the total power generated by a sound source, and it is more common to specify the magnitudes of sound in terms of their intensity, which is the sound power transmitted through a unit area in a sound field.
Power spectrum   A plot showing the relative levels of energy contained in bandwidths across the frequencies of a complex sound.
Pure tone   A tone resulting from a single sine wave.
Receptive field   The range of values of a parameter, such as space or frequency, for a stimulus which triggers a neuron's response that is significantly above its background or spontaneous activity of the neuron.
Residue pitch   The low pitch heard when a group of partials is perceived as a coherent whole. For a harmonic complex tone, the residue pitch is usually close to the pitch of the fundamental component, but that component does not have to be present for a residue pitch to be heard. Also known as virtual pitch, low pitch and periodicity pitch.
Response area   The set of combinations of two parameters that can elicit a response above threshold from a neuron. For the frequency–amplitude parameters, this includes the area within a frequency-tuning curve.
Rhythm   The inter-beat interval in a musical segment or the inter-syllable interval between repeated syllables in a call. A fast rhythm consists of short inter-beat or inter-syllable intervals. Intervals may also be created by repeated broadband sounds, such as the striking of cymbals within a musical piece.
Rise time   The time it takes for a sound-wave amplitude to rise from baseline to the maximum or peak level, also referred to as attack or the buildup of sound in the case of a complex tone.
Song   A sequence of same or different syllables repeated at certain intervals.
Sonogram   Another term for a sound spectrogram.
Sound pressure level   A term used to indicate that the reference sound pressure used for calculating a tone's decibel (dB) rating is set to $2 \times 10^{-5}\,\mu\text{Pa}$. Abbreviated as SPL.
Sound spectrogram   A display showing the pattern of frequencies and intensities over a period of time, usually calculated using Fourier analysis. The abscissa is time, the ordinate is frequency, and the amount of energy is shown either using colors or darkness of shading or as three-dimensional line plots.
Sound waves   Periodic pressure changes in a medium such as air.

Syllable A basic phonetic unit of speech and language that usually consists of a consonant or a vowel or a combination of the two. Part of or a whole animal communication call segment surrounded by silence intervals. A phoneme is more clearly defined from an acoustic viewpoint, whereas a syllable in human speech is based primarily on articulation.

Timbre The quality of sound arising from the spectral distribution of its acoustic energy. Different instruments or voices have different timbres when producing the same musical note at the same loudness levels.

Tonal A sound consisting of one or several harmonics.

Tone burst A constant frequency tone of a finite duration.

Tuning curve For a single neuron, this is a graph of the lowest sound levels, at which that neuron barely responds, plotted as a function of another parameter, e.g. frequency. The frequency tuning curve is the border line of the frequency-threshold curve or -response area.

Ultrasound Sound frequencies above the frequency range of human hearing ($f > 20\,\text{kHz}$).

White noise Broadband noise containing a random distribution of a theoretically infinite number of frequencies, all at the same amplitudes.

Index

A

AC, mustached bats
  echolocation, functional organization 162–163
  neural mechanisms
    call representation 164–165
    pulse–echo representation 163–164
acoustic meaning, perception 90–92
acoustic patterns continuums, categories 100–106
  frequency sweeps, direction 104–105
  information-bearing elements, duration 105–106
  spectral (formant) structure 102–103
acoustic signal design, mustached bats and call types 77–80
  motivation–structure hypothesis 79–81
active electrolocation 271, 276
adaptations
  behavioral
    meaning perception 117–119
  physiologic
    and anatomic 115–117
    audiovocal, blind mole rat 116
aggression, caged bats
  experimentally elicited 74–77
  natural 73–74
AI responses, vocalized sounds 195
amplitopy 164
amplitude modulation, vocal source 11–14
androgen expression 231–232
anterior forebrain pathway, song control system 234, 275, 277, 280

area, cortical
  anterior (A) *190*, *192*
  anterior primary auditory (AIa) 158
  CF/CF 158, *159*, 163, 165, 172, 178, 198
  DF *159*
  DIF *159*
  DM 158, *159*
  Doppler-shifted constant frequency (DSCF) 158, *159*, 162 165, 167, 168, 169, 170, 172, 178
  dorsal (D) *190*, *192*, 193, 195, 197, *198*, 200
  dorsoanterior (DA) *192*, 193, 199
  dorsocaudal (DC) *191*, *192*, 193, 194, 197, *198*, 200
  dorsocaudal belt (DCB) *192*, 193
  dorsoposterior (DP) *192*, 193, *197*, *198*, *199*
  dorsorostral belt (DRB) *192*, 193
  FM–FM 158, *159*, 163, 169, 172, 175, 178
  H_1–H_2 *159*
  posterior (P) *192*, 193, 197, *198*, 199, 202
  primary auditory (AI) 160, 173, 189, 191–200
    anterior (AIa) 158, *159*
    posterior (AIp) 158, *159*, 161, 172, 173, 175, 178
  rostral belt (RB) *192*
  small field (S) *192*
  TE 158, *159*
  ventroanterior (VA) *159*, *192*, 193, *197*, *198*, 199
  ventrocaudal belt (VCB) *192*, 193
  ventroposterior (VP) *159*, *192*, 193, 197, *198*, 199
  ventrorostral belt (VRB) *192*, 193

351

area, cortical (contd.)
  VF 159
arousal and emotions level, acoustically
    expressed 92–93
attraction, perceived meaning 96–98, 106
audible parameter space
  broadband group 96
  high-frequency group 96
  low-frequency group 96–97
audiograms 87, 95
auditory belt projections
  to prefrontal cortex
    rhesus monkeys 217–218
auditory brainstem and midbrain, bats
    structure
  brainstem pathways 136
  isofrequency contours 137
  tonotopic organization 135
auditory communication, cortical
    plasticity 294
  auditory cortex, experience-dependent
      plasticity
    adult animals 301–304
    during early development 297–301
  central auditory system, cortex-oriented
      plasticity
    corticofugal modulation 304–309
  cortical plasticity perspectives
    vocalization, cortical coding 309–313
auditory communication, in primates
  coding/decoding 205–206
  pattern recognition problem 205–206
auditory concept learning
  category learning
    Aristotelian view 319
    new animal model 323–3291
    theoretical underpinnings 319–320
auditory cortex, experience-dependent plasticity
  adult animals
    cortical plasticity, after localized hearing
        loss 303–304
    cortical representation, enhanced 301–302
    cortical representation, enhancement 301
  during early development 297–301
    cortical IBEs/IBPs, normal development
        298–299
    distorted acoustic environment, cortical
        development 299–300
    inner ear ablation impact, cortical
        development 300–301
    sensory experience, critical role 225,
        297–298

tonotopic maps, normal development
    298–299
auditory cortical fields 158
  tonotopic organization 191–193
auditory feedback 230, 234, 252
auditory nerve 132
auditory nerve fibers 135
auditory pathways
  brainstem 136
aversion, audible parameter space 97
avian communication, vocal mechanisms
  avian vocal systems
    integration with respiration 7–9
    labia 5–6
    syringeal membranes 6–7
    syrinx 3–5
  bipartite syrinx, lateral independence
      14–20
    acoustic specialization 19–20
    song lateralization, patterns 19–20
    two-voice vocalizations 17–19
    unilateral dominance versus bilateral
        parity 16–17
  future directions 29–30
  production, propagation, and perception
      28–29
  vocal learning 9–10
  vocal source, controlled and intrinsic
      modulation 10–14
    amplitude modulation 11–13
    frequency modulation 13–14
  vocal tract filtering 24–28
  vocalizations, nonlinear dynamics 20–24
avian vocal systems
  integration with respiration 7–9
  labia 5–6
  syringeal membranes 6–7
  syrinx 3–5

B

behavioral context, mustached bats
  simple syllabic calls 67–69
behavioral postures, captive mustached bats
  and associated calls 64–67
behavioral tests
  binaural deafening 43
    head drumming behavior 46
  blind mole rats 43–46
  vibratory stimuli effect 45
belt fields 199
bipartite syrinx, lateral independence

acoustic specialization 19–20
song lateralization 19–20
two-voice vocalizations 17–19
unilateral dominance versus bilateral
    parity 16–17
birdsong, learning and production
  anterior forebrain pathway
    learning pathway 234
  neural circuit
    song system 232–236
birdsong development 227
  sensitive periods, learned behavior 227
  sensorimotor learning 229
  song crystallization 231–232
  variability and plasticity 230–231
  zebra finch song development *228*
birth cry 96
blind mole rat
  behavioral tests 43–46
  electrophysiologic experiments 42–43
  lower jaw *49*
  middle ear *48*
  morphology 46–49
  seismic communication, via acoustic
      channels 36
  seismic signals, perception 40–41
  somatosensory or auditory system 41–42
  vibratory signals *38, 40, 45*
    properties 38–40
boxing and poking behavior, captive
    mustached bats 65
BPN, selectivity
  sound processing, neuronal responsiveness
      207–208
brain-derived neurotrophic factor (BDNF)
    244
  BDNF-induced variability 247
  developmental declines 247
  roles 245
  synapse, structure and function 248–249
brainstem auditory evoked responses
    (BAER) 40
broadband group 95, 96

C

caged bats
  aggression
    experimentally elicited 73–76
    natural 73–74
  social interaction
    recording procedures 71–73

call representation, neural mechanism
  basis functions 164–165
  single unit call response data 164
call responses, mustached bats 165–175
  CF/CF combinations, high-frequency
      domain 172, 178
  FM–CF combinations 167–169
  FM–FM combinations, time domain
      169–171
  harmonic complexity, low-frequency
      domain 172–175
call types, mustached bats 77–81
calls, cortical representation
  multiparametric distributed 175–180
  versus perception 180–182
Cape mole rat
  vibratory signals 49
captive free-flying bats
  experimental results
    behavioral postures and associated
        calls 64–68
    inspection and appeasement 69–71
    social dominance 70–71
    social interaction 63–64
  social behaviors
    maintenance and recording
        procedures 60
    scoring system 60–62
captive mustached bats 57–59
catbird song *18*
categorization
  eco-ethologic aspects 320–321
    vervet monkeys 321
  and generalization 321–323
    learning curves 321
category learning *322*
  new animal model
    frequency-modulated tones 324–326
    physiologic correlates 326–329
    theoretical underpinning 319–320
central auditory system, cortex-oriented
    plasticity
  corticofugal modulation 304–305, *306*
    combination-sensitive neurons, response
        properties 308
    species-specific differences 308–309
  frequency tuning, frequency-specific
      modulation
    and tonotopic 305–307
  frequency-specific neural plasticity
      304–305
  IBEs/IBPs modulation 307–308

central nucleus of the inferior colliculus (ICC) 50, 143
CF/CF combinations, high-frequency domain 172
characteristic frequencies (CFs) 50
    SAM tones 51–52
cochlear tonotopy 137
cohesion, audible parameter space 97
communication sound perception, common rules 85–86
    psychoacoustical measures and relations 87
        audiograms 87
        duration discrimination 88
        frequency discrimination 88
        intensity discrimination 88
        pitch, perception and discrimination 89–90
        spectral resolution 89
        spectral summation 89
        temporal summation 87–88
    six rules 90
        acoustic meaning, perception 90–92
        acoustic patterns continuums, categories 100–106
        arousal and emotions level, acoustically expressed 92–93
        audible parameter space 93–99
        individualized perception 90
        urgency of response 99–100
communication sounds
    lower auditory nuclei
        convolution process 139–141
        Tadarida 137
        waveforms *138*
    processing, inferior colliculus
        heterogeneity 143–145
        inhibition 147
        response selectivity 143–145
    spatio-temporal pattern
        functional advantage 148–149
        in ICC 147
        population response 150
communication sounds, in primates
    auditory belt projections
        to prefrontal cortex 217–218
    auditory communication
        pattern recognition problem 205–206
    auditory cortex
        early parallel processing 206–207
        hierarchic processing 205
        human imaging studies 218–219
        sound processing
            with intermediate complexity 207–214
        species-specific calls
            responses 214–217
complex sounds, acoustic structure
    in mustached bats
        echolocation signals, structure 161
        social communication calls, structure 161–162
constant frequency (CF) 156, 161, 163
corollary discharge signal 283, 284
cortical coding, of vocalizations
    cortical plasticity perspective 309–313
    three envelopes 310–311
cortical development
    distorted acoustic environment 299–300
    inner ear ablation, impact 300–301
cortical IBEs/IBPs, normal development and tonotopic 298–299
cortical plasticity, after localized hearing loss 303–304
cortical representation, enhancement
    acoustic signals induced 301
corticofugal modulation 304–305, *306*
    combination-sensitive neurons, response properties 308
    species-specific differences 308–309
critical bandwidth (CBW), properties 89
crouching behavior, captive mustached bats 64

D

direct motor pathway, song control system 277
distorted acoustic environment
    cortical development 299–300
distress calls, pups 96
Doppler-shifted constant frequency (DSCF) area *158*
    neurons 167
dorsal lateral geniculate nucleus (dLGN) 50
duration discrimination 88–89

E

early parallel processing, in primates
    auditory cortex 206–207
echolocation 134
    calls, common features 162
echolocation signals, mustached bats 61
    functional organization 162–163
    structure 161

electric fish sensory processing 273
electric organ discharge (EOD) 271–276, 284
electrocorticograms
　category learning, single trial analysis 328–329
electrophysiologic experiments
　blind mole rat 42–43
electrosensory lateral line (ELL) 273, 274
electrosensory processing 271–276, 284–285
enhanced cortical representation 301–302, 313
error-correction function
　LMAN activity patterns
　　song related 236
　supports
　　experimental directions 235
excitatory response areas 137, 160, 167, 169

F

fast Fourier transform (FFT) *38*
females, defensive calls 96
fly-by behavior, captive mustached bats 67
FM–CF combinations 167–169
　Doppler-shifted constant frequency, data 167
FM–FM combinations, time domain 169–171
frequency discrimination 88, 295, 302, 304
frequency jumps, bird vocalizations *23*
frequency modulated (FM) pulses 163
frequency modulation, vocal source 13–14
　air sac pressure 13, *14*
　songbirds 13, *15*
　syringeal muscles 13
frequency sweeps, direction
　acoustic patterns continuum 104–105
frequency sweeps parameters, selectivity 210–214
　DS index 210
　FM rates, preferred 213–214
frequency tuning, frequency-specific modulation
　and tonotopic maps 305–307
frequency-modulated tones 156, 193, 324–325, 328–332
　modulation direction, categorization
　　apparatus and training paradigm 324–325
　　behavioral results 325–326
　　stimuli 324

G

Georychus capensis see Cape mole rat
guinea pig
　auditory cortical
　　binaural response *199*
　　neural recording *190, 194, 197*
　　principal component analysis *201*
　　tonotopic map *191, 192*
Gulf toadfish, natural behavior
　steroids 125–126

H

hallmarks, sensorimotor learning
　song
　　plasticity 230
　　variability 230
harmonic complexity, low-frequency domain 172–175
　multi-peak response areas, tuning 173
hearing 133–134
　importance, bats
　　echolocation 134
　via lower jaw
　　bone conduction 46–47
higher auditory fields, neural activity spreads 197–198
hormone-dependent plasticity
　teleost fish 129
human imaging studies
　Heschl's gyri 218–219

I

IBEs/IBPs modulation 310
　in different domains 307–308
individualized perceptions, level of matched groups 90
inferior colliculus (ICC) 50, 133, 135, 137, 143, 145–150, 305–309
information-bearing elements (IBE) 85, 295, 307, 310
　duration, acoustic patterns continuum 105–106
information-bearing parameters (IBP) 296, 307, 310
inhibitory response areas 160
inner ear ablation, impact
　cortical development 300–301
　monocular dominance plasticity 300

inspection behavior, captive mustached bats 66–67
intensity discrimination 88
isofrequency contours 137

J

Japanese macaque monkeys 91
just meaningful differences (JMDs) 86, 90, 91, 100, 101
just noticeable difference (JND) 88, 91, 94, 96, 101
juvenile
  sensitive period 226
  -time window 224
juxtalobar nucleus (JLN) 273

K

kissing behavior, captive mustached bats
  arching back 66
kitten
  auditory cortical
    response to vocalizations 311, 312
    tonotopic plasticity 303
  auditory development 298
  vocalization 295

L

labia 5–6, 12
learned behavior
  birdsong development, sensitive period 227–232
  sensitive period 223
  transfer, to novel stimuli 323

M

mammalian AC 157–161
  belt, parabelt areas 158
  core-belt–parabelt organization, *Rhinolophus* 158
  cortical areas 157
  non-neurophysiologic strategy, complex sounds 157–158
  organization 158–160
  primary auditory cortical neurons, response properties 160–161
  role, understanding 157
  tonotopic organization 158

marking behavior, captive mustached bats 64, 69
mesoscopic neurodynamics, auditory cortex
  auditory concept learning
  categorization
    eco-ethologic aspects 320–321
    and generalization 321–323
  category learning
    new animal model 323–329
    theoretical underpinnings 319–320
Mexican free-tailed bats 80
  auditory brainstem and midbrain structure 135–137
  communication sounds
    inferior colliculus 143–147
    lower auditory nuclei 137–143
    spatio-temporal pattern, ICC 147–150
  importance of hearing 133–134
  neural response profiles 142, 144, 146
  neural responses 139, 149
  species-specific vocalizations
    processing 132
    vocalizations 138
middle latency responses (MLR) 40, 42, 44
midshipman fish
  auditory communication 334–335
  vocal-auditory coupling 127
  vocal communication 123
  vocalizations 124
molecular effectors
  molecular signals 244
  synaptic and vocal plasticity 244–249
monkey calls 199, 211, 214
mormyrid electric fish 265
  electrosensory processing 271–276
    active electrolocation 271
    AND-gate 272, 274
    corollary discharge signal 282–284
    electric organ corollary discharge (EOCD) 271, 272
    electric organ discharge (EOD) 271, 272
    electrosensory lateral line (ELL) 273, 274
    novelty response 272
  sensory processing, neuronal substrates 265
  sensory system 266
  songbirds, comparison 282–285
    sensorimotor processing systems 269
morphology, blind mole rat 46
  lower jaw articulation 49

middle ear, structure *48*
Morton's rule 94
  modified 97
motifs 229
mouse vocalizations *92, 95*
multiple auditory fields, functional differences 198–200
mustached bats
  AC 162–165
    neural responses *166, 168, 170, 174*
  agonistic vocalizations 75
  audiovocal communication and social behavior 57, 78, 79
    acoustic signal design 77–81
    behavioral context, simple syllabic calls, 67–69
    behavioral postures and associated calls 64–67
    caged bats, social interaction 71–73
    call representation, neural mechanisms 164–165
    call types 77–80
    captive free-flying bats, social behavior 60–62
    echolocation, functional organization 162–163
    experimentally elicited aggression 74–77
    inspection and appeasement 69
    natural aggression 73–74
    pulse–echo representation, neural mechanisms 163–164
    roosting structure, activity patterns and social interaction 63–64
    territoriality and social dominance 70–71
  auditory cortical fields *159*
  behavior measurement *71*
  call–emotion relationship *79*
  call responses 165
    CF/CF combinations 172
    FM–FM combinations 167–171
    harmonic complexity 172–175
  calls 58
  call-type representation *176*
  complex sounds, acoustic structure
    echolocation signal *159*
    echolocation signals, structure 161
    male–female interaction 73
    roosting behavior 62
    social communication calls 161–162

N

Nanospalax ehrenbergi see blind mole rat
neural circuit correlates, model *248*
neural circuit plasticity
  diverse systems
    common themes 225
  sensitive period 224
neural correlates
  nucleus RA 237
  song plasticity
    sensorimotor learning 236–237
  synaptic connectivity, projection neurons 237
  vocal plasticity
    nucleus RA 237
neural mechanism
  call representation 164–165
  pulse–echo mechanism 163–164
  vocal–communication 123–130
neuroendocrine mechanisms 123
  future prospects 129
  vocal–auditory coupling 126–129
  vocal communication, interfacing 123
  vocalization
    hormonal control 123–126
neuron's response region 140–141
neuronal sensitive periods
  and behavioral 224, 226, 241, 245
  common themes 224
nipping behavior, captive mustached bats 65
nonlinear dynamics
  in vocalizations 20–24
non-neurophysiologic strategy 157
notes 11, 12, 229, 239
nucleus RA 237
  cellular analysis
    sensorimotor learning 240–241
  functional synaptic analysis
    sensorimotor learning 241–244
  neuronal codes, for song 237–238
  subregion
    structure and function 239
  synaptic connectivity
    extrinsic and intrinsic patterns 238

O

optical recording methods 189–191
  voltage-sensitive dyes 189

optical signals, principal component analysis 200–202
oscillation 6, 22, 179, 336
oscine songbirds, vocal learning 9

P

parakeet syrinx 5
patterns, song lateralization and acoustic specialization 19–20
Pavlovian conditioning 326–327, 327
principal component analysis (PCA)
 advantages 202
 optical signals 200–202
perceptual processing, song 267–271, 276–282, 340
periodotopy 164
physiologic correlates, learning 323
 electrocorticograms
  single trial analysis 328–329
 general aspects 326–327
 stimuli utilized
  neuronal representation 327–328
pitch, perception and discrimination 89–90
Plexiglas tube
 mole rat, vibration 37
Porichthys notatus see midshipman fish
posterior primary auditory (AIp) 158
primary auditory cortex (AI) 160
primary auditory cortical neurons, response properties
 DSCF processing area 160
psychoacoustical measures and relations
 communication sound perception
  audiograms 87
  duration discrimination 88
  frequency discrimination 88
  intensity discrimination 88
  pitch, perception and discrimination 89–90
  spectral resolution 89
  spectral summation 89
  temporal summation 87–88
Pteronotus parnellii see mustached bats
pulse-echo representation, neural mechanism 163–164
 neural response, characteristics 163

R

RA subregions
 song patterning 239
 structure and function 239

Rayleigh and Love surface waves see seismic communication
response profile
 DNLL neuron 142–143
rhesus monkey
 auditory cortical
  connectivity *218*
  fields *208*
  responses *209, 212*
  response selectivity *215, 216*
  vocalizations *211*
ring dove *14, 23, 26*

S

scoring behaviors
 and calls
  captive bats 60–62
seismic communication
 via acoustic channels, blind mole rat 36
  auditory system 41–42
  behavioral tests 43–46
  electrophysiologic experiments 42–43
  morphology 46–49
  overview 36–38
  seismic signals, perception 40–41
  somatosensory system 41–42
  vibratory signals, properties 38–40
seismic signaling 37
 perception, blind mole rat 40–41
 see also vibratory signals
self-directed behaviors, captive mustached bats 64
sensitive periods
 learned behavior
  birdsong development 227–232
 neural circuit plasticity
  mechanisms 224–227
  NMDA 225–226
 neuronal and behavioral
  common themes 224
 zebra finch
  song development 228
sensory acquisition
 sensorimotor learning 229
sensory experience, critical role
 cross-modal plasticity studies 297
 early cortical development 297–298
sensory processing, neuronal substrates
 mormyrid electric fish 265
  electrosensory processing 271–276
  sensory system 266

song perception and learning
  perceptual processing and memorization 276–282
  songbirds 265, 267–271
  songbirds and electric fish, comparison 282–285
sensory systems
  auditory systems
    bone conduction 41, 43, 47, 48, 50
  somatosensory systems
    mechanoreceptors 41–42
sinusoidal amplitude modulated (SAM) tones 51–52
social communication calls
  cortical representation 156
  hypothesis 175–182
  mammalian AC 157–161
  mustached bats
    AC 158, 159, 162–165, 175, 182
    call responses 165, 338
    complex sounds, acoustic structure 161–162
    divisions 162
    structure 161–162
social interaction, captive bats
  roosting structure and activity patterns 63–64
somatosensory or auditory system
  seismic communication 41–42
song control system 277–278
  anterior forebrain pathway 277
  direct motor pathway 277
  features 278
song crystallization
  androgen expression 231, 232
  mechanisms 232
  vocal-paralysis study 231
song learning 267–271, 340
  nucleus RA
    cellular analysis 240–241
    functional synaptic analysis 241–244
    neuronal codes 237–238
    synaptic connectivity 238
    vocal plasticity, neural correlates 237
  learned behavior
    birdsong development 227–232
  molecular effectors
    synaptic and vocal plasticity 244–249
  RA subregions
    structure and function 239
  sensitive periods
    for learned behavior 223–224
    neural circuit plasticity 224–227

neuronal and behavioral 224
  and synaptic mechanisms 223
sensorimotor phase 268
sensory phase 267–268
song control system 277–278
  features 278
  pathways 277
song plasticity
  neural correlates 236–237
song system
  neural circuit for learning 232–236
  vocal imitation 267–268
song memorization 229, 230, 267, 276–282
song perceptual processing 267–271
song system
  anterior forebrain pathway (AFP), lesions
    learning pathway 234
  direct-motor pathway 233
  juvenile deafening 234
  LMAN activity 236
  neural circuit
    birdsong, learning and production 232
song variability 230
songbirds
  helium results mechanisms 25
  labia 5, 12
  mormyrid electric fish, comparison 282–285
  perceptual processing 276–282
  respiratory motor patterns 8
  sensory-motor coupling 269, 275
  song
    discrimination 278–279
    learning 267–268
    perceptual processing 266, 267–271, 280, 285
  song consolidation 248
    plasticiy 246
  song system of the brain 233
  syrinx 4, 13, 16
  vocal communication 266, 267
  vocal learning 9–10, 116, 227, 266, 271, 282
sound processing 207–214
  intermediate complexity 207
  selectivity
    BPN 207–210
    frequency sweeps parameters 210–214
Spalax ehrenbergi see blind mole rat
spatiotemporal processing, guinea pig auditory cortex
  activity spreads, higher auditory fields 197–198

AI responses, vocalized sounds 195
multiple auditory fields, functional
  differences 198–200
optical recording methods 189–191
optical signals, principal component
  analysis 200–202
spatiotemporal representation
  constant and frequency-modulated tones
    193–195
  functional significance 195–197
tonotopic organization, auditory cortical
  fields 191–193
spatiotemporal representation
  constant and frequency-modulated tones
    193–195
  functional significance 195–196
species-specific calls, responses
  MC
    preference index 216
    and spatial selectivity 216–217
  nonlinear integration mechanisms 214–216
    combination-sensitivity 215
species-specific communication sound 217
spectral (formant) structure, acoustic patterns
  continuum 102–103
spectral resolution 89
spectral summation 89
subsongs 229
subsyllables 57, 59
syllables
  mustached bat 57, 59
  songbird 229
syringeal membranes 6–7
syrinx 3–5

T

Tadarida brasiliensis see Mexican
  free-tailed bats
temporal summation 87–88, 216
territoriality and social dominance, mustached
  bats 70–71
tests, neural circuit model
  alternative possibilities 249
  developmental regulation 251
  selection, RA 252
  synaptic connectivity and vocal plasticity
    correlations 250
tonotopic organization
  auditory system, fundamental 135
  auditory cortical fields 191–193

tonotopic organization, guinea pig auditory
  cortex 191–193
  characteristic frequencies (CFs)
    contours 192
tonotopy 160
two-voice vocalizations 17–19

U

unilateral dominance versus bilateral parity
  bipartite syrinx 16–17
urgency of response, audible parameter space
  99–100
USVs, pups 96

V

vibratory dialogue 37, 38
vibratory signals
  properties 38–40
  temporal patterns, analysis 39
  see also seismic signals
vocal communication
  neural mechanisms
    future prospects 129
    neuroendocrine mechanism 123
    vocal–auditory coupling 126–129
    vocalization, hormonal
      control 123–126
  vocal signature, mothers and pups 134
vocal imitation
  sensorimotor phase 268
  sensory acquisition phase 267–268
vocal learning 9–10, 116, 265, 266,
  271, 282
vocal mechanisms, for avian
  communication 3
vocal source 10
  amplitude modulation 11–13
  frequency modulation
    air sac pressure 13, *14*
    songbirds 13, *15*, 16, 19, 29
    syringeal muscles 13
vocal tract filtering 24–28
vocal–auditory coupling
  steroid hormones 126–129
vocalizations 17, *23*, 26, 28, 86
  composite 57, 59
  frequency jumps, bird *23*
  hormonal control, steroid modulation
    124–125

nonlinear dynamics 20–24
simple syllables 57, 59
turtle-dove *21*
two-voice 17–19

W

wing-flicking behavior 65, 76
wrestling and biting behavior 66

wriggling calls, pups 96, 102, 104
spectral (formant) structure 102–103

Z

zebrafinch
HVC-RA plasticity *244*
inspiratory syllables 7
song development *228*